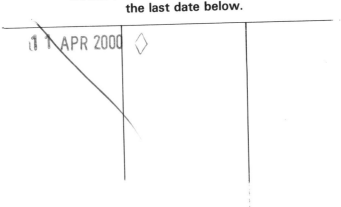

FACT–Vol. 18

# COMBUSTION MODELING, SCALING AND AIR TOXINS



**presented at**
The 1994 International Joint Power Generation Conference
Phoenix, Arizona
October 2–6, 1994

**sponsored by**
The Fuels and Combustion Technologies (FACT) Division, ASME

**edited by**
Ashwani K. Gupta
University of Maryland

Albert Moussa
Blazetech Corporation

Cary Presser
National Institute of Standards and Technology

Michael J. Rini
ABB Power Plant Laboratories

Roman Weber
International Flame Research Foundation

George Woodward
Wheelabrator Environmental Systems

THE AMERICAN SOCIETY OF MECHANICAL ENGINEERS
UNITED ENGINEERING CENTER / 345 EAST 47TH STREET / NEW YORK, NEW YORK 10017

Printed in U.S.A.



ISBN No. 0-7918-1383-5

Library of Congress Catalog Number 94-78220


Printed in U.S.A.

# FOREWORD

This volume contains papers presented at the 1994 International Joint Power Generation Conference, October 2–6, in Phoenix, Arizona. These papers represent contributions from sessions sponsored by the ASME Fuels and Combustion Technologies (FACT) Division, and included participants from university, industry, and government.

- Pollutants Emissions and Combustion Modeling
- Particle Transport and Combustion Phenomena
- Scaling and Mathematical Modeling of Combustion Systems
- Air Toxic Issues
- Recent Advances in Low $NO_X$ Combustion
- Corrosion and Fouling Effects of Alternative Fuel Combustion
- Advancements of Reburn Commercialization

The sessions shared much in common, particularly with regard to the strong emphasis on $NO_X$ control, and modeling and experiments relevant to combustion, pollutants emission, and air toxins. The sessions on pollutants emission and combustion modeling focused on the dissemination of information pertaining to current research on theoretical simulation and prediction of combustion systems and pollutants emission. Topics included modeling of furnaces and boilers, burners, combustors, gaseous flames, spray combustion, coal and coal slurry combustion. Both fundamental and applied studies in the area of pollution control, including $NO_X$ and efficiency were of interest. Systems of special interest included furnaces and boilers, gas turbine combustion, internal combustion engines and other power generation systems. All of the above-mentioned topics have provided a unique contribution to the theoretical and predictive capabilities of combustion systems, and are of particular interest to the scientific, research, and technical communities. Modeling aspects were also examined on the corrosion and fouling effects of alternative fuel combustion. The session on scaling and combustion modeling for the utility industry presented various modeling approaches, their potential uses in the design and operation of real combustion systems, and their limitations. The session on air toxic issues examined the environmental aspects of conventional fuels: coal, gas, oil, low-grade fuels, and MSW. The session on recent advances in low $NO_X$ combustion examined developments of fossil fuel fired turbulent diffusion flames at both the pilot plant- and full-scale. Advancements of reburn commercialization were examined in the final session.

The editors wish to express their sincere thanks to those individuals who contributed to the sessions and the publication of these proceedings. They are particularly thankful to the authors, abstract and manuscript reviewers, and the ASME Headquarters staff for their complete cooperation throughout the publication process.

Ashwani K. Gupta
University of Maryland

Albert Moussa
Blazetech Corporation

Cary Presser
National Institute of Standards and Technology

Michael J. Rini
ABB Power Plant Laboratories

Roman Weber
International Flame Research Foundation

George Woodward
Wheelabrator Environmental Systems

# CONTENTS

FACT-Vol. 18, Combustion Modeling, Scaling and Air Toxins
ASME 1994

# CHLOROBENZENE OUTPUTS FROM COMBUSTION OF CHLORINATED ORGANIC AND INORGANIC COMPOUNDS

**A. E. S. Green, J. A. Vitali, and T. L. Miller**
Department of Mechanical Engineering
Clean Combustion Technology Laboratory
University of Florida
Gainesville, Florida

## ABSTRACT

We consider the gas phase formation of chlorinated benzenes and phenols as precursors of chlorinated dioxins and furans from the combustion of solid fuels containing organically bound chlorine. The model investigated is intended to apply to the combustion of medical waste, municipal waste and coals containing chlorine. Assuming a temperature-time profile drawn from incinerator experiments, we use kinetic modeling with known reaction rates and simple assumptions on unknown reaction rates to further investigate four models of chlorinated benzene formation. Since reaction rates for most chlorination processes are not known, we choose simple systems of reaction rates that yield outputs that can be made approximately compatible with results of the Pittsfield-Vicon incinerator and Clean Combustion Technology Laboratory experiments. We also consider recent measurements of HCl emissions from crematoria and the implication of this work with respect to the benefits of material substitution in medical and municipal waste incineration. These benefits should also accompany the dechlorination of coals. We note the disparity between the prevailing USA position and the emerging position of Germany on the issue of halogenated plastics. We also note that Europe and Asia are beginning to address solid fuel issues as a consolidated discipline. This pattern should be helpful in broadening the understanding of solid fuels combustion processes and in ferreting out erroneous data and conclusions. This is important in view of the recent concern about the role of low dioxin exposure levels on fetal development and the immune system.

## (1) Introduction

Since 1988 the Clean Combustion Technology Laboratory (CCTL) has been engaged in the measurement of toxic products of institutional waste incineration and attempts to model the results for various controlled input waste and combustion conditions. This work is a continuation of studies on co-combustion of fuels which began in 1980 (Green, 1981; 1986; Green and Pamidimukkala, 1982; 1984; Green, et al.,1990a; 1990b). Our experimental work has indicated that reducing chlorinated plastics in the waste stream leads to reductions in emissions of monochlorobenzene, dichlorobenzene and other chlorinated organic compounds (Green, et al., 1990a; 1990b). A similar conclusion has been reached for PCDD/PCDF emissions (Bulley, 1990). The CCTL has also carried out phenomenological analyses of the Pittsfield-Vicon (PV) incinerator data and have reached a similar conclusion using their data (Green, et al., 1991a; 1991b; 1991c; 1992). However, the PV study itself came to the conclusion that there is no relationship between the amount of polyvinyl chloride (PVC) in the waste and PCDD/PCDF emission (Neulicht, 1987), a conclusion that is implicitly restated in a recent assessment (Solid Waste Assoc., 1993). Thus, we have a disagreement whose resolution is important to the understanding of clean combustion technology and the formulation of a sound public policy.

At this time the experimental data available on chlorinated aromatic emissions from combustors with controlled PVC inputs is exceedingly sparse and noisy. Hence it is difficult to resolve important issues on the basis of the available experimental data since the conclusions reached can depend upon the statistical method of analysis and the rejection rules for outlyers. Pending the availability of more complete and refined data sets, kinetic modeling of combustion processes can be helpful in exploring relationships between inputs to a combustor, its operating conditions, and its various emissions. While the results of kinetic modeling are suspect since they depend upon assumed reaction rates they should not have the signal to noise problems which currently plague experimental toxic emission data.

We focus here on gas phase production of chlorobenzenes at high temperatures in a modular incinerator up to the exit of the secondary combustion chamber. Chlorobenzenes and the closely related chlorophenols are considered to be the primary precursors of chlorinated dioxins and furans. These are probably mostly formed at lower temperatures in or after a boiler or in a baghouse (Altwicker and Milligan, 1993). The PV measurements (Neulicht, 1987) provide output concentrations of the various congeners of dichlorobenzene through hexachlorobenzene. These were obtained by modified method 5 sampling at the exit of the secondary combustion chamber. The CCTL measurements report emission factors for benzene, monochlorobenzene and the congeners of dichlorobenzene obtained with a volatile organic sampling train. Within the limitations of the sparse and noisy data sets and the different inputs and combustor sizes, it might be possible to reconcile these measurements (Wagner and Green, 1993). In the present work we seek kinetic models of the formation of benzenes with various degrees of chlorination (n) from 0 to 6. Thus, in effect we sum the congener concentrations for n= 2,3, and 4.

## (2) Kinetic Modeling

Main frame computers, minicomputers and even PC's can now readily solve large numbers of simultaneous differential equations (Hindmarsh, et al.,1980). The number of simultaneous equations that must be solved depend upon the number of species followed. For example, to completely follow H and $H_2$ concentrations when hydrogen passes through a high temperature-time cycle would require 2x2 = 4 reactions. For hydrogen and oxygen reactions, if we only consider the stable species $H_2$, $O_2$ and $H_2O$ and the intermediate free radicals H, O, OH, and $HO_2$, which play major roles in flame chemistry, we must follow 7x7 = 49 reactions. If we follow only the forward reactions, this reduces to (7x6/2) + 7 = 28 reactions.

If one goes to hydrogen, oxygen and carbon chemistry, the number of reactions that could play a significant role increases greatly. For example if, as in this study, we simply follow the fate of 14 additional species, C, CH, $C_2$, $CH_2$, $CH_3$, $CH_4$, $C_2H_2$ and $C_6H_6$, CO, $CO_2$, CHO and $CH_2O$, we must, in principle, contend with a total of 21x21 = 441 reactions. If, for simplification, we follow only the forward reactions we still should consider (21x20/2) + 21 = 231 reactions. If we add C,H,O compounds to the reacting pool including all $C_2$, $C_3$, $C_4$, $C_5$, $C_6$, etc the number of species to be followed increases accordingly and the number of reactions proliferates tremendously.

Suppose now one wishes to also follow the evolution of simple chlorinated hydrocarbon species including, for example, Cl, $Cl_2$, HCl, ClO, CCl, CHCL, $CH_2Cl_2$, $CHCl_3$, $CCl_4$, $C_2HCl$, $C_2Cl_2$, $C_2ClO$, $CHCl_2$. The full reaction set will then total 34x34=1150 and the forward reaction set (34x33/2) + 34 = 595. Clearly this can only be carried out with large computers. If we begin to include aromatic compounds such as chlorobenzene the number of reactions proliferates again. If we also allow for all three body reactions the number of reactions that, in principle, must be followed becomes a monumental computational task requiring the use of supercomputers. We see that it would be easy to get lost in going from one dimension (H) to two dimensions (H and O), to three dimensions (H, O and C) and four dimensions (H,O,C and Cl) in the application of kinetic theory.

The traditional approach in the application of kinetics to toxic's generation has been to await the measurement of individual reaction rates before undertaking kinetic calculational experiments that include the species involved. Unfortunately, this tradition postpones to the distant future the day that kinetics can contribute to our understanding of the generation of organic toxics. An alternative approach is to take a very selective path in four dimensional kinetics space based upon experimental evidence and current understanding as to the key species of interest. This type of venture into H, O, C, Cl space has been an endeavor of the CCTL in several past studies (Green et al., 1991b, c; 1992, 1993, 1994). In these studies we have attempted to develop simple phenomenological kinetics models that are compatible with the results of CCTL experiments and experiments reported in the literature. Such phenomenological modeling must be carried out with great care since it is easy to get lost even in 2 space (H and O). As evidence we note that most engineering treatments of combustion still describe the combustion of $H_2$ by $2H_2+O_2 \rightarrow 2H_2O$ and ignore the essential role of free radicals.

## (3) The CCTL Kinetic Modeling Efforts

In the CCTL kinetic studies, we have taken a standard (relatively old) collection of H, O and C reaction rates (Green et al., 1993) and considered very selective sets of additional reactions when PVC is added to the input. These selective sets define our models. We have focused our attention on simple models leading to the formation of chlorobenzenes. Since in some of our previous studies we became bogged down in the multiplicity of reaction rate choices, in our recent studies we have confined ourselves to very simplistic sets of chlorinated species. Our hope has been that we might gain some insight into the pattern of reaction rates needed to achieve order-of-magnitude agreement with the available experimental data. In this study we seek simple models that can describe the evolution of $C_6H_6$, $C_6H_5Cl$, $C_6H_4Cl_2$, $C_6H_3Cl_3$, $C_6H_2Cl_4$, $C_6HCl_5$ and $C_6Cl_6$. These are important aromatic toxics in themselves and also act as precursors to chlorinated dioxins and furans.

A chemical pyrolysis/kinetics code, originally developed for modeling coal-water-gas combustion (Green and Pamidimukkala, 1982; 1984), has been adapted for studying emissions from modular incinerators by providing synthetic data pertinent to the present application (Green , et al., 1994). In the present work, we continue to focus on the gas phase production of benzene ($B_o$) and chlorobenzene with various degrees of chlorination ($B_n$).

Following models of well stirred reactors, we assume that the hot gases in the incinerator are well mixed so that gas phase kinetics, rather than diffusion, limit the rates of reactions. Furthermore, to simplify the calculations, we impose a temperature-time profile roughly compatible with the PV and CCTL experimental data sets. The standard profile chosen here conforms to

$$T = T_0 \exp [t/t_1) - (t/t_2)] / [g + \exp (t/t_1)] \quad (1)$$

where $t_1$ = 0.6 sec, $t_2$ = 70 sec, g = 0.4 and $T_0$ =1900°F. This temperature-time schedule is followed for 4.7 seconds which on the average will take a gas parcel through the exit of the secondary chamber in the PV incinerator.

Nineteen basic molecules or free radicals were followed in the reactions, including atomic and molecular hydrogen, carbon and six hydrocarbons: H, $H_2$, C, $C_2$, CH, $CH_2$, $CH_3$, $CH_4$, $C_2H_2$ and $C_6H_6$, and nine molecules generated in combination with oxygen, including: O, OH, $HO_2$, $O_2$, $H_2O$, CO, $CO_2$, CHO and $CH_2O$. Sixty reaction rates between these species were included in the code for the basic H-C-O combustion reactions as compiled from various sources (Green, et al., 1993). Where reaction rates are not known, we have chosen order of magnitude estimates of reaction rates for complete classes of reactions. Thus the models we have investigated consist of the assembled chemical mechanisms and the system of reaction rates for the chlorinated species which are assumed.

To go from the solid, liquid and gaseous inputs into what has essentially been developed as a gas phase reaction code we have made several simplifying assumptions that have evolved from our prior work (Green and Pamidimukkala, 1982; 1984; Green, et al.,1991c). Thus, using representative ultimate C, H and O analyses we postulate that the input garbage may be taken as molecules that pyrolyze to tar and char in accord with a simple scheme:

$$\begin{aligned} &\text{garb} \rightarrow \text{tar} + 3CO_2 + 13H_20 \\ &\text{tar} \rightarrow \text{char} + 10CH_2 \\ &\text{char} \rightarrow 4C_2 + \text{Ash} \end{aligned} \quad (2)$$

where the Ash "atom" has a weight of 60. In effect, we assume that our garbage consists of $C_{21}H_{46}0_{19}Ash$ molecules with molecular weight 662. We also assume that liquid water input simply evaporates according to

$$H_2O(l) \rightarrow H_20(g)$$

## The Fate of PVC

In our earlier studies (Green, et al., 1991b,c) we reported the behavior of several kinetic models at a single peak temperature parameter and various levels of PVC input. In our simplistic models, PVC plastic is assumed to depolymerize and pyrolyze according to

$$PVC \rightarrow HCl + C_2H_2 = X + A_0 \quad (3)$$
$$PVC \rightarrow H_2 + C_2HCl = H_2 + A_1 \quad (4)$$

In this work we assume that $A_2$ ($C_2Cl_2$) is made by the reaction

$$A_1 + X \rightarrow A_2 + H_2 \quad (5)$$

We define for convenience

$$\begin{aligned} P_n &= C_3H_{3-n}Cl_n \\ F_n &= C_4 H_{4-n} Cl_n \\ B_n &= C_6 H_{6-n} Cl_n \end{aligned} \quad (6)$$

In three models considered, benzene and its chlorinated counterparts ($B_n$) are basically produced by the chemical combination of acetylene and its chlorinated counterparts ($A_n$) via propargyl ($P_n$) or four body ($F_n$) intermediates, or directly by 3-body capture reactions.

## (4) Prior Kinetic Phenomenology

In our prior kinetic studies of incineration (Green, et al., 1991b, c, 1994; Wagner and Green, 1993) the reaction rates involving chlorinated components are varied from model to model. In assigning these reaction rates, we first varied the parameters by trial and error until we found a set of reaction rates that gave reasonable $B_n$ magnitudes at a standard $T_o$. Here each Arrhenius activation parameter is placed in the form $E= \beta T_s = \beta 1255\,°K$ where $\beta$ can be positive or negative. Thus

$$k = A \exp(-E/RT) = A \exp(-\beta T_S/T) \quad (7)$$

We first used the code to obtain noise-free "data" for the variation of $B_n$ ($C_6H_{n-6}Cl_n$) vs x (=PVC) input. Then we formed the sum

$$B_t = \sum B_n \quad \text{where } n = 1 \text{ to } 6 \quad (8)$$

and found that $B_t(x,T_S)$ for all models generally increased rapidly with x input (Green, et al., 1993). We next launched a more ambitious effort to simultaneously fit all $B_n$ (x,T) and $B_t$(x, T) where target values of $B_n$ and $B_t$ were inferred from the PV data. While we succeeded with several of the models by choosing families of reaction rates we did not bring all of our models into conformity with reasonable $B_n$(x,T) and $B_t$(x,T) behaviors. After considerable puzzlement as to the sources of our difficulties, we concluded the following:

(1) The congener data available are too sparse and noisy to reliably infer seven functions of two variables.

(2) When the objective is to determine the influence of PVC upon chlorinated benzene production the choice of $B_t$ given by (8) as used in the PV study is not a good one. A better choice would be the chlorine weighted sum

$$B_W = \sum n B_n \quad \text{where } n = 1 \text{ to } 6 \quad (9)$$

which should track the amount of PVC or chlorine in a more reasonable way.

(3) The reaction sets used in our models (Green, et al., 1991b,c, 1993) were rather complex so that too many choices of unknown Arrhenius parameters were necessary. We essentially got bogged down in these choices.

As a result of these problems we decided recently to take a new approach in which we first chose a simplistic set of reactions for achieving reasonable $B_w(x, T)$ values with minimal adjustment of Arrhenius parameters (Green, et al., 1994). The results of this approach showed $B_w$ increasing with increasing PVC and decreasing with increasing temperature. In addition, plots of $B_w$ vs PVC showed a linear relationship while plots of $B_t$ vs PVC showed more complex behavior. This can be explained by the fact that $B_t$ does not maintain the chlorine atom balance.

### (5) Present Kinetics Analysis

In this study, we have summed the PV chlorobenzene congener data to obtain values for $B_2$, $B_3$ and $B_4$. For the 15 runs which gave such data we calculated median values for the di- through hexachlorobenzenes. The PV study contained no data on benzene ($B_o$) and monochlorobenzene ($B_1$). On the other hand the CCTL data (Green, 1992; Wagner and Green, 1993) contains measured emission factors for $B_o$, $B_1$ and $B_2$. By normalizing the CCTL $B_2$ to the PV $B_2$, the two data sets can be combined in what we refer to as our base case. These base case components $B_0$ through $B_6$ have reasonable magnitudes and reasonable relative concentrations.

Each of the four models produces benzene and its chlorinated counterparts via different reaction pathways. In the **Propargyl Model** it is assumed that the $C_3$ hydrocarbons are first formed by two-body reactions which then undergo a capture reaction leading to benzene ring formation (Miller and Melius, 1992). The output concentration of $B_n$ with input PVC's are assumed to result from the reactions

$$A_i + A_j = P_q + CH \qquad q = i + j = 0 - 3 \qquad (10)$$

and the capture reactions

$$P_q + P_r \dashrightarrow B_n \qquad n = q + r = 0 - 6 \qquad (11)$$

In the **Two-Body Acetylene Model** chlorobenzenes are assumed to be formed via 2-body reactions from acetylene like $C_2$ compounds and intermediate $C_4$ compounds via :

$$A_i + A_j \dashrightarrow F_m \qquad m = i + j = 0 \text{ to } 4 \qquad (12)$$

$$A_p + F_m \dashrightarrow B_n \qquad n = p + m = 0 \text{ to } 6 \qquad (13)$$

The **Three-Body Acetylene Model** assumes the chlorobenzenes are formed directly from acetylene-like molecules under low oxygen (sooting) conditions. From the acetylene-like compound we obtain chlorobenzenes via

$$A_i + A_j + A_k = B_n \,;\; n = i + j + k = 0 \text{ to } 6 \qquad (14)$$

In the **Successive Chlorination Model**, we assume that benzene enters the system as a pyrolysis product of styrene (ST) which comes from polystyrene (PST), via the reactions

$$PST \dashrightarrow ST \qquad (15)$$

$$ST \dashrightarrow B_0 + A_0 \qquad (16)$$

We also allow for the pyrolysis of styrene via

$$ST \dashrightarrow 4\,A_0 \qquad (17)$$

We next assume that a system of successive chlorinations occurs via

$$B_{n-1} + HCl \dashrightarrow B_n + H_2 \qquad (18)$$

This is the only model where PST was included to produce chlorinated benzenes. HCl, again, is a product of the breakdown of PVC.

### (6) Results of The Simplistic Kinetic Models

In our most recent kinetic modelling effort, (Green, et al., 1994) we finally found simplistic models that gave reasonable behaviors for $B_w(x,T)$. In the course of this effort, we found another possible explanation of the questionable conclusion that chloro-organic outputs are not correlated with PVC input. Essentially (in our notation) the PV analysis for the case of chlorobenzene used the sums $B_t = \sum B_n$, n = 2 to 6 (since they did not measure monochlorine, n=1). They found poor correlations between PVC and $B_t$. While carrying out our analysis of the Propargyl model results it became obvious that when there is substantial production of the higher congeners ($B_4$, $B_5$ and $B_6$) the correlations between PVC (or its surrogate HCl) and $B_t$ can be poor. On the other hand if one seeks correlations between the chlorine weighted sum of chlorobenzenes, i.e.,

$$B_w(x,T) = \sum nB_n(x,T)$$

one should expect a good correlation since chlorine atoms are preserved in $B_w$. We observed this behavior in our modeling efforts and also found that we could achieve reasonable dependence upon temperature with all of our four models. However, only the Propargyl model gave a promising distribution of $B_n$ with respect to n, the degree of chlorination. The other three models gave results that declined orders of magnitude with n. In the present work we have applied additional effort to this issue with the hope of finding a unique model that not only generated a reasonable $B_w(x,T)$ but also reasonable $B_n(x,T)$ for n=0 to 6.

In connection with this effort we first used the PV congener concentrations together with the CCTL emission factors to arrive at a base case set of $B_n$ concentrations for a standard temperature time pattern. The target base distribution arrived at was $B_o = 6$, $B_1 = 2$, $B_2 = 1$, $B_3 = 0.5$, $B_4 = 0.5$, $B_5 = 0.2$ and $B_6 = 0.1$ (all in units of $10^{-12}$ kilomoles per dry standard cubic meter). By trial and error adjustment, we found, as we suspected, that we could bring the constrained propargyl models within an order of magnitude of these target numbers. However, to our surprise, by applying what we learned on sensitivity of output-to-parameter adjustment, we found that we could also bring the other three models into order-of-magnitude agreement. Table 1 to Table 4 give parameters that accomplished this. Where parentheses are shown they represent the numbers of reactions assumed to have the same reaction rate. Table 5 gives the concentrations generated with these parameters at 4.7 seconds.

Table 1. Propargyl Model
Reaction Rates

| | A | β |
|---|---|---|
| PVC --> X + $A_o$ | 1D+2 | +2 |
| --> $H_2$ + $A_1$ | 5D-4 | -2 |
| A+X --> $H_2$+$A_2$ | 1D+5 | +2 |
| A+A --> P +CH(5) | 5D+7 | +2 |
| P+A --> B(10) | 1D+8 | +2 |

Table 2. Two-body Model
Reaction Rates

| | A | β |
|---|---|---|
| PVC --> X + $A_o$ | 1D+2 | +2 |
| --> $H_2$ + $A_1$ | 2D-4 | -2 |
| A+X --> $H_2$+$A_2$ | 2D+5 | +2 |
| A+A --> F (6) | 5D+7 | +2 |
| A+F --> B (15) | 1D+8 | +2 |

Table 3. Three-body Model
Reaction Rates

| | A | β |
|---|---|---|
| PVC -->X+$A_o$ | 1D+2 | +2 |
| PVC -->$H_2$+$A_1$ | 4D-4 | -2 |
| $A_1$+X-->$H_2$+$A_2$ | 1D-5 | +2 |
| A+A+A-->B(10) | 1D13 | +2 |

Table 4. Successive Chlorination Model
Reaction Rates

| | A | β |
|---|---|---|
| PVC -->X+$A_o$ | 1D+2 | +2 |
| PST --> ST | 1D+2 | +2 |
| ST --> 4$A_o$ | 1D+4 | +2 |
| ST -->$B_o$+$A_o$ | 1D+0 | +2 |
| $B_{n-1}$+X -->$B_n$(6) | 1D+4 | -1 |

Table 5. Chlorobenzene Concentrations
($10^{-12}$kmole/dscm)

| | Base | P | 2B | 3B | SC |
|---|---|---|---|---|---|
| $B_o$ | 6 | 4.4 | 4.3 | 15 | 1.4 |
| $B_1$ | 2 | 0.2 | 3.0 | 0.2 | 2.8 |
| $B_2$ | 1 | 1.8 | 2.1 | 0.1 | 2.7 |
| $B_3$ | 0.5 | 0.5 | 0.4 | 2.6 | 1.8 |
| $B_4$ | 0.5 | 0.9 | 0.3 | 0.9 | 0.9 |
| $B_5$ | 0.2 | 0.2 | 0.1 | 0.6 | 0.3 |
| $B_6$ | 0.1 | 0.08 | 0.04 | 0.5 | 0.2 |

We have not made a serious effort to tune the parameters to achieve optimum matches between calculated and base case numbers. From the great sensitivity of the calculated values to parameter changes, we have concluded that such optimization could be quite productive but had best be carried out with an automatic search routine as the senior author has done on several other scientific problems. This would entail working with a larger computer and certainly should be a worthwhile effort when more comprehensive and accurate data become available. At that point it would be purposeful to compare the inferred reaction rates with reaction rates projected by quantum chemical modeling, and any reaction rates involving organic chlorination processes that might be measured.

Figure 1A illustrates the time dependence of the major species concentrations and Figure 1B the free radicals when the reaction rate for Eq. (4) is set to zero (Universal Model). This eliminates the channels for $A_n$, $P_n$, $F_n$ and $B_n$ production in the Propargyl, Two-body and Three-body models and hence mainly gives the big picture on H, C, and O chemistry. This chemistry pattern does not change appreciably in going to any of our models since all of them produce only trace quantities of organo-chlorines. It should be noted that our CO output is high which we suspect is due to our use of a limited H, C, and O reaction rate set and stoichiometric conditions. Figure 2A shows the intermediate chlorinated species and 2B the chlorobenzenes for the Propargyl model. Figures 3A and 3B show the corresponding chlorinated species for the Two-body model , Figure 4 for the Three-body model and Figure 5 for the Successive Chlorination model. The intermediate chlorinated compounds $A_n$, $P_n$ and $F_n$ should be viewed as surrogates for all other chlorinated hydrocarbon species and the $B_n$ concentrations the surrogates for all aromatic species. The Three-body and Successive Chlorination models have fewer intermediate chlorinated species and to this extent are less realistic.

Studies of benzene and soot formation in hydrocarbon combustion give additional support to the propargyl model approach and possibly the Two-body model (Miller and Melius, 1991).The best $B_n$ pattern we have obtained with the Three-body model is not as satisfying as patterns from the other three models. This pattern might be improved by using a fifth set of adjusted reaction rate parameters as is used in the other three models. However, we tend to believe that three body capture reactions between trace species $A_o$, $A_1$, and $A_2$ are not very likely

## (7) The Fate of Inorganic Chlorides

We have recently investigated the possibility of inorganic Cl forming HCl under typical medical waste type combustion conditions with the help of HCl measurements from crematoria (Green, et al., 1994). Granting the chlorine content of bodies is approximately 0.18%, one would approximately expect an HCl emission factor of 1.8 gr/kg if inorganic chlorine were completely converted to HCl. However a tabulation of HCl emissions data from crematoria in California and Florida reveals that all but one test shows that less than 20% of the Cl available is emitted as HCl (Green, et al.,1994). Thus the data suggests that only a small percentage of inorganic forms of Cl are converted to HCl.

More compelling evidence as to the release of HCl from combustors containing sources of inorganic chlorine comes from tests of HCl emissions from kraft mills (Someshwar, 1994). Results from many mills show that a small percentage of the Cl in the liquor was emitted as HCl. In addition, continued studies of these kraft mills indicate that the HCl is produced through secondary reactions involving $SO_2$, gas phase alkali chlorides, water vapor and $O_2$ forming alkali sulfates and HCl (g) below about 900°C. Someshwar (1994) concluded that under normal operating conditions, the HCl produced is less than 5% of the total inorganic chloride input to a furnace.

## (8) Discussion

The senior author (AESG) would like to cite past experience to illustrate the goal of this effort. After World War II, with its dramatic nuclear ending, a large scale experimental and theoretical effort was mounted to reach an understanding of the basic interaction between nucleons (neutrons and protons). As abundant and reliable experimental data with high energy accelerators accumulated it became evident that the N-N interaction is very complex, showing strong dependencies on the distances between two nucleons, the relative orientation of their spins, the orientation of the spins with respect to the radius joining the two nucleons as well as the angular momentum and velocity of relative motion. Even with some theoretical understanding of the interaction at long ranges it was necessary to use models with 40 adjusted parameters to organize the experimental data. Then in the 1963-1966 time-frame several research groups in America and Japan developed physical models that could approximately represent the data with only one to five parameters. Soon afterwards an international conference on this topic (Green, et al., 1967) revealed that all of these successful models explained the complex behavior of the N-N interaction in terms of the approximate cancellation of major static interactions which left relativistic interactions as major residues. Thus the ability to represent complex data with only a few adjusted parameters signalled a major breakthrough in our physical understanding.

In kinetic modeling of toxics produced in combustion, unfortunately, we do not now have a large and reliable data set to attempt to distill into phenomenological models with only a modest number of adjustable parameters. If we get to that point then, in principle, it should be possible to relate the adjustable

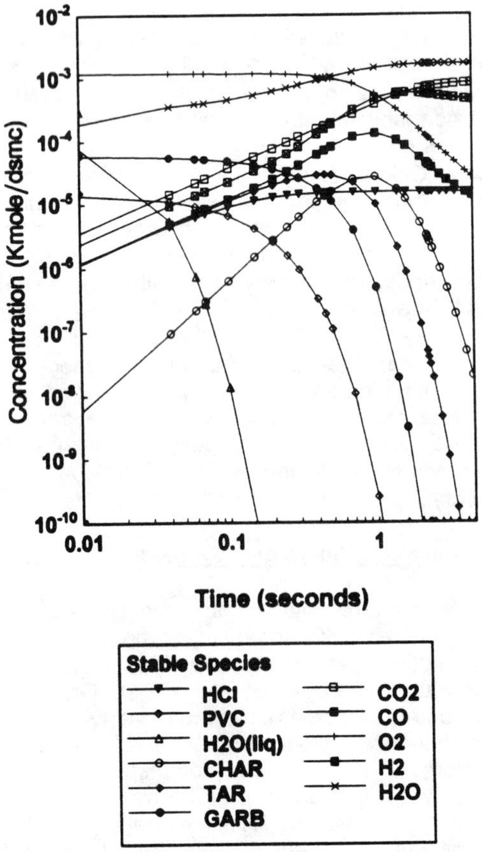


**Figure 1A. Universal Model time dependence of stable species .**

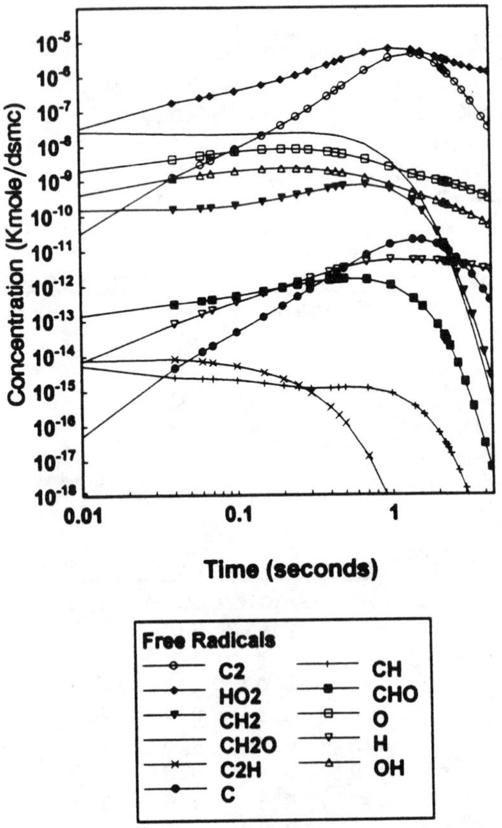


**Figure 1B. Universal Model time dependence of free radicals.**

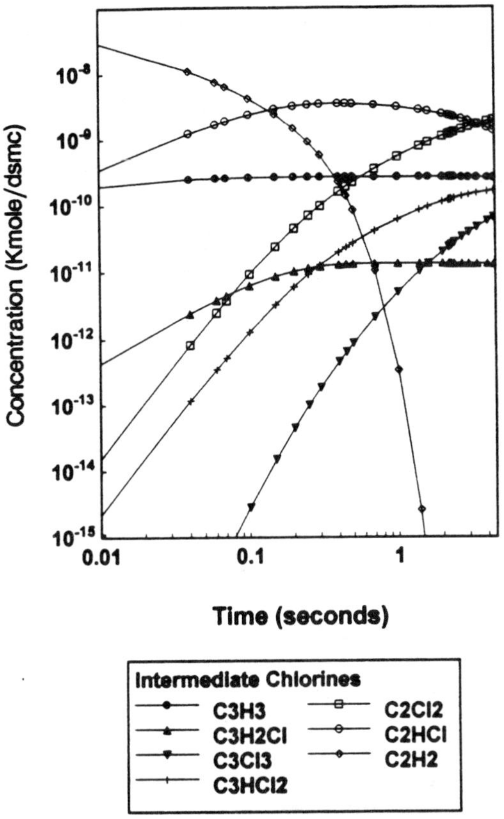


**Figure 2A. Propargyl Model time dependence of intermediate chlorinated species.**

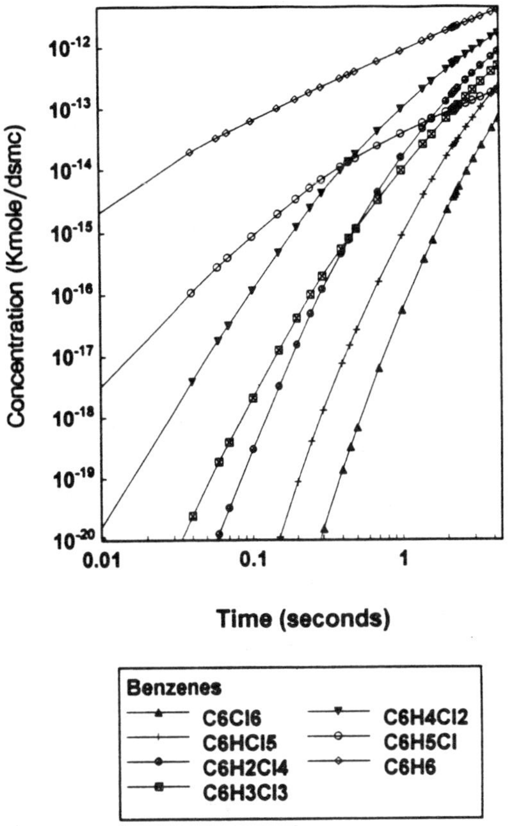


**Figure 2B. Propargyl Model time dependence of benzenes.**

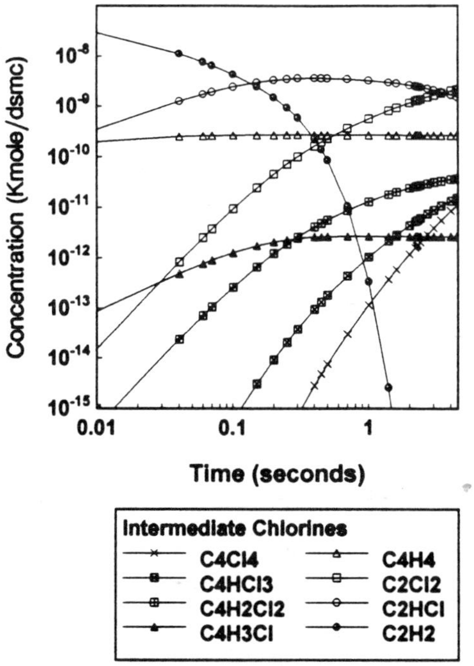


**Figure 3A. Two-Body Model time dependence of intermediate chlorinated species.**

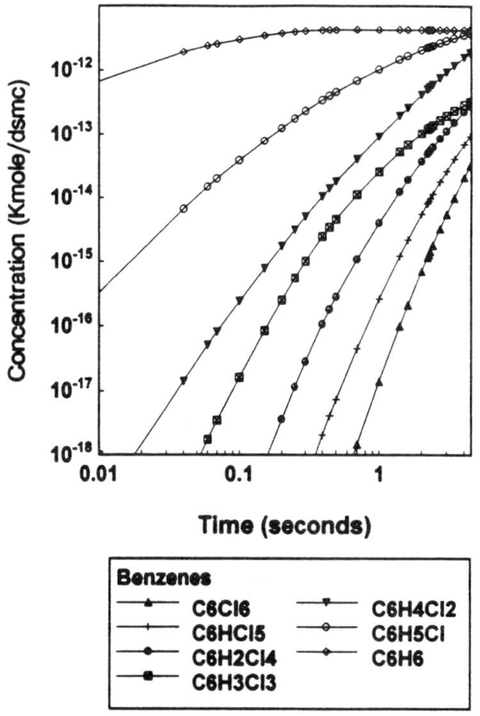


**Figure 3B. Two-Body Model time dependence of benzenes.**

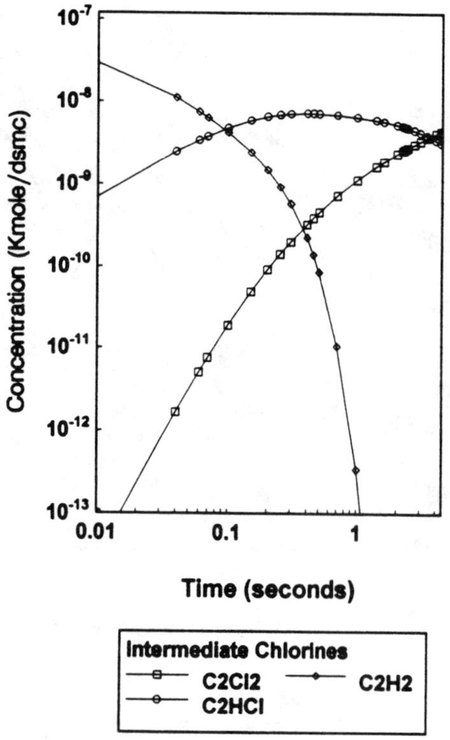


Figure 4A. Three-Body Model time dependence of intermediate chlorinated species.

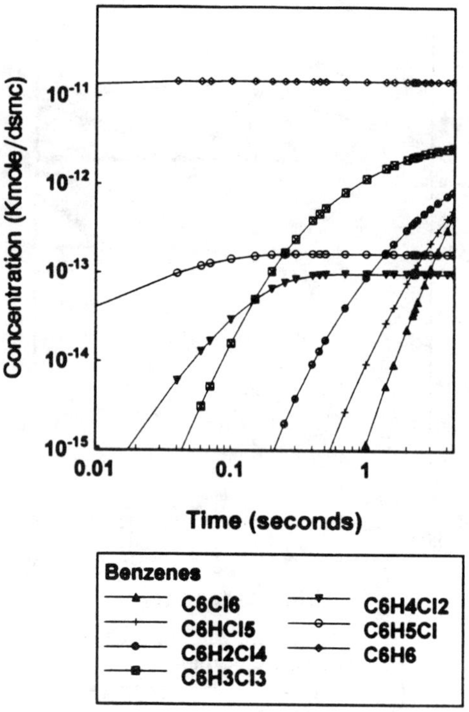


Figure 4B. Three-Body Model time dependence of benzenes.

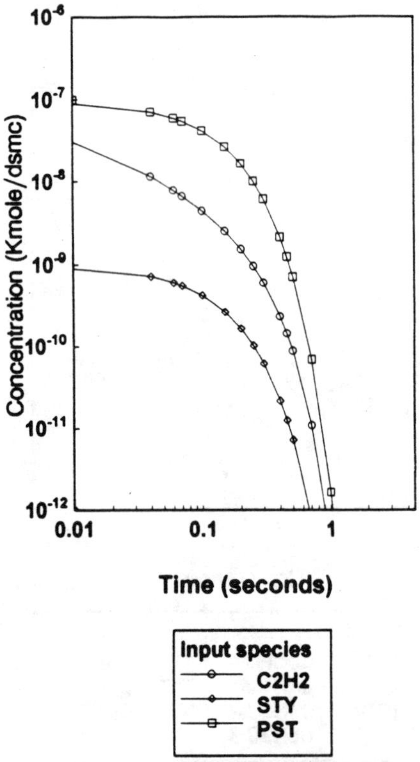


Figure 5A. Successive Chlorination Model time dependence of PST, Styrene, and Acetylene

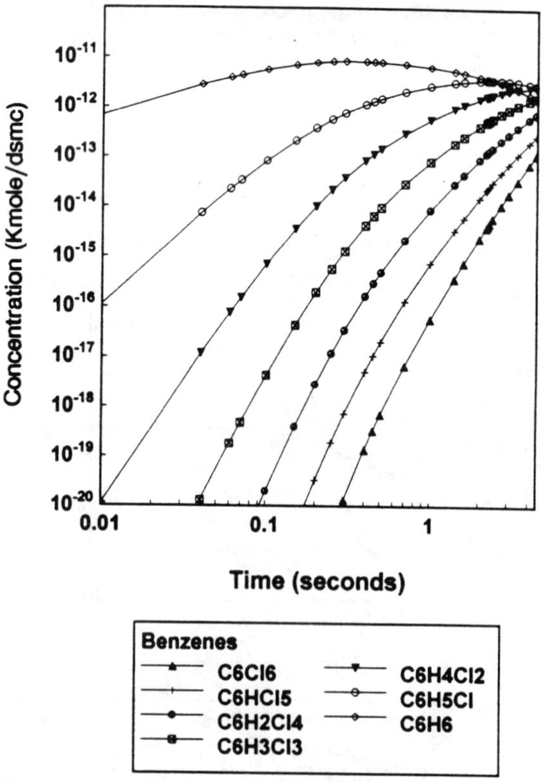


Figure 5B. Succesive Chlorination Model time dependence of benzenes.

parameters to the fundamental makeup of the atomic and molecular species involved. In the meantime some useful clarifications might be achieved with the aid of simplistic models which are economical in adjusted parameters that can account for general patterns of experimental observations. It is in this spirit that we have continued our kinetic modeling efforts. We are gratified that we are now close to defining models that give reasonable $B_n(x,T)$, seven functions of the variables x = PVC and T. All of the models we have found indicate that an increase in organic chlorine input leads to an increase in chlorinated organic output just as is intuitively expected.

## Policy Implications

The main focus of our study of chlorobenzene formation relates to these aromatic compounds as precursors to dioxins and furans. However, chlorinated benzenes themselves induce a wide range and severity of health effects (Peirano,1991; Hileman, 1993). Animal studies indicate a trend of increasing toxicity or carcinogenicity with increased chlorination of the benzene ring. In addition evidence that organochlorines have adverse effects on endocrine, immune and nervous systems of exposed fetuses has grown during the past few years (Hileman, 1993). The Chlorine Institute acknowledges problems with a few selected organo-chlorines but believe these problems can be dealt with by selective phase-outs. However, environmental groups tend to favor a broader-based phase-out of organo-chlorines.

Recently the German Federal Government took far-reaching legislative action to further reduce dioxin releases into the environment. Among these are strict limit values and bans on certain chemicals or restrictions on their use. Among the further recommendations proposed by the Federal Health Office (BGA), the Federal Environmental Agency (UBA) and the Joint Working Group of the Federation and the Lander on Dioxin are limitations on the use of halogenated plastics (Schultz, 1993).

It is interesting to note the contrast between the current positions of Germany and the United States on the halogenated plastics, particularly PVC. Commercial production of PVC began in 1933 in both Germany and the United States. The closest recent "official" position on PVC in the US comes out of the Department of Energy via the National Renewable Energy Laboratory (NREL) in a report (Solid Waste Assoc., 1993) prepared by the Solid Waste Association of North America (SWANA). The report, "developed through a synthesis of available information," leads to the conclusion that: "the presence or absence of PVC in the MSW stream will not reduce the need to employ control measures" .This conclusion essentially implies that there is no point to phasing out halogenated plastics or separating inputs to incinerators.

The fact that a synthesis of the literature by Germany (Schultz, 1993), where PVC was first synthesized, comes essentially to a conclusion contrary to that of SWANA (Solid Waste Assoc., 1993) is another example of the well known fact that a synthesis is often dependent upon the synthesizer. Since no scientific author or authors for the SWANA report are identified, no new data or reanalysis of old data is presented and only a selective set of literature is cited, it would appear to be unwise to use the SWANA report as a basis for public policy decisions in the United States. Instead it would be prudent to use pollution prevention or front end measures in incineration rather than complete reliance on end of pipe control measures.

The Declaration of the International Union of Air Pollution Prevention Association (IUAPPA) strongly supports greater reliance on the front end approach (Green, 1992c; Freeman, et al., 1992; Green, 1992a; Green, 1992b) as do Green Products or Green Technologies legislation now under consideration by the United States Congress. If Germany phases out medical-grade PVC products whereas the US does not and Germany's synthesis holds up better than the SWANA synthesis, as appears likely, the United States would again lose out on an important environmental market.

The May 1994 decision of the United States Supreme Court on ash from municipal waste incinerators provides another reason to encourage the Pollution Prevention or Green Technologies approach. Essentially this decision disallows rules exempting municipal waste ash from regulation as a hazardous waste when tests show that it fails to meet non-hazardous waste standards. To avoid substantial increases in ash disposal costs will require much greater attention to the avoidance of toxic producing species that enter the municipal waste stream (or the medical waste stream).

Some of the considerations of this chlorobenzene study might also be relevant to the use of coal with high chlorine content. The combustion of coal is usually carried out at higher temperatures which will influence the chlorinated organic outputs. Nevertheless it would be well to apply knowledge based upon incineration studies to the combustion of other solid fuels. It might be noted that European Asean combustion circles are now viewing the technologies for the combustion of all solid fuels as closely related technologies. On the other hand, in the US, the combustion of coal, biomass and solid waste are mainly practiced by separate - often competing industries. The continuation of this tradition could also place the US in a weakened competitive position.

Finally we note that a forthcoming 2000-page report from scientists at the Environmental Protection Agency has concluded that cancer is not the most serious health hazard at common exposure levels. Of greater concern are subtle effects on fetal development and the immune system, that may be the result of very low levels of exposure.

## Acknowledgments

The authors wish to thank the Ferro Corporation, the Mick A. Naulin Foundation, the University of Florida Gatorade Fund, Supelco, Tekmar, The Eye Research Laboratory, $ECO^2$, and an anonymous donor for direct or in-kind project support. We also thank Gene Hemp, Gerald Schaffer, Winfred Phillips, and Weilin Chang for other forms of help and encouragement and Akiba Green for his technical assistance.

## REFERENCES

Altwicker , E.R. and Milligan, M.S., 1993, "Formation of Dioxins: Competing Rates Between Chemically Similar Precursors and DeNovo Reactions", *Chemosphere*, 27:301-307.

Bulley, M., 1990, "Medical Waste Incineration in Australia," in *Proc. of the 83rd Annual Meeting of Air & Waste Manage. Assoc.*, Paper 90-27.5, Pittsburgh.

Freeman, et al., 1992, "Industrial Pollution Prevention", *J. Air Waste Management Assoc.*,42: 618.

Green, A., Ed., 1981, *Alternative to Oil, Burning Coal with Gas*, University Presses of Florida, Gainesville, FL, pp. 1-140.

Green, A., 1986, "Clean Combustion Technology" in *Proceedings of Conf. 'There is No Away'.* University of Florida Law School, Gainesville, FL. EPA-NTIS.

Green, A. , 1992a, "Reconciliation of Divergent Views on Municipal Solid Waste by Pollution Prevention", Air and Waste Management Association Meeting, Paper 92-5709.

Green, A. 1992b, "Response: Industrial Pollution Prevention", *J. Air Waste Management Assoc.*, 42: 1660.

Green, A. and Pamidimukkala, K., 1982, "Kinetic Simulation of Combustion of Gas/Coal and Coal Water Mixtures," in *Proceedings of the 1982 1st Conf. on Combined Combustion of Coal and Gas*, Cleveland, Ohio.

Green, A. and Pamidimukkala, K., 1984, "Synergistic Combustion of Coal and Natural Gas," *Energy, the International Journal*, 9, 477-484.

Green, A., MacGregor, M. H., Wilson, R., 1967, "International Conference on the Nucleon-Nucleon Interaction" Reviews of Modern Physics, 39-3:495-717.

Green, A., Batich, C., Powell, D., et al., 1990, "Toxic Products From Co-Combustion of Institutional Waste, " in *Proc. of 83rd Annual Meeting of the Air & Waste Management Association*, Pittsburgh, PA, paper 90-38.4.

Green, A. Batich, C, Wagner, J., et al., 1990, "Advances in Uses of Modular Waste to Energy Systems," Presented at the Joint Power Gen. Conf. ASME-IEEE, and published in A. Green, and W. Lear, Eds., 1990, *Advances in Solid Fuels Technologies*, Fuels and Combustion Technology (FACT) Division of ASME, New York, NY

Green, A., Wagner, J. and Lin, K., 1991a, "Phenomenological Models of Chlorinated Hydrocarbons," *Chemosphere* Vol.22, Nos. 1-2, pp. 121-135.

Green, A., Wagner, J., Saltiel, C., et al., 1991b, "Medical Waste Incineration with a Toxic Prevention Protocol," in *Proc. of the 84th Annual Meeting of the Air & Waste Management Association*, Vancouver, B. C., June 16-21, , Paper 91-33.5.

Green, A., Wagner, J., Saltiel, C., et al., 1991c, "Pollution Prevention and Institutional Incineration," in *Proceedings of the ASME Solid Waste Processing Conference*, Detroit.

Green, A., Ed., 1992, *Medical Waste Incineration and Pollution Prevention*, Van Nostrand Reinhold, New York, NY.

Green, A., Mahadevan, S., and Wagner, J., 1993, "Chlorinated Benzene Formation in Incinerators", in *Combustion Modeling, Cofiring and NOx Control*, Fuels and Combustion Technology-FACT Vol 17 A. Gupta et al., Eds. published by American Society of Mechanical Engineers, New York, NY..

Green, A., Miller, T., Baruch, S., et al. 1994, "Incinerator Inputs and Chlorobenzene Outputs", in *Proceedings of the 87th Annual Meeting of Air & Waste Manage. Assoc.*, 94-FA 154.01, Cincinnati, OH.

Hileman, B., "Concern Broadens Over Chlorine and Chlorinated Hydrocarbons",*Chemical and Engineering News*, April, 19:11-20.

Hindmarsh, A.C., LSODE and LSODI, 1980 "Two Initial Value Ordinary Differential Equation Solvers", ACM SIGNUM Newsletters, Vol.15, 10-11.

Miller, J. and Melius, C., 1992, "Formation of aromatic compounds in flames" Combustion Research Facility, Sandia National Lab, Livermore, CA.

Neulicht, R., 1987, "Results of the Combustion and Emissions Research Project at the Vicon Incinerator Facility in Pittsfield, Mass.", New York State Energy Research and Development Authority Report 87-16, June.

Peirano, B., 1991, Health Assessment of Chlorobenzenes, EPA/600/58-841015F.

Schultz, D.,1993, "PCDD-PCDF - German Policy and Measures to Protect Man and the Environment", *Chemosphere*, 27: 501-507.

Solid Waste Assoc. of North America, 1993, "Polyvinyl Chloride Plastics in Municipal Solid Waste Combustion: Impact Upon Dioxin Emissions", prepared for the National Renewable Energy Laboratory and US Department of Energy, NREL/TP-430-5518, April.

Someshwar, A. V., 1994, "A Study of Kraft Recovery Furnace HCl Acid Emissions" , NCASI Technical Bulletin, to be published.

Wagner, J., and Green, A., 1993, "Correlation of Chlorinated Organic Compound Emissions from Incineration with Chlorinated Organic Input", *Chemosphere*, 26: 2039-2054.

FACT-Vol. 18, Combustion Modeling, Scaling and Air Toxins
ASME 1994

# STUDIES OF POLLUTANT EMISSIONS IN A WELL STIRRED REACTOR

**Joseph Zelina**
Department of Mechanical and Aerospace Engineering
University of Dayton
Dayton, Ohio

**Dilip R. Ballal**
Department of Mechanical and Aerospace Engineering
University of Dayton
Dayton, Ohio

## ABSTRACT

Emissions of CO and $NO_x$ from coal-fired and/or the combined steam-gas turbine cycle power plants pose a threat to our environment. To achieve a pollution-free environment, CO and $NO_x$ emissions from a well stirred reactor (WSR) were investigated. The results of the investigation were applied to predict emissions and to explore performance improvements of combustion burners employed in power generation systems.

The combustion efficiency of a WSR represents the upper range of values which the burners used in power generation combustion equipment should approach. Existing prediction methods (devised for high combustor pressure) require only slight modifications to provide good predictions of CO and $NO_x$ emissions. Predicted values of $NO_x$ were lower than the WSR measurements because the intense mixing in the WSR eliminates cool spots in the combustion zone which are prevalent in practical combustion devices.

## NOMENCLATURE

| | |
|---|---|
| A | = area ($m^3$) |
| $C_p$ | = specific heat |
| f | = fraction of total air for combustion |
| $D_o$ | = fuel drop diameter |
| $m_a$ | = mass flow rate (kg/sec) |
| P | = pressure (kPa) |
| $\Delta P$ | = pressure drop (Pa) |
| Re | = Reynold's Number |
| T | = temperature (K) |
| $V_c$ | = reactor volume ($m^3$) |
| $\phi$ | = equivalence ratio |
| $\lambda_{eff}$ | = effective value of evaporation ($m^2/s$) |
| $\eta_c$ | = combustion efficiency |
| $\rho$ | = density |
| $\tau$ | = residence time |

**Subscripts**

| | |
|---|---|
| c | = combustion |
| f | = fuel, flame |
| m | = mixing |
| r | = reaction |
| e | = evaporation |
| st | = stoichiometric |
| $\theta$ | = reaction rate-controlled |

## INTRODUCTION

Emissions of CO and $NO_x$ from coal-fired and/or the combined steam-gas turbine cycle power plants are a major contributor to atmospheric pollution, acid rain, and ozone depletion. To decrease CO emissions, combustion efficiency should be as high as possible; to decrease $NO_x$ emissions, natural gas reburning is often employed in pulverized coal-fired boilers. Also, many industrial gas turbine combustors today have a high-output low-volume design because they are aero-engine derivatives. In these combustors, the overall kinetics of chemical reactions is controlling. A well stirred reactor (WSR) is a laboratory idealization of such a gas turbine combustor and, hence, its study is of current and future importance. Further, in low-output combustion equipment, such as a furnace, boiler, or incinerator, the actual chemical reaction time is negligible in comparison to fuel evaporation and mixing processes. However, the Clean-Air Act of 1991 is forcing the burning of natural gas fuel and the application of sophisticated aerospace-derived technology to improve these ground-based stationary combustion systems. Thus, the study of combustion in a WSR is relevant to current and future power generation systems.

In this paper, we describe the semi-empirical equations for practical gas turbine combustors derived by Lefebvre (1984) and adapted for power-generation systems by Ballal (1993) to predict combustion efficiency, CO, and $NO_x$, emissions. Next, we evaluate how well the theory and WSR measurements agree. Finally, we discuss reasons for any discrepancies and suggest future improvements in combustor modeling.

## THEORY

### Combustion Efficiency

In a continuous-burn combustion system, the main factors affecting the level of combustion efficiency are evaporation (or vaporization) rate (ER), mixing rate (MR), and reaction rate (RR). In practical combustion burners, the maximum heat release is rarely controlled by all three rates. Rather, *one* of the three key rates participates in determining overall combustion efficiency. Lefebvre (1984) has shown that:

***RR controlled***

$$\eta_\theta = f\left(\frac{P^{1.75} V_c \exp\left(\frac{T}{300}\right)}{m_a}\right) \tag{1}$$

***MR controlled***

$$\eta_m = f\left(\frac{P_3 A_c}{m_a T^{0.5}}\right)\left(\frac{\Delta P}{P_3}\right)^{0.5} \tag{2}$$

***ER controlled***

$$\eta_e = \left(\frac{\lambda_{eff} \rho_g V_c}{f_c m_a D_o^2}\right) \tag{3}$$

$$\lambda_{eff} = \left(\frac{1.76 k \ln(1+B) Re_{D_o}^{0.5}}{C_p \rho_f}\right). \tag{4}$$

Eq. (1) for the *RR-controlled* system (such as a WSR) is used only when the combustion zone is supplied with natural gas or well-atomized, refined, light distillate fuel intensely premixed with air. For most natural-gas-fired boilers and furnaces, the combustion system is *mixing controlled* and Eq. (2) would be appropriate. When residual oil or pulverized solid waste is burned, the combustion process is *evaporation controlled* and Eq.(3) will apply.

Examination of Eqs. (1) through (4) shows that combustion efficiency will reach a theoretical maximum for RR-controlled systems and will remain minimum for ER-controlled systems. Eq. (3) states that combustion efficiency is improved by increasing the fuel volatility, turbulence intensity, combustion volume, and gas pressure, and is impaired by increases in air mass flow rate and mean drop size of the fuel slurry. Therefore, improvements in the combustion efficiency of an ER-controlled system can be achieved primarily by decreasing the fuel droplet or pulverized fuel particle size to as low a value as the costs will allow. For example, Ballal (1993) has estimated that increasing the mean drop size from 20 percent to 60 percent decreases the combustion efficiency from 40 percent to 25 percent. Likewise, Eq. (2) illustrates that increasing the pressure drop will facilitate an increase in the combustion efficiency of a MR-controlled combustion system.

### CO Emissions

CO and UHC (unburned hydrocarbons) emissions are important because they represent a direct manifestation of combustion *inefficiency*. If the combustion zone in a boiler, furnace, or ground-based combustion burner operates fuel-rich, has inadequate mixing, or is prematurely quenched by cold air, large quantities of CO (which is relatively resistant to oxidation) will be formed due to the lack of oxygen needed to complete the reaction to $CO_2$. UHC emissions include fuel which emerges at the combustor exit in the form of droplets or vapor, and the products of the thermal degradation of the parent fuel into species of lower molecular weight (such as methane and acetylene). UHC emissions are associated with poor atomization, inadequate burning rates, and premature quenching. In general, UHC emissions parallel those of CO. Any modification that decreases CO emissions will also decrease UHC emissions.

In deriving an empirical model of CO emissions, it is assumed that the CO concentration is proportional to the product of three parameters,( i.e., $[CO] \propto \tau_r^{-1}\, RR.\, MR$). By definition, we have the residence time $\tau_r \propto (P\, V/m_a\, T)$, $RR \propto P^n \exp(cT)$, and $MR \propto (\Delta P/P)^n$. The above relationships embody the main variables of combustion volume, pressure loss (or turbulence energy), inlet pressure, combustion temperature, and air mass flow. Using these relationships, Lefebvre (1984) derived a semi-empirical correlation for CO that fits the experimental data from a large number of gas turbine combustors. This relationship is:

$$CO = \left(\frac{C_1 m_a T_c \exp(c T_c)}{(V_c - V_e)\left(\frac{\Delta P}{P}\right)^{0.5} P^{1.5}}\right) \tag{5}$$

where

$$V_e = \left(\frac{0.55 f_c m_a D_o^2}{\rho_c \lambda_{eff}}\right). \tag{6}$$

In Eq. (5), $C_1$ = 86 and c = -0.00345 are empirical constants. Note that for a RR-controlled system, $V_e$ = 0. Therefore, any improvement in fuel evaporation either via a decrease in the mean drop size or an increase in fuel volatility, both of which cause a decrease in $V_e$, will produce decreased CO emission. For example, Ballal (1993) estimated a 50 percent increase in normalized CO emissions when $V_e$ is increased from 25 percent to 50 percent. Also, the inclusion of a pressure-loss in Eq. (5) suggests that the higher turbulence created by an increase in pressure drop promotes better mixing and eliminates CO formation. For example, an increase in pressure drop from 25 percent to 75 percent causes the normalized CO emissions to decrease from 90 percent to 50 percent. In this manner, the importance of rapid fuel evaporation (volatility), fine fuel slurry drop size, and high level of turbulent mixing to decreasing the CO and UHC emissions is made clear.

## $NO_x$ Emissions

For a combustion zone, a key factor controlling $NO_x$ is the flame temperature. At low power or off-load conditions, $NO_x$ emissions are quite small. At high power, peak-load conditions, $NO_x$ emissions are prominent. For practical combustors, Lefebvre (1984) has derived that:

$$NO_x = \left( \frac{C_2 P^{1.25} V_c \exp(0.01 T_{st})}{m_a T_c} \right). \tag{7}$$

In Eq. (7), $C_2 = 9 \times 10^{-8}$ and $T_{st}$ is important for heterogeneous fuel-air mixtures, whereas the bulk temperature $T_c$ takes into account the residence time in the combustion zone. Ballal (1993) has calculated that increasing the temperature by 30 percent (from 1000K to 1300K) increases $NO_x$ emissions 15-fold. Almost a twofold decrease in $NO_x$ emissions can be achieved, either by diluting the combustion zone (i.e., increasing the air supplied) or by decreasing residence time (i.e., decreasing volume). Experimental data on practical combustion burners used for power generation are required either to verify or to refute these predictions.

## WSR DESIGN

### WSR Test Facility

The first WSR was developed by Longwell and Weiss [1955] at Exxon Corporation ,Linden, NJ. Other researchers have designed reactors of various geometries; this past work is summarized by Zelina and Ballal (1994). We capitalized on the findings of previous researchers and designed a 250-ml toroidal WSR. Some distinguishing design features of our reactor are: (1) high-pressure operation up to 5 atm. and use of zirconia-coated refractory materials capable of operation up to 2300K; (2) intense mixing caused by multiple-jet injection to ensure kinetically controlled combustion even at high temperatures; (3) large reactor volume and short residence time which eliminate possible surface reactions; and (4) use of a flow straightener at the exit to the WSR to produce an accurate PFR (plug flow reactor). The design details of this toroidal WSR are given by Zelina and Ballal (1994).

Fig. 1 illustrates the WSR test facility and its associated instrumentation: thermocouples, gas sampling probes, emissions analyzers, and gas chromatography-mass spectrometry (GC-MS) and Fourier Transform Infrared Spectroscopy (FTIR) equipment. Two Type B, Pt-Rh thermocouples (0.01" dia. bare wire) rated at 2100K and two Type K, Cr-Al thermocouples rated at 1660K were installed in the WSR and PFR, respectively. Air-cooled quartz gas-sampling probes were used to withdraw pollutant samples from the WSR. These gas samples pass through a water trap and a dehumidifier, then are fed into various emissions analyzers.

The HORIBA emissions analyzers comprise the following units: model MPA-510 oxygen analyzer (0 to 50 percent), model FIA-510 total hydrocarbon analyzer (0 to 10,000 ppm carbon), model VIA-510 CO (0 to 5000 ppmV), model VIA-510 CO (0 to 20 percent) and $CO_2$ (0 to 100 percent) analyzer, and model CLA-510 SS NO and NOx analyzer (0 to 2000 ppmV). A gas sample of 500 ml/min is drawn in by each HORIBA unit and analyzed for various pollutant species.

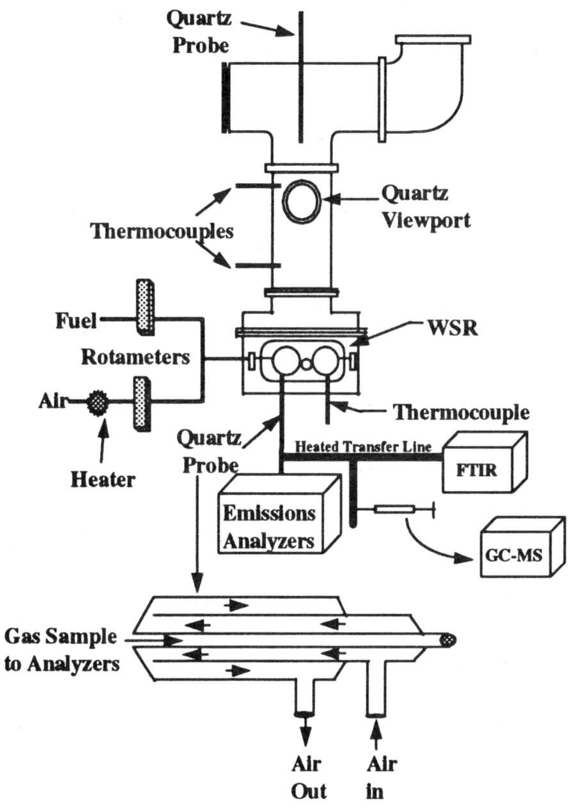


Fig. 1: WSR test facility with instrumentation.

### Test Conditions

As is normally the case for burners employed for power generation, the WSR test facility was operated slightly above atmospheric pressure (20 psig) and at room temperature (300K). In this facility, the air rotameter is rated at 0 to 800 SLPM and fuel supply is in the 0 to 100 SLPM range. The above flow conditions permit the operation of the reactor over a broad range of equivalence ratios $\phi = 0.4$ to 2.5 (paraffinic hydrocarbon fuels), residence time $\tau_c = 3$ to 12 ms, a cold mixture velocity in the toroid $\cong$ 100 m/s, mixing time $\tau_m \cong 0.1$ ms, and reactor temperature $T_f = 1200$ to 2100K.

### Error Analysis

Both the fuel flow and airflow were monitored to within 2 percent. The combined error produced an uncertainty of 5 percent in equivalence ratio and 2 percent in mean velocity measurement. The thermocouple temperature measurements, after correcting for heat loss were accurate to within 50K. The HORIBA emissions analyzers quote an accuracy to within 1 percent of full scale.

## RESULTS AND DISCUSSION

### Combustion Efficiency

Fig. 2 shows the measured combustion efficiency of the toroidal WSR as a function of reactor mixture ratio. The combustion efficiency in the WSR is more than 99 percent over a wide operating mixture ratio ranging from near-lean blowout to stoichiometric ratio. No corresponding data were available for practical boiler and furnace conditions to compare against these measured values. Nevertheless, the WSR data represent the upper range of values which the burners used in power generation combustion equipment should approach given adequate pressure drop (rapid mixing) and good atomization or pulverization of the fuel (rapid evaporation).

### CO Emissions

Fig. 3 shows the measured CO and $NO_x$ concentrations in WSR for methane, propane, and ethylene fuels as a function of flame temperature. It is seen that CO concentration reaches a minimum in the temperature range of 1650K to 1800K. CO emissions are important because they are directly related to combustion inefficiency. In combustion zones of furnaces, boilers, or ground-based gas turbines, it is known (Lefebvre 1984, Gupta et al.,1993) that for fuel-rich combustion, CO is formed because of the lack of oxygen, while for a slightly fuel-lean or stoichiometric combustion, dissociation of $CO_2$ produces CO. Also, quenching by cold air and inadequate mixing can lead to increased CO formation. A net result of these processes yields lowest values of CO in the range of $\phi = 0.7$ to 0.9, corresponding to the flame temperatures of 1650K to 1800K.

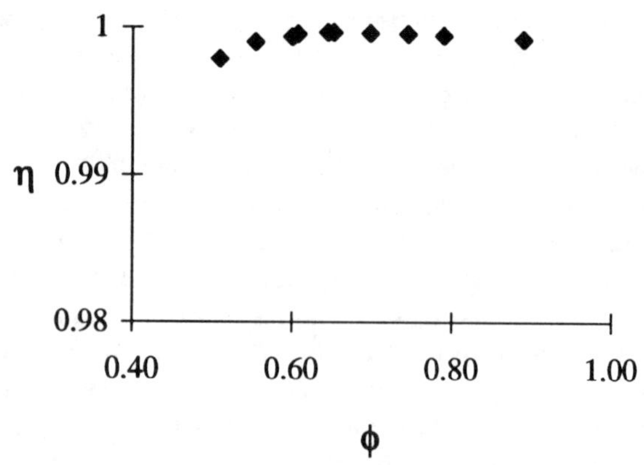


**Fig 2: Toroidal WSR combustion efficiency plotted as a function of reactor equivalence ratio.**

Fig. 4 shows actual CO production in the WSR as compared with the predicted values obtained from Eq. (5). As expected, the agreement is reasonable but not perfect. A main reason for the discrepancy is that Eq. (5) is especially accurate at high combustor pressure when CO emissions are lower. At slightly above-atmospheric conditions where boilers and furnaces normally operate, it underpredicts the level of CO emission.

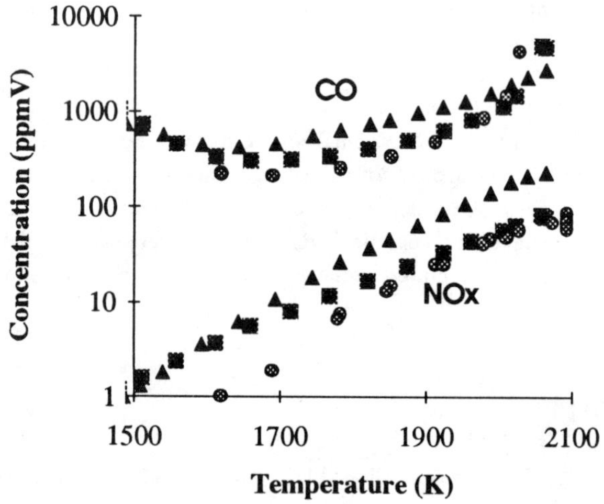


**Fig. 3: Emissions of CO and $NO_x$ as a function of flame temperature. (•=methane, ▲= ethylene,■= propane).**

To improve agreement, a new value for the constant *c* (c = 0.0031) was used. The modified equation gave good agreement over a range of WSR flow conditions from 3 g/s to 19 g/s; values obtained for 19 g/s are shown. Also, it was observed that the modified predictions were in good agreement with measured values between 1600 and 1900 K, the operating temperature range for most practical combustion systems. However, discrepancies occurred outside this temperature range. At the lower temperatures, WSR flame instabilities and resulting lower combustion efficiencies produce a large amount of CO which the model was not able to accurately predict. At temperatures near stoichiometric conditions, dissociation of $CO_2$ increases the amount of CO, which is also not accounted for by the simple fluid dynamic model.

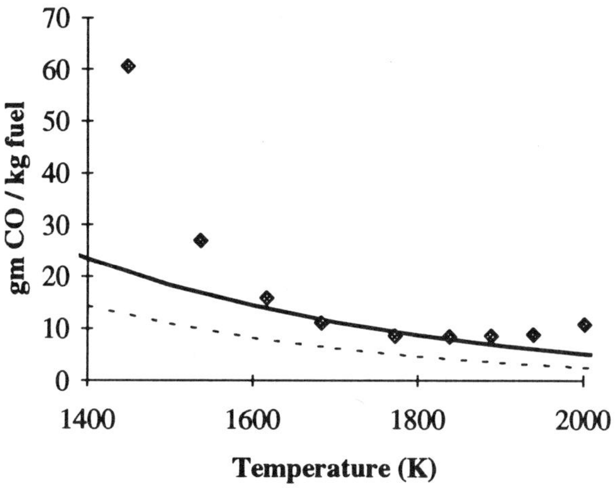


**Fig 4: CO emissions from WSR plotted with predicted values. ( ♦ = WSR data, ---- = Eq. 2 results, and ——— = Modified Eq. 2 results).**

## $NO_x$ Emissions

Fig. 3 shows the measured $NO_x$ concentrations in WSR for methane, propane, and ethylene fuels as a function of flame temperature. An increase in flame temperature produced by an increase in equivalence ratio towards the stoichiometric value will increase $NO_x$ emissions significantly, with maximum values of $NO_x$ occurring at an equivalence ratio of about one.

Fig. 5 shows $NO_x$ production from WSR tests and values obtained from Eq. (7). Predicted values of $NO_x$ were considerably lower than the WSR measurements because intense mixing in the WSR eliminates cool spots in the combustor which are prevalent in practical combustion devices. Therefore, the overall $NO_x$ concentrations in the WSR exhaust is higher than the emission values for the practical combustor. In fact, $NO_x$ emissions rise moderately with high pressure, rapidly with increasing residence time, and exponentially with flame temperature. Eq. (7) was modified by changing the value of the constant $a$ to 0.0103 and, as seen in Fig. 5, the modified predictions agree quite well with experimental data for temperatures between 1500K and 2100K at a loading of 19 g/s as shown in Fig. 5. The modified predictions were also tested at a loading of 3 g/s and agreed quite well with experimental data.

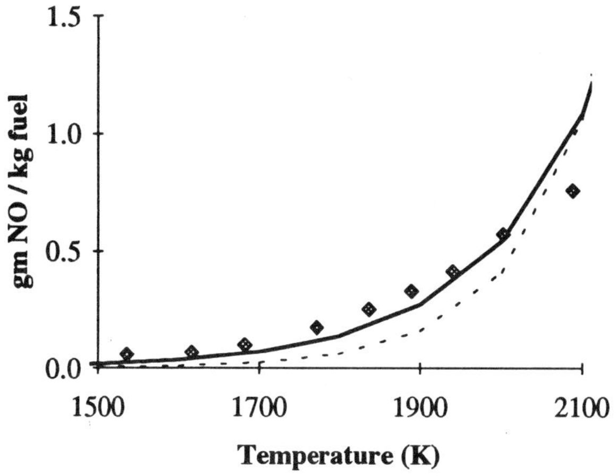


**Fig 5: $NO_x$ emissions from WSR plotted with predicted values ( ♦ = WSR data, ---- = Eq. 7 results, and — = Modified Eq. 7 results).**

## SUMMARY AND CONCLUSIONS

Emissions of CO, $NO_x$, and combustion efficiency of a WSR were studied and the application of this sophisticated aerospace-derived technology to improve power-generation combustion systems was explored. Our findings are summarized in the following paragraphs.

1. The combustion efficiency in the WSR is above 99 percent over a wide operating mixture ratio. The WSR data represent the upper range of values which the burners used in power generation combustion equipment should approach given adequate pressure drop (rapid mixing) and good atomization or pulverization of the fuel (rapid evaporation).

2. CO emissions are underpredicted by Eq. (5), which was originally developed for high combustor pressure where CO emissions are lower. At slightly above-atmospheric conditions where boilers and furnaces normally operate, a slight modification is required to provide good predictions of the level of CO emissions. At temperatures near stoichiometric conditions, dissociation of $CO_2$ increases the amount of CO, which also is not accounted for by the simple fluid dynamic model prediction.

3. Predicted values of $NO_x$ using Eq. (7) were lower than the WSR measurements, because intense mixing in the WSR eliminates cool spots in the combustor which are prevalent in practical combustion devices. Therefore, the overall $NO_x$

concentrations in the WSR exhaust is higher than the $NO_x$ emission values for the practical combustor. A slightly modified constant provides accurate $NO_x$ predictions for temperatures between 1500K and 2100K at two different air loadings of 3 g/s and 19 g/s.

## ACKNOWLEDGMENT

This work was supported by the Air Force Wright Laboratory, Aeropropulsion and Power Directorate, Wright-Patterson Air Force Base, OH, under contract F33615-92-C-2207 with Mr. Charles W. Frayne serving as the Technical Monitor.

## REFERENCES

Ballal D. R., 1993, "Combustion of Residual Fuel in Power Generation System," *ASME-FACT*-Vol. 17, pp. 11-16.

Gupta, A. K., Mehta A., Moussa, N. A., Presser C., Rini M. J., Saltiel, C., Warchol J., and Whaley H, 1993, "Combustion Modeling, Cofiring, and $NO_x$ Control," *ASME FACT*-Vol. 17.

Lefebvre A. H., 1984, "Fuel Effects on Gas Turbine Combustor-Liner Temperature, Pattern Factor, and Pollutant Formation," *AIAA J. Aircraft*, Vol. 21, pp. 887-898.

Longwell, J. P., and Weiss, M. A, 1955, "High Temperature Reaction Rates in Hydrocarbon Combustion," *Industrial and Engineering Chemistry,* Vol. 47, pp. 1634-1642.

Zelina J., and Ballal, D. R., 1994, "Combustion Studies in a Well Stirred Reactor" *AIAA Paper No.* 94-0114.

FACT-Vol. 18, Combustion Modeling, Scaling and Air Toxins
ASME 1994

# STUDIES OF COMBUSTION AND EMISSIONS IN A MODEL STEP SWIRL COMBUSTOR

**Mark D. Durbin**
Department of Mechanical & Aerospace Engineering
University of Dayton
Dayton, Ohio

**Dilip R. Ballal**
Department of Mechanical & Aerospace Engineering
University of Dayton
Dayton, Ohio

## ABSTRACT

Combustion and pollutant emissions from fossile-fired and gas turbine cycle power plants are increasingly receiving close scrutiny because of the Clean Air Act of 1991. In this paper, an investigation of combustion and pollutant emissions in a model step swirl combustor (SSC) is described, specifically flame characteristics, lean blowout (LBO), reactive flow field, and CO and $NO_x$ emissions. The SSC was designed to simulate a practical gas turbine combustor.

Flame photographs and LBO measurements illustrate that the co-swirl provides more stable combustion and yields slightly lower values of LBO than the counter-swirl arrangement. The co-swirl arrangement also contributes to high velocity gradients, intense mixing, and temperature uniformity. Finally, CO and $NO_x$ emissions follow opposite trends as expected, i.e., CO formation is a minimum in the region of maximum $NO_x$ emissions. Also, the recirculation of the outer flow appears to produce a strong radial gradient in the CO and $NO_x$ emissions.

## NOMENCLATURE

| | |
|---|---|
| D | = inner air nozzle diameter |
| LBO | = lean blowout |
| P | = pressure |
| Re | = Reynolds number |
| U | = axial mean velocity |
| T | = mean temperature |
| Y | = radial direction |
| Z | = axial direction |
| $\phi$ | = equivalence ratio |
| $\theta$ | = swirl vane angle |

### Subscripts

| | |
|---|---|
| i | = inner |
| o | = outer |

## INTRODUCTION

The Clean Air Act of 1991 is forcing the re-examination of combustion and pollutant emission strategies in fossile-fired power plants. Compliance with stringent CO and $NO_x$ emissions standards, and the need to improve thermal efficiency and meet the peak-power demands of the industry has led to the rapid use of gas turbines in power generation, usually in a combined-cycle configuration. Gas turbine combustors fire natural gas, light distillate oil, or, with proper prefiring treatment, a wide range of residual and crude oils or pulverized coal fuel. Proper design of the combustor and fuel nozzle (natural gas nozzle, airblast atomizer, or pulverized coal firing nozzle) is required to achieve good combustion performance and low emissions.

In this paper, a model step swirl combustor (SSC) was designed to simulate a practical gas turbine combustor. Gaseous propane fuel was injected in the form of an annular jet, coaxially sandwiched between co-swirling or counter-swirling airstreams. This fuel injection pattern simulates the fuel-air mixing just downstream of a natural gas nozzle, airblast atomizer, or pulverized coal-firing arrangement located in the dome region of an industrial gas turbine combustor employed in power generation. The specific areas investigated are: (i) flame characteristics such as length, shape, mixedness and LBO, (ii) combustor flow field patterns such as velocity and temperature, and (iii) emissions.

## TEST FACILITY

### The SSC

Fig. 1 shows a schematic diagram of the SSC which has a 150-x150-mm cross section with rounded corners, length of 754 mm, and a step height of 55 mm. The SSC provides a geometrically simple, optically accessible research combustor capable of reproducing the fuel-air mixing pattern in the dome region of a practical industrial gas turbine combustor. The SSC also offers independent control over inner and outer airstreams.

The SSC was mounted on a vertical combustion tunnel with a three-axis traversing mechanism. Measurements of velocity, temperature, and emissions were made to investigate combustion and pollutant emissions.

Fuel was supplied to the combustor by the annular fuel tube (20 mm i.d. and 29 mm o.d.) which is coaxially sandwiched between swirling airstreams; the inner air jet (20 mm dia.) and the outer annular air jet (29 mm i.d. and 40 mm o.d.). The combustor exit has a 45% blockage orifice plate on top which simulates the back pressure exerted by the dilution jets in a practical gas turbine combustor (see Sturgess et al., 1990). The SSC has quartz windows on all four sides to permit visual observations and laser diagnostics measurements.

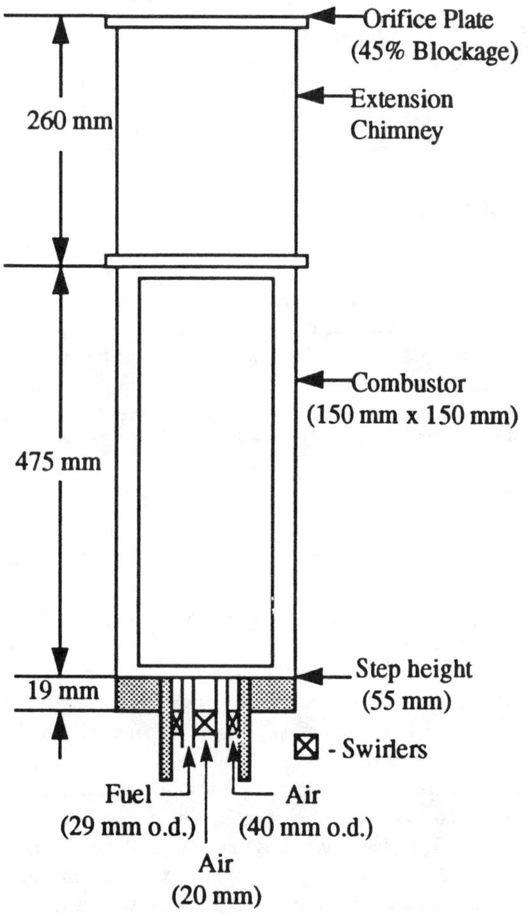


**Fig. 1: Schematic Diagram of a step swirl combustor.**

Stationary helical vane swirlers were located 25 mm upstream from the burner tube exit in each of the air passages. The inner swirler had six vanes with a central 1.4 mm dia. hole to prevent the flame from anchoring to the swirler. The outer swirler had twelve vanes. Inner swirler lengths are 25, 19, and 19 mm, respectively, for 30°, 45°, and 60° swirlers; outer swirler lengths are 32, 25, and 19 mm, respectively, for 30°, 45°, and 60° swirlers. The swirlers were precision-fabricated in a rapid prototype manufacturing process known as stereolithography. These swirlers performed satisfactorily at high temperatures in our combustor.

## Instrumentation

Durbin and Ballal (1994) have described the LDA system used here for velocity measurements. Essentially, this is a three-beam two component (axial and radial) set using a 514.5 nm line of an 18 W argon-ion laser with a component separation based on polarization. A two-beam third component (tangential) set uses a 488.0 nm line with separation by color. Since the third component is normal to the first and second components, the measurement volume had a quasi-spherical shape of 100 μm diameter and the calculated fringe spacing was 3.6 μm. The LDA system has Bragg cell frequency shifting (10 MHz for the first and second channels and 30 MHz for the third channel) for measurements in recirculatory flows, 4-$\sigma$ filtering software for spurious signals, for example, due to seed agglomeration, and a correction subroutine to account for the LDA signal biasing effects in combusting flows. A fluidized-bed seeder was used to inject sub micron-sized (0.1 μm) $ZrO_2$ particles into each passage. Typical LDA sampling rates exceeded 1 kHz for both isothermal and combusting flow measurements.

The CARS optics layout described by Pan et al. (1991) was used for temperature measurements. The laser source is provided by a Nd:Yag pulse laser with a 10-ns time resolution and a Boxcars configuration is used. The probe volume is approximately 25-μm by 250-μm. The CARS signal is collected by a Spex 1702 spectrometer, 2-D charge coupled diode (CCD) camera from Princeton instruments, and Tracor-Northern multichannel analyzer. The raw data is processed by in-house software on a personal computer.

Both fuel and airflow were monitored by separate electronic flow control units to within ±0.5% and ±1.5%, respectively. The combined error produced an uncertainty of ±1.5% in equivalence ratio, or ±50 K in mean temperature. A total of 500 samples were taken for each CARS measurement to ensure that the error in the rms temperature was less than 10 K.

An Ecom-AC portable microprocessor controlled emissions analyzer was used for emission sampling. The Ecom-AC was developed as a test instrument to ensure optimal fuel and air mixtures in burners and boilers. It is used by industrial, service, and installation contractors. The Ecom-AC uses electrochemical sensors with an accuracy of five percent of the total reading.

## Test Conditions

The experiments focused on one configuration for both co- and counter-swirl vane angles. This configuration was chosen because previous work of Durbin and Ballal (1994), revealed dramatic differences between co- and counter-swirl. The configuration was inner vane angle = outer vane angle = 30°. The average axial velocity of the inner airstream was 16 m/s and the average axial velocity of the outer airstream was 8 m/s. Equivalence ratio was stoichiometric, which gave average axial fuel velocity of 2 m/s. Mass flow was equal for both inner and outer airstreams.

## RESULTS

### Flame Characteristics and LBO

Figs. 2a and 2b are long exposure photographs, taken with a 35-mm camera, of the flame structure for the co- and counter-swirl arrangements. The co-swirl flame has a dimple shape at the centerline which is evidence of a strong inner recirculation zone. The counter-swirl flame has a bulbish shape at the centerline which indicates no inner recirculation zone. The counter-swirl flame is anchored in the region just above the fuel stream exit, indicating low velocities in this region. These assessments were verified with mean velocity measurements shown later in Figs. 4a and 4b.

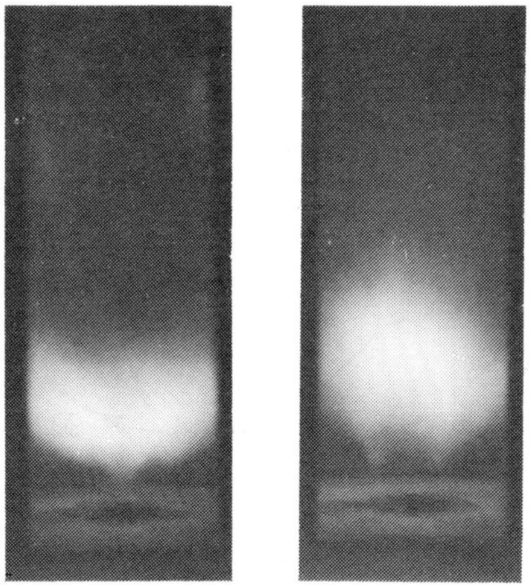

(a) (b)

**Fig. 2: Photographs illustrating the flame structure for (a) co-swirl and (b) counter-swirl arrangements. Test conditions were $\phi$ = 1.0, $U_i$ = 16 m/s, $U_o$ = 8 m/s, $\theta_i = \theta_o$ = 30°.**

Fig. 3 shows LBO data for co- and counter-swirl. When inner air velocity is significantly higher than the outer air velocity, co-swirl provides more stable combustion and always yields *slightly* lower values of LBO than the counter-swirl condition. These lower values occur because the co-swirl flow direction produces less uniform mixing in the combustor than the counter-swirl flow direction. An imperfectly-mixed diffusion flame blows out at a lower overall equivalence ratio because combustion is sustained in the locally rich mixture regions. This is good for stability but it should be recognized that the reactants are not uniformly mixed, which leads to low combustion efficiency and high CO emissions. Thus, there is an important *tradeoff* to be considered when assessing burning characteristics of co- and counter-swirl flows. No detectable differences in LBO values were found for co- and counter-swirl flow conditions when inner and outer air velocities were equal.

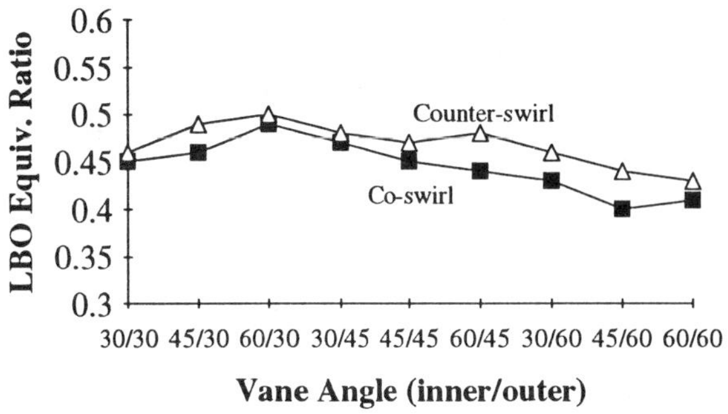


**Fig. 3: A comparison of LBO data for co- and counter-swirl arrangements ($U_i$ = 32 m/s, $U_o$ = 16 m/s).**

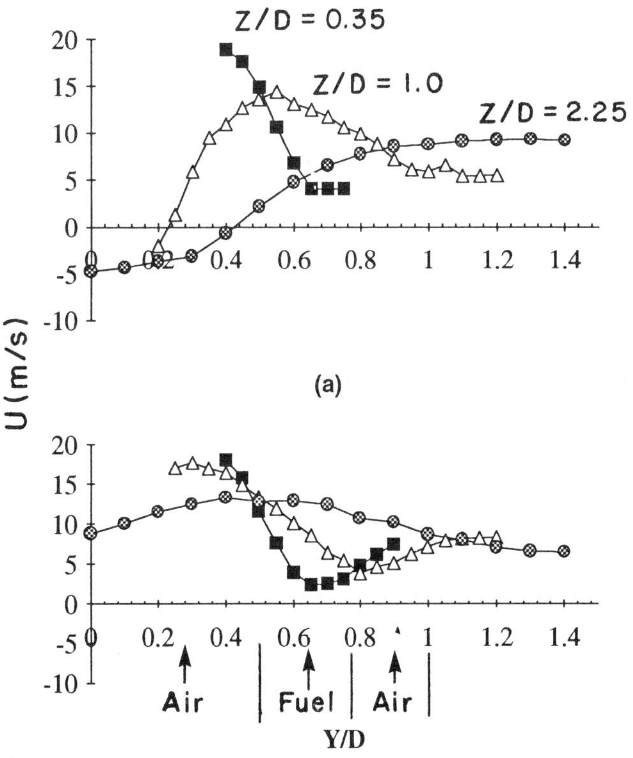


**Fig. 4: Radial variation of mean axial velocity at three different locations downstream of the fuel nozzle: (a) co-swirl, (b) counter-swirl. Same test conditions as Fig. 3.**

### Velocity Data

Figs. 4a and 4b show differences in the combusting flow field due to co- vs. counter-swirl effects. These LDA measurements were made by seeding the fuel passage.

As seen in Fig. 4a, at Z/D = 1.0, negative velocities are apparent for Y/D < 0.2. At Z/D = 2.25, the negative velocities extend to Y/D = 0.4. This indicates the presence of an inner recirculation zone which is growing in width with downstream position. The negative velocities associated with the inner recirculation zone provide an excellent flame stabilizing region. These results are consistent with the dimple-shaped central flame structure observed in the photograph in Fig. 2a.

Fig. 4b illustrates the axial mean velocity profiles for the counter-swirl direction. These results show no evidence of any inner recirculation zone. Moreover, in the annular gap corresponding to the fuel tube location, the velocity profile has a minimum. Presumably, this arises due to the flow velocity cancellation effect produced by the counter-rotating swirl. Also, the annular fuel jet is subjected to strong shearing action of counter-rotating braids along its inner and outer boundaries. As evident in Fig. 2b, this shearing action produces a very thin annular film near the nozzle exit (Z/D up to 1) which supports an attached flame structure in the combustor.

It should be noted that in our experiments, the co- and counter-swirl arrangement had identical overall pressure drop. Yet there is a difference in their mixing characteristics and, hence, also in the flame structures. Thus, expressing the mixing quality only in terms of the overall pressure drop can be deceptive.

## Temperature Data

Figs. 5a and 5b show CARS mean temperature measurements for the co- and counter-swirl configurations, respectively. The temperature measurements were made radially outward from the combustor centerline for four axial distances downstream from the fuel inlet to the combustor at 7, 20, 45, and 100 mm, but only up to Y = 30 mm (Y/D = 1.5). As such these data represent the reactive flow field around the center of the combustor. Fig. 5a shows that, at Z = 7 mm (Z/D = 0.35), the combustor center contains ambient temperature air; no burning or recirculation occurs this close to the fuel nozzle. Beyond a radial distance of 20 mm (which corresponds with the outer radius of the outer air passage), the temperature shoots up to 1200K, indicating that combustion in the outer recirculation zone provides hot products back to the flame zone. Further downstream (Z = 20 mm), high centerline temperature (1250K) indicates an inner combusting recirculation zone, whereas relatively cool temperatures are apparent radially outward. This cool temperature location is fed by the inner airstream. The mean temperatures past Y = 20 mm increase because of the outer recirculation zone, but the real strength of the outer recirculation zone is beyond Y = 30 mm. At Z = 45 mm, the central zone is at the flame temperature of nearly 2000K, and the temperatures fall off slightly in the radial direction. At Z = 100 mm, which is above the visible flame, combustion product temperatures are measured which are less than the flame temperature.

Fig. 5b shows the counter-swirl temperature measurements. At Z = 7 mm, the centerline temperature is the same room temperature measurement as the co-swirl configuration from the inner airstream, but at Y = 13 mm, there is a peak temperature. This peak temperature is the location of a thin flame. The flame is stabilized at this location because of reduced velocities from the counter-rotating braids. The temperatures reduce to near room temperature at Y = 19 mm, where the outer airstream would influence the temperature. Moving farther in the radial direction, temperatures increase due to the outer recirculation zone. At Z = 20 mm, this peak temperature where the flame sits is enhanced as the flame front becomes wider, then temperatures are reduced due to the outer airstream and then they increase due to the outer recirculation zone. Comparing co- and counter-swirl at Z = 20 mm, counter-swirl still has room temperature air at its centerline, whereas co-swirl has elevated temperatures at its centerline. This finding verifies that co-swirl has a strong central recirculation zone at this location and counter-swirl does not. At Z = 45 mm, the inner core has high temperatures for co-swirl, while for counter-swirl the inner core temperatures are relatively low. This indicates that co-swirl has a greater heat release per unit volume in the vicinity of the fuel nozzle than the counter-swirl configuration. At Z = 100 mm, the temperatures for co- and counter-swirl are similar.

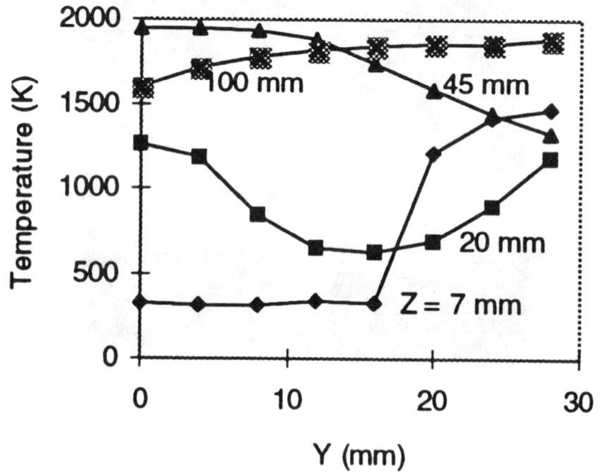


(a)

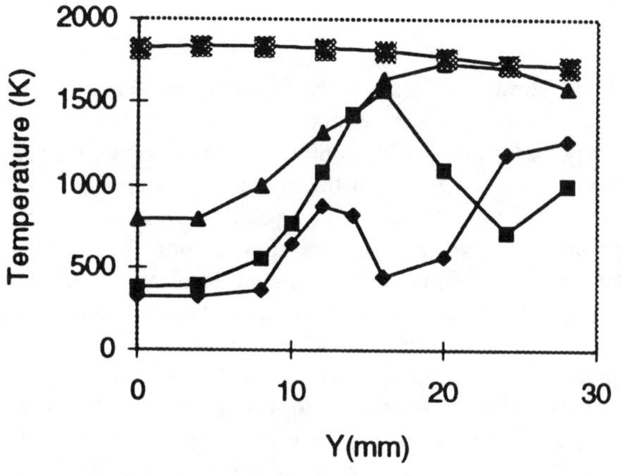


(b)

**Fig. 5: Radial variation of mean temperatures at four different locations downstream of the fuel nozzle: (a) co-swirl, (b) counter-swirl.**

## Emissions

CO and $NO_x$ measurements were taken in the SSC with an Ecom-AC portable emissions analyzer. A 1/4-in. quartz probe surrounded by a 3/8-in.-o.d. quartz air cooling passage was used to take the samples. Zelina and Ballal (1994) have described the probe design in detail. The probe was inserted through 3/8-in. holes in the wall and emissions measurements in the near-field and far-field were made in the outer recirculation zone. Figs. 6a and 6b show, for the co-swirling configuration, CO and $NO_x$ measurements respectively (arbitrary units).

Figs. 6a and 6b show (and flame photographs confirm) that, in the flame region located around Z = 20 mm, CO emission is minimum and the corresponding $NO_x$ emission is maximum. However, downstream of this flame region, dilution and quenching of the combustion products by cold air increases CO and decreases $NO_x$ emissions. Also, the recirculation of the outer flow appears to produce a strong radial gradient in the formation of CO and $NO_x$ measurements. More measurements are in progress to examine these interesting results in detail.

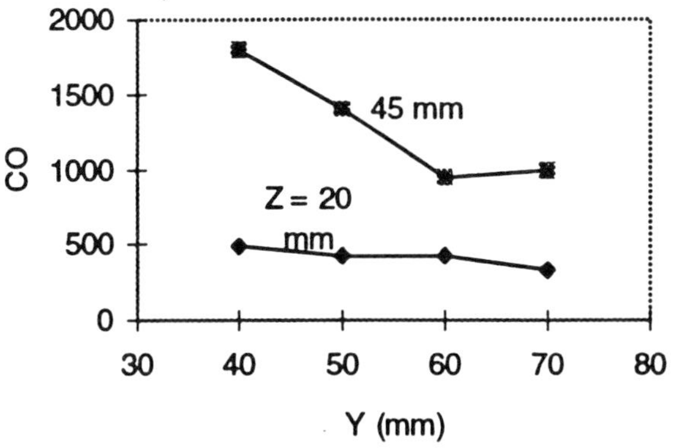


(a)

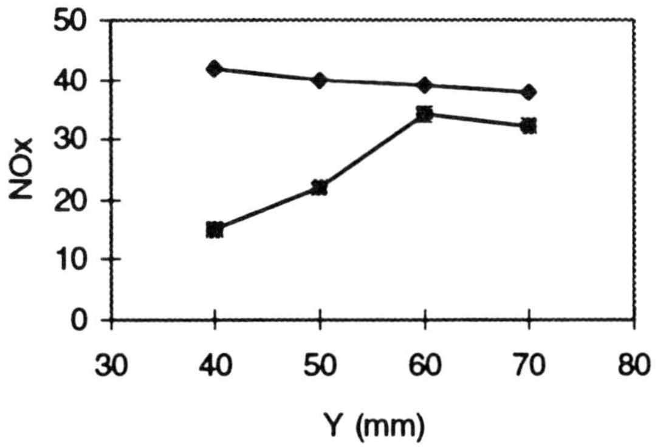


(b)

**Fig. 6: Measurements of (a) carbon monoxide, and (b) nitric oxide emissions in the co-swirling outer recirculation zone (arbitrary units).**

## SUMMARY AND CONCLUSIONS

An investigation of flame characteristics, LBO, reactive flow field, and pollutant emission was performed in a model SSC designed to simulate a practical gas turbine combustor. The following conclusions were reached.

1. Flame photographs and LBO measurements illustrate that co- and counter-swirl arrangements yield different flame characteristics. For example, the co-swirl provides more stable combustion and yields slightly lower values of LBO than the counterswirl arrangement.
2. For these experiments, the co-swirl configuration was found to have a greater heat release per unit volume in the vicinity of the fuel nozzle than the counter-swirl configuration.
3. Emissions of CO and $NO_x$ follow opposite trends, i.e., CO formation is a minimum in the region of maximum $NO_x$ emission. Also, the recirculation of the outer flow appears to produce a strong radial gradient in the emissions of CO and $NO_x$.

## ACKNOWLEDGMENTS

This work was supported by the U.S. Air Force, Wright Laboratory, Fuels and Lubrications Division, Aero-Propulsion and Power Directorate, Wright-Patterson Air Force Base, Dayton, OH under contract No. F33615-92-C-2207, with Mr. Charles W. Frayne serving as the Air Force Technical Monitor.

The authors wish to thank Mr. Marlin D. Vangsness for the CARS system alignment and data processing.

## REFERENCES

Ballal, D.R., 1993, "Combustion of Residual Fuel in Power Generation System," ASME FACT-Vol. 17, pp. 11-16.

Durbin, M.D., and Ballal, D.R., 1994, "Studies of Lean Blowout in a Step Swirl Combustor," ASME Paper No. 94-GT-216, to appear in *Transactions of the ASME, Journal of Engineering for Gas Turbine and Power.*

Pan, J.C., Vangsness, M.D., Heneghan, S.P., and Ballal, D.R., 1991, "Scalar Measurements in Bluff Body Stabilized Flames Using CARS Diagnostics," ASME Paper No. 91-GT-302.

Sturgess, G.J., Lesmerises, A.L., Heneghan, S.P., and Ballal, D.R., 1990, "Design and Development of a Research Combustor for Lean Blowout Research," *Transaction of ASME Journal of Engineering for Gas Turbines and Power*, Vol. 114, pp. 13-19.

Zelina J. and Ballal, D.R., 1994, "Combustion Studies in a Well Stirred Reactor," AIAA paper No. 94-0114.

FACT-Vol. 18, Combustion Modeling, Scaling and Air Toxins
ASME 1994

# EFFECT OF JET VELOCITY ON RADIATION CHARACTERISTICS OF A LAMINAR NATURAL GAS JET FLAME

**Michael J. Hubbard**
Wayne Home Equipment-Burner Division
Fort Wayne, Indiana

**S. R. Gollahalli**
School of Aerospace and Mechanical Engineering
University of Oklahoma
Norman, Oklahoma

## ABSTRACT

An experimental study of the temperature profiles, radiative characteristics, and emission index of nitrogen oxides in natural gas diffusion flames at jet exit Reynolds numbers of 625, 1240, and 1850 in a concentric air stream is presented. A scanning infrared camera was used to measure brightness temperature field and a fine-wire thermocouple was used to measure gas temperature in the flame. A pyrheliometer was used to measure total radiation flux. Results show that the radiative fraction of heat release and soot concentration decrease slightly at higher Reynolds numbers. Flame emissivity determined using the infrared camera agrees well with the predictions of two theoretical models.

## NOMENCLATURE

| | |
|---|---|
| d | Burner exit diameter |
| F | Radiative fraction of heat release rate |
| H | Flame Height |
| r | Radial distance |
| Rej | Burner exit Reynolds number |
| x | Axial distance |

## INTRODUCTION

Natural gas is recognized today across the world as the cleanest burning fuel in terms of emitted pollutants. As well, it is an abundant source of chemical energy. The government and industry generally recognize natural gas as the most viable source of energy for controlling environmental pollution in the near future. The composition of natural gas varies locally depending on reservoir conditions, but is mainly composed of methane with trace amounts of carbon oxides, nitrogen and other hydrocarbons. The chief constituent of natural gas, methane, is the smallest molecule of all the hydrocarbons. Its flame is generally the lowest emitter of unburned hydrocarbons, unburned carbon, and carbon monoxide. The amount of emitted radiation is also the lowest of all the hydrocarbon flames. This is primarily due to the low emissions of soot, which is due, at least in part, to the low carbon-to-hydrogen ratio of methane.

Since the methane flame ranks lower than all other hydrocarbon flames in terms of its power to radiate, its application is somewhat limited. Methane, and hence natural gas, is certainly well suited for situations where radiative heat transfer is not desired, as in gas turbine applications. However, it is not necessarily as good in situations such as furnaces, boilers, and domestic space heating devices since all these devices are reliant on high levels radiation emission for efficient heat exchange. Although techniques such as using a porous matrix to enhance radiation have been employed they are considerably expensive to manufacture due to material costs. If the flame itself might somehow be manipulated to increase the amount of radiative heat transfer, simpler and more economical heating units might be realized.

A number of publications on radiation and soot emission characteristics of hydrocarbon flames (Tien and Lee, 1982; Tien and Sohrab, 1990; Markstein, 1974; Olson et al., 1985; D'Alessio et al., 1974; Gomez et al., 1987; Saito et al., 1986; Hura and Glassman, 1987) exist in the literature. Since most of these studies have focused on highly luminous hydrocarbon fuels, systematic studies on the radiative characteristics of natural gas flames, which are on the borderline of nonluminous and luminous flames depending on the combustion conditions, are very limited.

This paper describes a part of a comprehensive investigation aimed at understanding the specific effects of operating variables on the radiative heat transfer characteristics from natural gas diffusion flames. The variable discussed in this study is the fuel jet velocity. In order to eliminate the effects of secondary air entrainment the structure of the flame in this study was limited to laminar case only.

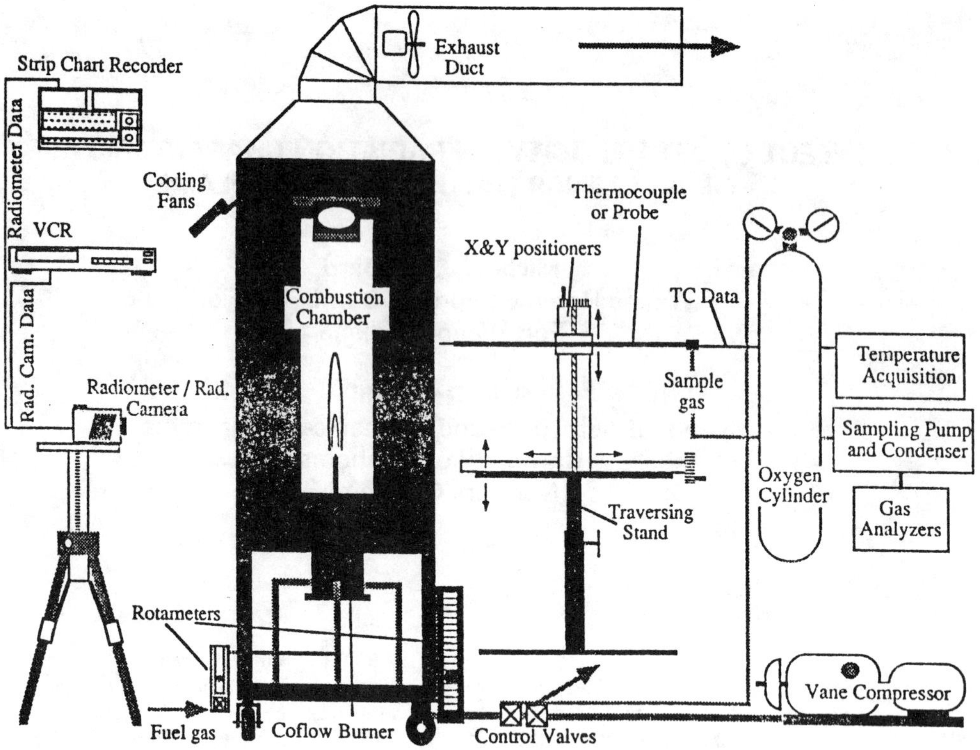


FIG. 1. Schematic Diagram of the Experiemtal Apparatus

## EXPERIMENTAL APPARATUS

The combustion facility consists of a laboratory-scale chamber with exhaust discharging and filtering capabilities. The combustion chamber was constructed of plate steel of 4.8 mm (0.1875") thickness. The bottom plate of the chamber is 76 cm (30") long and 76 cm wide. The chamber height is 164 cm (64"). The chamber itself has fan-cooled Pyrex windows on three of the four side walls which facilitate flame visualization. The chamber may be closed so that quiescent conditions are obtainable or opened for fast servicing by using one of the walls as a door. Another useful feature of the chamber is a small access door in that side wall for igniting the fuel jet when the chamber is closed. Exhaust gases are ducted through a 30.5 cm (12") diameter channel with a fan. A manually operated damper is used to control the amount of exhaust draft. All walls are painted flat black so that a black body emissivity of the chamber walls might be simulated and to avoid reflections for optical investigations. A schematic of the burner is shown in Figure 1.

The burner assembly, shown in Figure 2, is comprised of an outer burner body which supplies secondary air flow to the flame in the co-flow arrangement and a burner tube which supplies fuel gas. The co-flow burner is equipped so that flow velocity and composition of secondary air may be varied. Air flow is supplied with a vane-type rotary compressor and is metered with a high capacity rotameter. Air is ducted from the compressor to the outer burner body through 15.9 mm (0.625") outer diameter thin walled copper tubing. The fuel tube assembly consists of a stainless steel tube (3.7 mm I.D.) and flow metering equipment. This facility is capable of providing variable fuel flow to the burner tube, by means of a rotameter/valve assembly with which flow is also metered. All rotameters are calibrated with a standard wet test meter. The burner tube can be moved in the vertical direction so that radiation may be measured at different heights above the burner without moving the radiometer/detector. This burner is capable of generating a wide range of laminar to turbulent flames. Natural gas was used as the fuel and its average composition is given in Table 1.

### Instrumentation

A type R thermocouple (Pt/Pt-13% Rh) was used in conjunction with a digital data acquisition system to obtain temperature readings. The thermocouple had a wire diameter of 0.25 mm, and a bead diameter of 0.5 mm. The wire was insulated from the flame with a ceramic sleeve and protected from physical damage with a stainless steel tube as a sheath. The thermocouple bead was silica-coated to negate the possibility of catalytic effects. The data acquisition system consists of a PC with analog-to-digital conversion and data acquisition boards. A software written in BASIC was used to control sample rate, to correct data for convective and radiative losses from the thermocouple, and to provide statistical information of fluctuations in the data.

A radiometer (pyrheliometer type) mounted on a tripod stand was used with a strip-chart recorder to measure total radiative heat flux and axial profiles of radiative heat flux. This unit has a sensitivity of 29.8 $W/m^2/(mV)$ and is capable of response to broad band thermal radiation from the very near to the very far infrared regions of the electromagnetic spectrum.

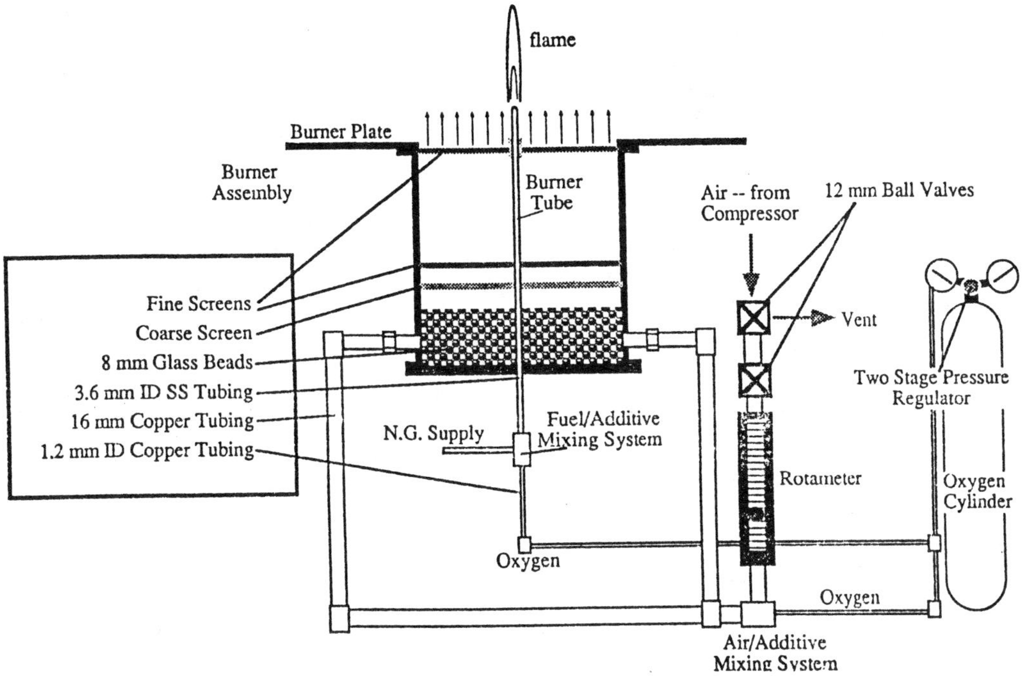


FIG. 2. Details of the Coflow Burner Apparatus

To obtain axial profiles of emitted radiation, a view limiting set-up was constructed to control the solid angle of view of the radiometer.

Direct photographs of the flames were taken using a 35 mm SLR camera with ASA 100 print film and ASA 64 slide film. Two images were obtained of each flame; one at an exposure of 1 second to provide flame length information and a second image at 1/30 second exposure to provide indications of the fluctuating flame structure. A graduated scale was photographed at the same position as of the flames to provide a true scale for determination of flame length.

In addition to the photographic method described, a *thermal* imaging technique was used to provide isotherms which were used to calculate overall emissivity at certain locations in the flame. The equipment used was a scanning infrared imaging camera equipped with a mercury-cadmium-telluride sensor which operates at 77 K within an evacuated metal cryogenic dewar for maximum thermal sensitivity. A vertical and horizontal scanner are used to scan over the field of view and obtain a spatial distribution of emitted radiation. The detector is sensitive to radiation in the 3-14 μm range which corresponds to broad-band spectral response. The camera also calculates apparent or brightness temperature information based upon measured radiative flux and an assumed emissivity.

A sampling system consisting of a moisture condenser and pump, along with sampling probes and Teflon tubing for flow directioning was used for exhaust-gas sampling. The condenser was a glass tube immersed in an ice bath to remove any liquid from the sample. Glass wool and a filter element with fine pores were used to remove particulates from gas samples. Both the condenser and filter were used to protect equipment and ensure proper readings. A rotameter was also employed to regulate flow to the analytical equipment. A conical Pyrex gas collector was placed above the flame to obtain a thoroughly mixed gas sample for determining emission indices. An uncooled quartz probe (of orifice diameter 1 mm) was used to draw the gas samples. A traversing mechanism was used to move the probe. A polarographic catalyst type oxygen analyzer was used to measure the volumetric concentration of $O_2$ in samples of the secondary air supply and the exhaust gases. Volumetric concentrations of CO and $CO_2$ were measured using nondispersive infrared (NDIR) analyzers. A chemiluminescent analyzer was used to detect the concentrations of oxides of nitrogen within the sampled exhaust gases. Following the methods described by Yagi and Iino (1962), soot concentration in the flames was calculated using laser light attenuation measurements. Soot volume fraction was calculated knowing the optical path length which the laser traversed (the width of the flame) and the wavelength of the laser used (0.635 μm). The laser used was a 0.95 mW Helium-Neon optical laser. The laser was directed through the center of a flame and it was detected using an optical phototube power meter.

Some experimental runs were repeated over ten times. The uncertainty intervals at a confidence level of 95% using Student's t-test were estimated from the repeated readings. The estimated uncertainty intervals expressed as a percent of their mean values in temperature, volumetric concentrations of CO, NO, NOx, emissivity based on brightness temperature, and radiative fraction of heat release were ± 1.5, 1, 1.5, 2.5, 5, and 2 respectively.

Table 1
AVERAGE COMPOSITION OF NATURAL GAS

| Species | Mole % |
|---|---|
| Methane | 92.66 |
| Ethane | 3.46 |
| Propane | 1.10 |
| Nitrogen | 1.62 |
| Oxygen | 0 |
| Carbon dioxide | 0.45 |
| Iso-butane | 0.14 |
| n-butane | 0.26 |
| iso-pentane | 0.07 |
| n-pentane | 0.08 |
| hexane+ | 0.15 |

## RESULTS AND DISCUSSION

### Effects of Secondary Air Velocity on the Radiative Heat Transfer from Flames

Flames of three fuel flow rates corresponding to the jet exit Reynolds numbers, $Re_j$, 625, 1240, and 1850 were examined in pure secondary air of varying velocity, the upper limits of which was determined by the flow rate capability of the vane compressor used. The Reynolds numbers are based on the average fuel jet velocity which was calculated from the measured volume flow rates. The effect of co-flow air velocity on the radiative heat release fraction (F) of the flame, where F is defined as the radiative heat power normalized by the total heat release rate computed assuming complete combustion, is presented in Fig. 3.

There was a minimal effect of the co-flow air stream on the radiative fraction of heat release ($\approx$2 % ) over the flow range of the compressor for all three fuel flow rates. This indicates that in the experiments described below any variation in F should be independent of the small fluctuations in secondary air flow velocity.

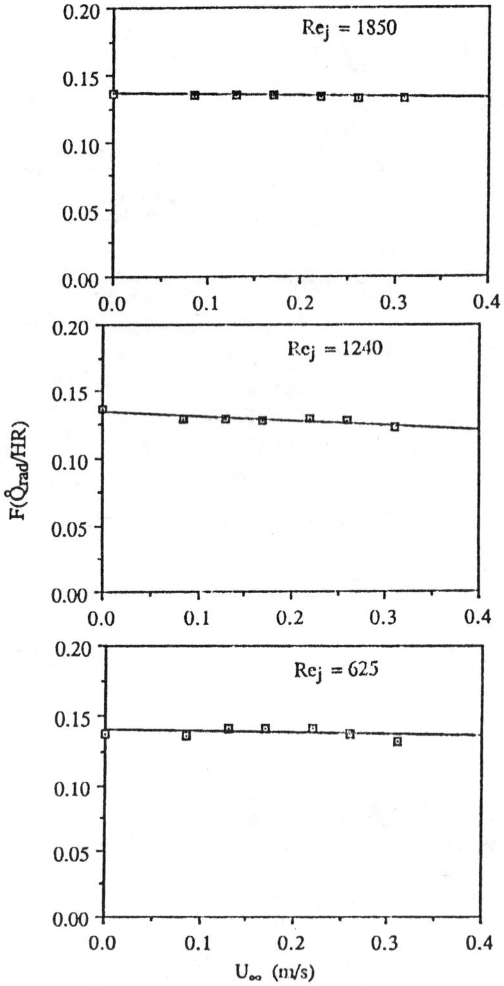


FIG. 3 Effects of Coflow Velocity on the Radiative Heat Release Fraction At Three Burner Exit Reynolds Numbers

### Effect of Jet Reynolds Number on the Radiative Fraction of Heat Release

Figure 4 shows the fraction of heat release F for the three flames. It is seen that the flame with the lowest fuel flow rate (Rej = 625) displayed the highest radiative fraction of heat release, and the highest fuel flow rate (Rej = 1850) yielded the lowest F. These values of F ($0.1 < F < 0.2$) agree with the values of Burgess and Zabetakis (1962). The lower radiative fraction of heat release values for the flames of higher Reynolds numbers are due to the higher degree of mixing and air entrainment for the higher jet exit velocities and consequent lower soot generation in the near-burner region.

### Axial Profiles of Radiation

Axial profiles of radiative heat transfer emitted from a slice of flame were taken using the conventional radiometer and a view limiting device for flames in pure air. As the angle of view of the radiometer was limited, readings were taken at such a distance to prevent any overlap in measurements. This distance was a function of solid angle of view of the radiometer.

Radiative power is plotted in Fig. 5 as a function of normalized axial location for the three flames in pure air with Reynolds numbers of 625, 1240, and 1850. Flame heights for these flames vary; however, a constant area and not a constant fraction of flame is viewed, as the distance from radiometer-to-flame was held constant. The radiation profiles were of similar shape for all flames, lowest in magnitude in the near nozzle region, increasing to a maximum in the far-flame region, and dropping at the flame-end. For the flame with the Reynolds number 625 the radiative power was the lowest and for Reynolds number 1850 the highest.

### Radial Temperature Profiles

Figure 6 shows the radial temperature profiles at three axial locations of the flames with jet Reynolds numbers of 625, 1240, and 1850 respectively. The temperature profiles for the flames in pure secondary air exhibit similar behavior as flow field was laminar and the fuel-air mixing is molecular diffusion governed. For all flames, the radial temperature profiles exhibit a recognizable double off-axis hump structure

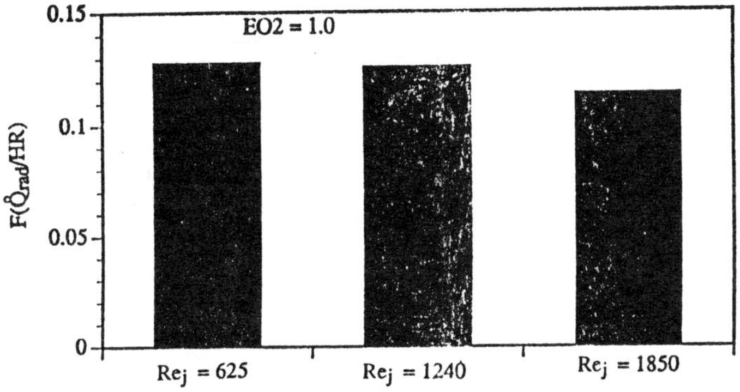


FIG. 4 Radiative Heat Release Fraction of Three Flames in Pure Air

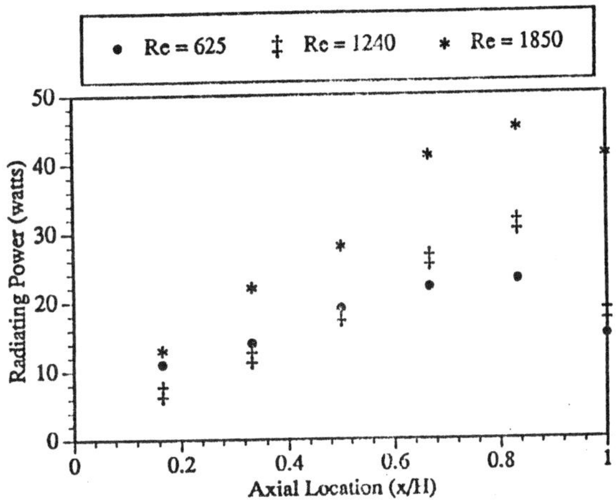


FIG. 5 Axial Distribution of Emitted Radiation of Three Flames in Pure Air - Emitted Radiation as a Function of Axial Location

in the near nozzle region. The strength of this peak diminishes downstream until x/H ≈ 0.5 and in all flames this structure vanishes at a location of x/H ≈ 0.67. Beyond that point, the radial temperature profile peaks on the central axis and decreases monotonically with an increase in r/d. The double off-axis hump structure is characteristic of diffusion-governed gas phase combustion where maximum temperatures occur in the zone of reaction viz., the outer flame edge. The axis peak structure is indicative of the solid phase soot combustion since particulate matter does not diffuse out.

All flames display similarly shaped temperature profiles which indicate similar behavior in terms of the relative dominance of gas- or soot-dominated combustion within the flame. With respect to temperature values, the two flames of lower flow rates ($Re_j$ = 625, 1240) have very similar values of temperature at identical positions in the flame. However, the flame of $Re_j$ = 1850 displays slightly higher temperatures.

## Brightness (Apparent) Temperature Measured from Thermal Imaging of Flames

Images were made from a videotape of the movie recorded with the scanning infrared camera and a videocassette recorder using a thermal video transfer technique. Each image was of a 1/60 second exposure. The camera images were created on a color scale which associated a color to a prescribed brightness (apparent ) temperature level. The brightness temperatures were calculated from levels of irradiation assuming black body emission.

The plots of brightness temperature as measured by the scanning radiometer are shown in Fig. 7. These plots were made by viewing the movies of the flames which contain a temperature distribution along the central axis of the flame. These plots display a similar tendency of increasing apparent temperature along the axial distance to a maximum and then decreasing in the flame end region. The location of maximum apparent temperature occurs in the region 0.5<x/H< 0.833. The brightness temperatures are substantially lower than temperatures measured by the thermocouple because the flames were optically thin and their emissivity was far less than unity. It is seen that the brightness temperature, in general, increases with temperature, although the increase is more marked between Rej=1240 and 1850. This trend is in accordance with the actual temperatures measured by thermocouple (Fig. 6). The increase in brightness temperature indicates a higher radiation level which can be attributed to higher energy input rate or the higher radiative fraction of heat release F. As the data in Fig. 5 have shown that F indeed decreases with the increase in Rej, the increase in temperature suggests that the increase in dilution due to secondary air diffusion is masked by the increase in heat release rate, at least in the case of laminar flames.

## Flame Length and Appearance

Flame lengths are determined from one-second exposure photographs. The visible flame length of the contiguous part of the flame on these photographs was 330 cm, 500 cm, and 570 cm for Rej equal to 625, 1240, and 1850 respectively. All the flames have a similar appearance. In all cases, there exists a very small black region directly atop the nozzle. This is, in all cases, approximately, of 1 mm length with little noticeable variation with Reynolds number. There is a larger bright blue band (4 to 10 mm in height) perpendicular to the axis of symmetry of the flame which lies directly above the black region. This band is the portion of the flame where only gas-phase reactions take place. A yellow luminous region is connected to the blue zone and continues over the entire length of the flame. This is the region where soot is formed and burned. Interior to the luminous zone of the flame is a dark region where no combustion takes place; only fuel pyrolysis occurs. Surrounding the lower half of the flame is a thin blue region which is duller and less well-defined than is the brighter blue band. This is again a region of gaseous reactions, and is separated from the rest of the flame by an extremely thin black region (≈0.5 mm in width).

When the flame is traversed radially in the very near-nozzle region, a dull blue band is crossed. Then an extremely thin, well-defined, bright blue band, is followed by a dark region as the central axis of the flame is approached. In the near- to mid-flame region, the dull blue band is first encountered, followed by a very thin dark region, then a bright yellow region, then a dark region. At approximately one half of the flame length, the outer dull blue zone disappears, but the luminous zone and the inner dark zone persist. Beyond one half of the flame length, the flame radiation is dominated by the coagulation and combustion of soot.

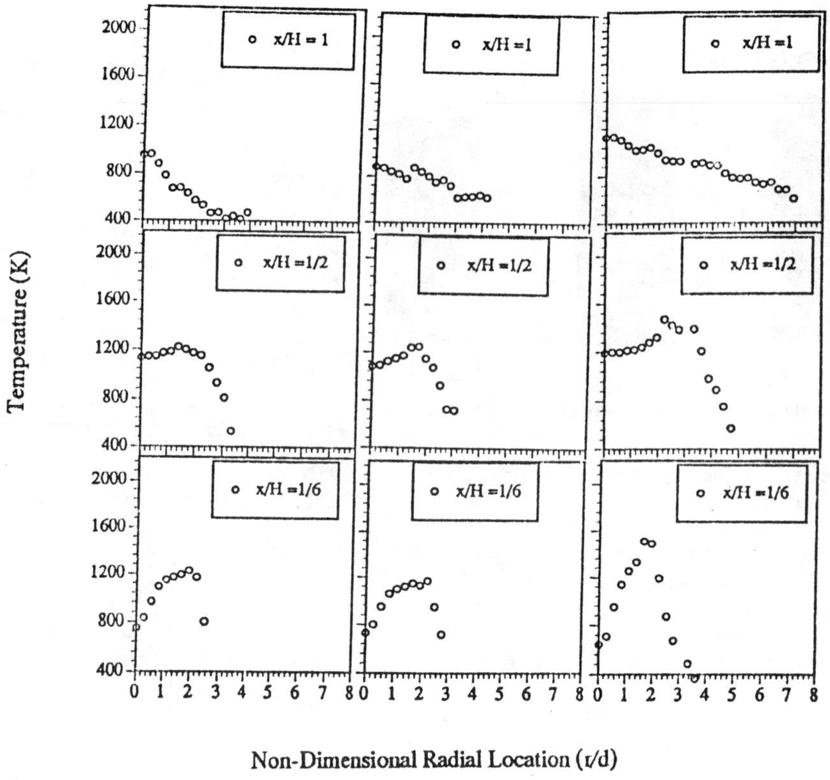


FIG. 6 Radial Temperature Profiles in the Natural Gas Diffusion Flames (Left: Rej=625, Middle: Rej=1240, Right: Rej=1850).

## Emission Indices of NO and $NO_x$

The emission indices are plotted as bar graphs in Figure 8. To enhance the radiative characteristics, the secondary air of all three flames was enriched with oxygen to have 50% more oxygen than air. It is interesting to note that the ratio of $NO_x$ emission to that of NO is approximately two, agreeing with observations of Ramavajjala and Gupta (1991). In general, we notice a very slight change of the emission indices of both NO and NOx with the increase in $Re_j$ at least for the laminar flame regime tested. This result is surprising in view of the increased temperature levels, particularly at Rej=1850, which suggests the presence of a counteracting factor that reduces the formation of NO or NOx. Since the emission index is based on the energy input rate, it appears that the formation rate of NO/NOx could not keep pace with the increase in energy input rate at higher Rej, perhaps due to competition for oxygen atom pool by the increase in CO formation that accompanies the increased fuel flow rate.

## Soot Concentration

To calculate emissivity for comparison with the emissivity values determined using the scanning radiometer, it was necessary to know the gas composition of the flame, average flame temperatures, the combined width of flame jet and shear layer, and also the concentration of particulates in the flame. Soot concentration was measured along the central axis of the flame due to the larger optical path length for the laser traverse at this location which results in a higher absorptivity. Soot concentration was measured at six axial locations and the distribution in the radial direction was observed but not recorded due to the inherently high level of uncertainty in the off-axis measurements. For both flames examined, the lowest amount of soot occurred in the near-nozzle region, increasing to a maximum in the far-flame region, and decreasing at the flame-end region. The average value of the volumetric soot concentration in the flame at $Re_j$=625 was 0.85 µl/ml, whereas in the flame at $Re_j$=1850 the corresponding value was 0.81 µl/ml. The slight decrease in soot concentration is in agreement with the changes in the value of radiative fraction of heat release F. The maximum values of volumetric soot concentration recorded in these flames were 1.33 and 0.99 µl/ml. The decrease in the maximum value of soot concentration when $Re_j$ was increased from 625 to 1850 also explains the increase in the maximum temperatures caused by soot combustion in the far-nozzle region at higher $Re_j$. The larger difference in the maximum soot concentration compared to the average concentration also suggests that the oxidation rate of soot is more affected than the formation rate of soot by the increased $Re_j$. That also explains the reason for higher formation of species like CO and the influence of oxygen atom depletion reactions counteracting the temperature effect on NO/NOx emission indices.

## Flame Emissivity

In this study, the flame emissivity was computed using three different methods. The first method of calculation was based on the data provided by the scanning radiometer. In this

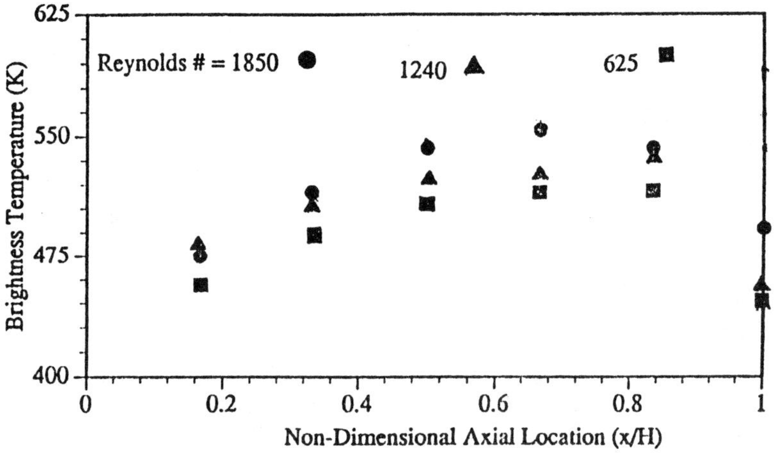


FIG. 7 Effect of Jet Reynolds Number on the Axial Profiles of Brightness Temperature.

method, emissivity was calculated as the ratio of apparent temperature to actual temperature raised to the fourth power using the scanning radiometer data. The apparent temperature is measured by the scanning radiometer based on the radiation which is incident on the detector and an assumed black body emissivity of unity. Actual temperatures were determined from the temperature profiles as the average temperature in the flame region and the shear layer. It was not possible to infer separate soot and gas emissivity terms from the results of the scanning infrared camera. The second method employed measured species concentrations to determine gaseous radiation (Hottel and Sarofim, 1967) and flame absorptance as an estimate of soot emissivity (Yagi and Iino, 1960). The third method of determining emissivity was based upon the model proposed by Yuen and Tien (1976), which has two terms representative of soot emissivity and of gas emissivity. Some modifications of this method were required. The method of Yuen and Tien applies to flames dominated by soot radiation or luminous flames. However, a considerable amount of the radiation in these flames came from gaseous species contribution. The emissivity equation in the third method has two terms, one is representative of soot emissivity and the other is representative of gas radiation. In the original method, the same value for path length of gas and soot matter is used. This method was modified in the present study as follows: the soot emissivity term was calculated based upon flame width, while the gas emissivity term was based upon the width of the flame plus the width of the shear layer containing combustion products. In addition, soot temperatures were averaged over the flame width, and the gas temperatures were averaged over the width of the flame plus the shear layer. This was done since the particulate matter does not diffuse as the gaseous products do. This differs slightly in numerical value but the results agree more closely with the results of the two other models.

The computed emissivity results are presented in Table 2. It is seen that the predictions of the models agree well with that calculated from the infrared camera measurements. For the case of the flame in pure air, the average emissivity over the flame length agrees with the values from the three models within 10%.

Table 2
Comparison of Emissivity Values
of the Flame at Rej = 625

| | |
|---|---|
| IR Camera | 0.126 |
| Model 1 | 0.120 |
| Model 2 | 0.122 |

## DISCUSSION

For the three flames in pure air (Rej = 625, 1240, 1850), the radiative fraction of release was seen to slightly decrease at the higher values of Reynolds number. This is due to the greater degree of air entrainment for the flames at higher burner exit Reynolds number, resulting in a lower degree of soot formation and an increase in the rate of soot combustion . This is visually confirmed by the brighter flames and by the measurement of soot concentration. The flames of lower fuel flow rates displayed slightly higher soot concentrations.

For the three flames of different Reynolds number it is not possible to make direct comparisons of emitted radiation as opposed to the case of F factor because the heat release rates are different. However, the flames display similar behavior in a qualitative sense; the amount of radiation increases to a maximum value which occurs in the far-flame region (x/flame height = 0.833), and the profiles drop off at the flame end.

The temperature profiles of flames in pure air display a double off-axis hump structure in the near-nozzle region and single axis peak structures in the far-nozzle region. The double off-axis humps indicate the dominance of gas phase combustion, and the axial peaks indicate the dominance of combustion of soot. For the flames in pure secondary air, soot combustion dominates the entire upper half of the flames.

Emissivity was calculated from the data provided by the scanning radiometer and compared to the values determined with conventional methods which were based upon species concentration and soot absorption. The predicted values of emissivity agree with the value determined from infrared camera measurement..

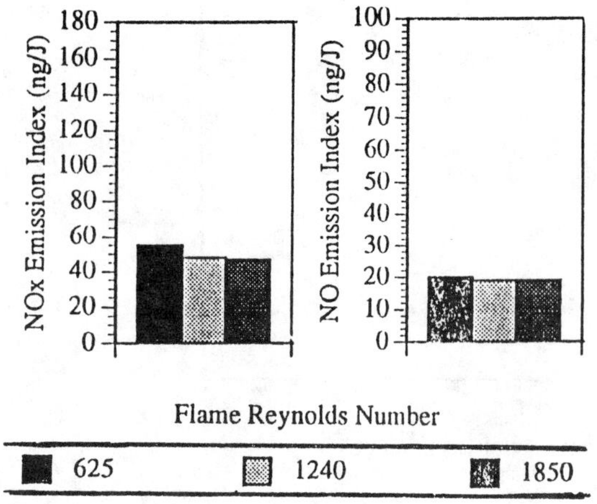


FIG. 8 Effect of Jet Reynolds Number on the Emission Indices of Nitrogen Oxides

## ACKNOWLEDGMENTS

This research was partly supported by the United States Department of Energy in the form of a grant for natural gas combustion research

## CONCLUSIONS

The following conclusions are drawn from this study on the radiative characteristics of a burner rim stabilized laminar (Rej= 625, 1240, and 1850) natural gas flame in a concentric stream of air: (i) The effect of co-flow air stream on the radiative fraction of heat release is small, (ii) The radiative fraction of heat release decreases with the increase in jet Reynolds number; the higher difference occurs for Rej between 1240 and 1850, (iii) A major part of the radiative power is emitted from the far-nozzle region of the flames, (iv) Temperature profiles in the corresponding regions of all three flames are similar; the maximum temperature in the flame at Rej=1850 is the highest, (v) The brightness temperature profiles determined with the scanning infrared camera exhibit trends similar to those of radiation power profiles, (vi) The maximum value of soot concentration decreases with the increase in Rej, (vii) The emission indices of NO and NOx are affected very little by Rej, (viii) The flame emissivity calculated using brightness temperature measured with scanning infrared camera and the gas temperature measured with thermocouple agrees well with the predictions of two theoretical models .

## REFERENCES


D'Alessio, A., Beretta, F., and Venitozzi, C., 1972, "Optical Investigations on Soot Forming Methane-Oxygen Flames," Combustion Science and Technology, Vol. 5, pp. 263-272.

Burgess, D. S. and Zabetakis, M. G., 1962, "Fire and Explosion Hazards Associated with Liquefied Natural Gas," U. S. Bureau of Mines, Report R1, 1099.

Gomez, A. Littman, M. G., and Glassman, I., 1987, "Comparative Study of Soot Formation on the Centerline of Axisymmetric laminar Diffusion Flames: Fuel and Temperature Effects," Combustion and Flame, Vol. 70, pp. 225-241.

Hottel, H. C., and Sarofim, A. C., 1967, Radiative transfer, McGraw-Hill Publications, New York, NY.

Hura, H. S. and Glassman, I., 1986, "Fuel Oxygen Effects on Soot Formation in Counterflow Diffusion Flames," Combustion Science and Technology, Vol. 53, pp. 1-21.

Markstein, G. H., 1974, "Radiative Energy Transfer from Gaseous Diffusion Flames," Fifteenth Symposium (international) on Combustion, The Combustion Institute, Pittsburgh, PA. pp. 1285-1294.

Olson, D. B., Pickens, J. C., and Gill, R. J., 1985, "The Effects of Molecular Structure on Soot Formation II, Diffusion Flames," Combustion and Flame, Vol. 62, pp. 43-60.

Ramavajjala, M., and Gupta, A.,1991, "Swirl Effect on NO and NO2 Emissions in a Variable Geometry Combustor," AIAA Paper No. 91-0643.

Saito, K. Williams, F. A., and Gordon, A. S., 1986, "Effects of Oxygen on Soot Formation in Methane Diffusion Flames," Combustion Science and Technology, Vol. 47, pp. 117-138.

Spiegel, M. R., 1961, Theory and Problems of Statistics, Schaum Publishing Co., New York, NY.

Tien, C. L., and Lee, S. C., 1982, "Flame Radiation," Progress in Energy and Combustion Science," Vol. 8, pp. 41-59.

Tien, J. H. and Sohrab, S. H., 1990, "Effects of Air-Side Oxygen Addition on Soot Formation in Methane Co-Flow Diffusion Flame," Combustion Science and Technology, Vol. 73, pp. 617-623.

Yagi, S. and Iino, H., 1960, "Radiation From Soot particles in Luminous Flames," Eighth Symposium (International) on Combustion, The Combustion Institute, Pittsburgh, PA, pp. 288-293.

Yuen, W. W., and Tien, C. L., 1976, "A Simple Calculation Scheme for the Luminous Flame Emissivity," Sixteenth Symposium (International) on Combustion, The Combustion Institute, Pittsburgh, PA, pp. 1481-1487.

**FACT-Vol. 18, Combustion Modeling, Scaling and Air Toxins**
**ASME 1994**

# CHARACTERISTICS OF VAPORIZING CRYOGENIC SPRAYS FOR ROCKET COMBUSTION MODELING

**Robert D. Ingebo**
NASA Lewis Research Center
Cleveland, Ohio

## ABSTRACT

Experimental measurements of the volume-median drop diameter, $D_{v.5e}$, of vaporizing cryogenic sprays were obtained with a drop size measuring instrument developed at NASA Lewis Research Center. To demonstrate the effect of atomizing-gas properties on characteristic drop size, a two-fluid fuel nozzle was used to break up liquid-nitrogen, $LN_2$, jets in high-velocity gasflows of helium argon and gaseous nitrogen, $GN_2$. Also, in order to determine the effect of atomizing-gas temperature on specific surface-areas of $LN_2$ sprays, drop size measurements were made at gas temperatures of 111 and 293 K.

## NOMENCLATURE

a — acceleration, cm/sec$^2$

Cd — drag coefficient

Do — liquid jet diameter, cm

$D_{v.5}$ — volume median drop diameter, cm

$k_c$ — correlation coefficient for Eq. (6)

$k_e$ — correlation coefficient for Eq. (1)

Nu — Nusselt number, based on $D_{v.5e}$

n — exponent for Eqs. (1) and (6)

Re — Reynolds number, based on $D_{v.5e}$

To — ambient airflow temperature, K

t — vaporization time, sec

Vc — acoustic velocity, cm/sec

W — weight flow of fluid, g/sec

We — Weber number, based on $D_{v.5e}$

x — axial downstream sampling distance

$\mu$ — absolute viscosity, g/cm sec

$\rho$ — fluid density, g/cm

Subscripts

c — calculated

e — experimental

g — gaseous nitrogen, $GN_2$

l — liquid nitrogen, $LN_2$

o — orifice

## INTRODUCTION

Highly volatile $LN_2$ sprays, with relatively large surface area per unit volume, were produced with a multiphase flow fuel nozzle. Such fuel sprays are desirable since they produce rapid fuel vaporization and efficient combustion in gas-turbine and rocket combustors. How atomizing-gas and liquid-jet properties effect the spray characteristic dropsize needs to be determined mathematically in order to compute changes in spray surface-area with residence time in the combustor. Mathematical expressions of this nature were investigated in the present study and used to correlate volume-median diameter, $D_{v.5c}$, with atomizing-gas mass flux and temperature. Such information is needed in the development of fuel spray combustion models that are based on characteristic drop-size expressions and cover wide ranges of liquid fuel and atomizing-gas properties. The resulting combustion models can then be used to develop more efficient fuel nozzles that will improve combustor performance and reduce exhaust emissions.

In the present investigation of vaporizing $LN_2$ sprays, measurements of $D_{v.5e}$ were made in the presence of relatively high thermal gradients. Atomizing-gas temperatures were 111 and 293 K and $LN_2$ droplet surface temperatures were near the boiling point of $LN_2$, 77 K. As a result, relatively high heat-transfer rates occurred across the gas film and they were considerably higher than those encountered in the study of water sprays, as described in Ref. 1. In that study, the effect of vaporization on drop size measurements was negligible when measurements were made very close to the fuel nozzle orifice. However, in the present study of $LN_2$ sprays, small droplets vaporized very fast and it was necessary to compute the effect of droplet vaporization on drop size measurements. Thereby, it was possible to determine values of $D_{v.5c}$ for unvaporized cryogenic sprays produced at the fuel nozzle orifice.

Multiphase flow atomization of $LN_2$ jets in sonic velocity He, Ar and $GN_2$ gasflows was experimentally investigated with the scattered-light scanner shown in Fig. 1. Sonic velocity gasflows were used so that liquid jet breakup occurred primarily in the regime of aerodynamic stripping. As a result of using high-velocity gasflow, very small $LN_2$ droplets were rapidly and completely formed at the fuel nozzle orifice. If the low gas-velocity regime of capillary wave breakup had been studied, then ligaments would have been formed and only partial breakup of the $LN_2$ jets would have occurred at the fuel nozzle orifice. Heat-transfer and drag coefficients obtained from earlier studies of droplet vaporization in

**Figure 1.—Apparatus and auxiliary equipment.**

high-velocity gasflow, Ref. 2, were used in the present study to calculate $LN_2$ droplet acceleration and vaporization rates.

Numerous investigators, as reported in Refs. 1 to 7, have studied the effects of atomizing-gas mass flux and gas velocity on spray drop size. As a result, different effects of atomizing-gas velocity on the characteristic drop size, Dc, have been observed as shown in table I. For the expression $Dc \sim V^n$, the exponent n determined by various investigators varied from −1.00 to −1.33. Atomization theory predicts n = −1.33 for liquid jet breakup in high-velocity gasflow. The disagreement between various investigators could to some extent be attributed to the effect of droplet vaporization on drop size measurements, especially when measurements were taken at distances relatively far downstream of the atomizer. This occurred, as reported in Ref. 1, when sampling distances downstream of the atomizer were increased from 2.2 to 6.7 cm. In that case, the exponent for atomizing-gas velocity decreased from −1.33 to 1.00.

A scattered-light scanner developed at NASA Lewis Research Center by Buchele, Ref. 8, was used to measure the volume-median diameter, $D_{v.5e}$, of $LN_2$ sprays formed in high-velocity He, $GN_2$ and Ar gasflows at atomizing-gas temperatures of 111 and 293 K. Sprays were sampled at a distance of 2 mm downstream of the fuel nozzle to minimize losses of very small droplets due to vaporization. Values of $D_{v.5e}$ varied from 3 to 30 μm.

## APPARATUS AND PROCEDURE

Multiphase flow atomization of $LN_2$ jets in high-velocity gasflows of He, $GN_2$ and Ar was investigated at atomizing-gas temperatures of 111 and 293 K. A two-fluid fuel nozzle was used as shown in Fig. 1. It was mounted at the center line of a 24 cm diameter duct and operated over pressure ranges of 0.2 to 1.0 MPa for both fluids, i.e., $LN_2$ and the atomizing gases. The $LN_2$ sprays were injected downstream into the atmosphere at the duct exit. The center of the 4.4 × 1.9 cm laser beam was located at a distance of 1.2 cm downstream of the atomizer orifice. Figure 2 shows details of the fuel nozzle. At a temperature of 77 K, $LN_2$ was injected into the airflow by gradually opening a control valve until the desired flowrate of 51 g/sec was obtained, as indicated by a turbine flow meter. To breakup the jet, an atomizing gas was turned on and the mass flowrate was measured with a 0.51 cm diameter sharp-edge orifice.

The scattered-light scanner optical system is shown in Fig. 3. It consisted of a laser beam expander with a spatial filter, rotating

TABLE I.—ATOMIZING-GAS VELOCITY EXPONENT, n, FOR LIQUID JET BREAKUP IN HIGH-VELCOITY GASFLOW

| Source | Exponent, -n |
|---|---|
| Adelbert, Theory[13] | 1.33 |
| Present study, x = 2.2 cm | 1.33 |
| Kim and Marshall[3] | 1.14 |
| Lorenzetto and Lefevbvre[4] | 1.00 |
| Nukiyama and Tanasawa[5] | 1.00 |
| Weiss and Worsham[6] | 1.33 |
| Wolf and Anderson[7] | 1.33 |

scanning-slit and a detector. The instrument measures scattered light intensity as a function of scattering angle by repeatedly sweeping a variable-length slit in the focal plane of the collecting lens. Scattered-light energy data were obtained as a function of scattering angle relative to the laser beam axis. The method is similar to that described in Ref. 9. As recommended by studies made in Ref. 8, measurements of scattered-light energy were normalized by the maximum energy and plotted against scattering angle, in order to determine the volume median diameter, $D_{v.5e}$. Also, it should be noted that this method can be used independent of particle size distribution function. In making a typical drop size measurement, the scan is repeated 60 times per second. This averages out any temporal variations in the energy curve. By measuring $D_{v.5e}$ for the entire droplet cloud, spray pattern effects were minimized.

The scattered-light scanner was calibrated with five sets of monosized polystyrene spheres having diameters of 8, 12, 25, 50 and 100 μm. The sprays were sampled very close to the fuel nozzle orifice, i.e., 2 mm downstream of the nozzle. As a result,

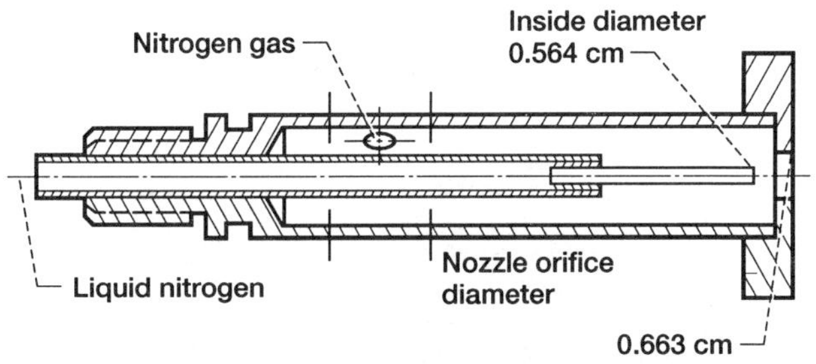


Figure 2.—Diagram of pneumatic two-fluid atomizer.

the $LN_2$ sprays contained relatively high number-densities of very small droplets and drop size measurements required correction for multiple scattering as described in Ref. 10. Also, the measurements were corrected to include Mie scattering theory, when very small drop diameters, i.e., 10 μm or less were measured. Reproducibility tests of the drop size data gave an agreement of ±5 percent. The effect of gas-density gradients on characteristic drop-size measurements was avoided by obtaining new background readings at each gasflow condition when the atomizing-gas temperature, Tg, was 111 K and 422 K. At Tg = 293 K, the background reading remained constant when gas flowrate was varied, since the ambient airstream temperature was also 293 K.

## EXPERIMENTAL RESULTS

Effects of atomizing-gas properties on experimental measurements of $D_{v.5e}$ were investigated. The entire spray cross section was sampled with the downstream edge of the laser beam located at a distance of 2.2 cm downstream of the fuel nozzle orifice, as shown in Fig. 3. Small $LN_2$ droplets were partially vaporized as they passed through the 2 cm scattered-light scanner laser beam and some of the very small droplets were completely vaporized before they could exit the beam. As a result, experimental measurements of drop size could only be obtained for partially vaporized $LN_2$ sprays. Therefore, the change in dropsize, $\Delta D_{v.5c}$, was calculated by using previously obtained, Ref. 2, vaporization rate expressions. It was then used to compute the initially unvaporized spray drop size, $D_{v.5c}$, that had been formed at the fuel nozzle orifice.

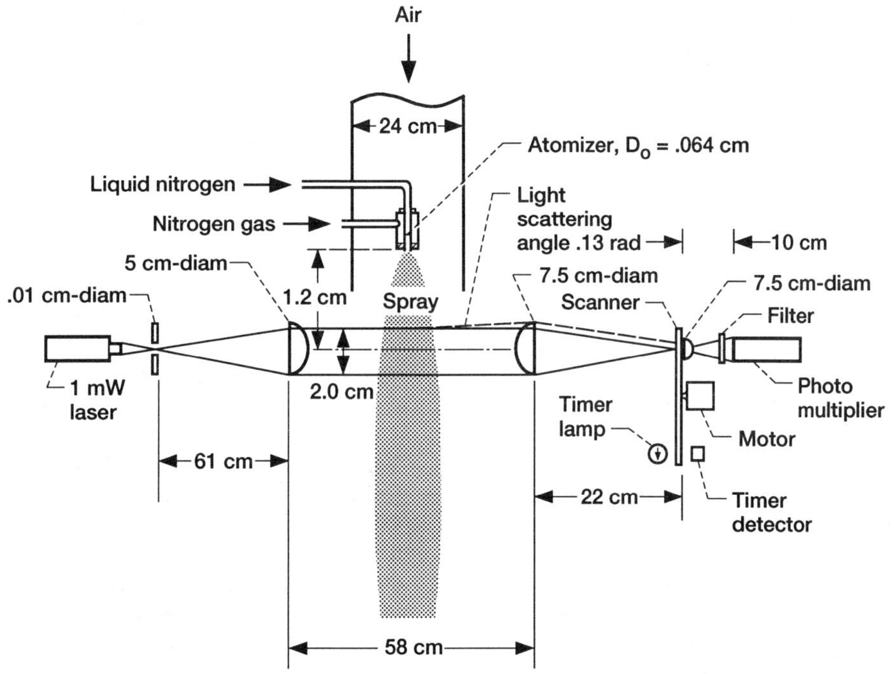


Figure 3.—Atmospheric pressure test section and optical path of scattered-light scanner.

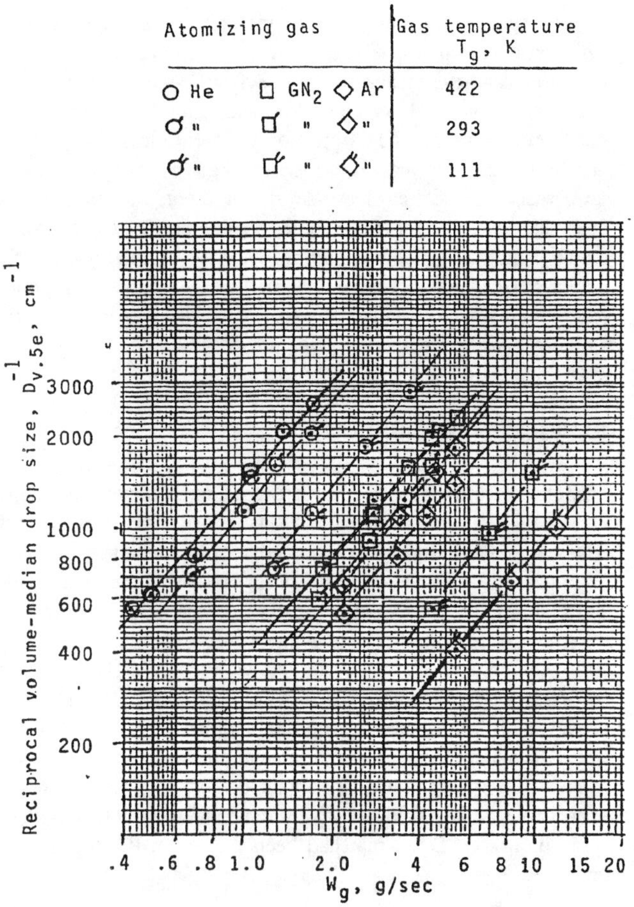


Figure 4.—Variation of experimentally determined volume-median drop size, $D_{v.5e}$, with atomizing-gas flowrate, $W_g$.

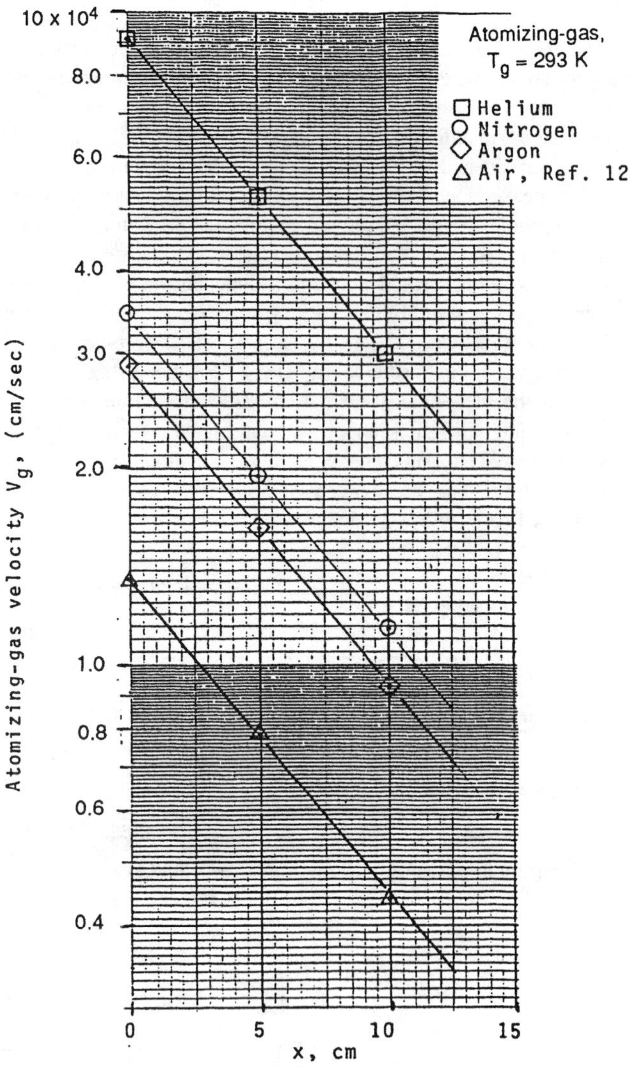


Figure 5.—Deceleration of atomizing-gases downstream of fuel-nozzle orifice. At x = 0, $V_g = V_c$.

## Atomizing-Gas Flowrate Effects On $D_{v.5e}$

Effects of atomizing-gas flowrates and the physical properties of He, $GN_2$ and Ar on $D_{v.5e}$ were investigated. Sonic velocity gasflows were used to atomize $LN_2$ jets. A plot of atomizing gasflow, Wg, against reciprocal values of $D_{v.5e}$ is shown in Fig. 2. From this plot, the following general expression was obtained:

$$D_{v.53}^{-1} = k_e W_g^n \qquad (1)$$

where $k_e$ = 1125, 275 and 222 and n = 1.10, 1.11 and 1.08 for the atomizing gases He, $GN_2$ and Ar, respectively, at a gas temperature of 293 K. Values of $D_{v.5e}$ and $W_g$ are given in cm and g/sec, respectively.

Experimental values of the exponent n varied slightly, from 1.08 to 1.11 and they were considerably below the value of n = 1.33 as predicted by atomization theory and reported in Ref. 13, for the breakup of liquid jets in high-velocity gasflow. This discrepancy between experimental and theoretical values of exponent n is attributed to the complete vaporization of very small and highly volatile $LN_2$ droplets and incomplete vaporization of the relatively large $LN_2$ drops. This occurred before dropsize measurements could be completed with the scattered-light scanner. Therefore, in the present study, the effect of droplet vaporization on drop size measurements of highly volatile $LN_2$ sprays is recognized. In Ref. 11, this effect was not accounted for when the drop size data were analyzed. As a result, even though effect of atomizing-gas flowrate on $D_{v.5e}$ did appear to agree with theory, it is apparent now that the proportionality constant $k_e$ was too low to be used in determining the characteristic dropsize of an initially unvaporized cryogenic spray.

## Droplet Acceleration and Vaporization Time

Effects of $LN_2$ droplet vaporization rates on $D_{v.5e}$, in high-velocity gasflows, were determined by calculating vaporization time, t, as based on drop velocity, Vd, for a given characteristic dropsize. Time, t, was calculated over a distance of 2.2 cm, i.e., from nozzle orifice to the downstream edge of the laser beam, as shown in Fig. 3. Values of Vd and acceleration, a, for $LN_2$ droplets were calculated according to the following momentum balance, as given in Ref. 8:

$$Md\ a = 1/2\ Ad(Vg - Vd)^2 Cd \quad (2)$$

where Md and Ad are mass and area of drop size $D_{v.5e}$, respectively. Also, $Md = \rho_2 \pi D^3_{v.5e/6}$ and Cd is the drag coefficient based on characteristic length, $D_{v.5e}$.

Rewriting Eq. (2) in terms of incremental changes in Vd over a distance $\Delta x$, gives the following relationship:

$$\frac{\Delta V^2 d}{\Delta x} = \frac{3(Vg - Vd)Cd}{2\,D_{v.5e}} \quad (3)$$

where $Cd = 27\ Re^{0.84}$, as given in Ref. 2 and Re is based on the characteristic drop size, $D_{v.5e}$.

The atomizing gases He, $GN_2$ and Ar decelerated into low-velocity airflow and gas velocity at the nozzle orifice was equal to the acoustic velocity of the gas. Values of Vg used to solve Eq. (3) were calculated at downstream distances of 5 and 10 cm, respectively, and are plotted in Fig. 5. Values of Vg based on data given in Ref. 12 are plotted in Fig. 5 for comparison. The percent deceleration of the atomizing gases was assumed to be approximately the same as that observed in Ref. 12, since similar two-fluid fuel nozzles were used in both studies.

Initial droplet velocity, $V_d$, was 2.55 m/sec., i.e., the injection velocity of the $LN_2$ jet. Acceleration of $LN_2$ droplets characterized by $D_{v.5e}$ were determined by numerically integrating Eq. (3) and plotting $V_d$ against downstream distance, x, as shown in Fig. 6. From this plot, vaporization time t was calculated by means of the expression $t = x/V_d$. Values of t are given in table II, for Tg = 293 K, along with Reynolds numbers averaged over the distance x and values of $D_{v.5e}$. Atomizing-gas transport properties used in calculating vaporization times are given in table III, for Tg = 293 K.

TABLE II.—VAPORIZATION TIME, Δt, AND REYNOLDS NUMBER

| Atomizing gas at Tg = 293 K | Wg, g/sec | $D^{-1}_{v.5e}$, $cm^{-1}$ | $\Delta t \times 10^4$ sec | Re |
|---|---|---|---|---|
| Helium | 1.00 | 1125 | 1.35 | 15.2 |
| Nitrogen | 4.54 | 1650 | 1.44 | 35.3 |
| Argon | 5.43 | 1370 | 1.52 | 31.4 |

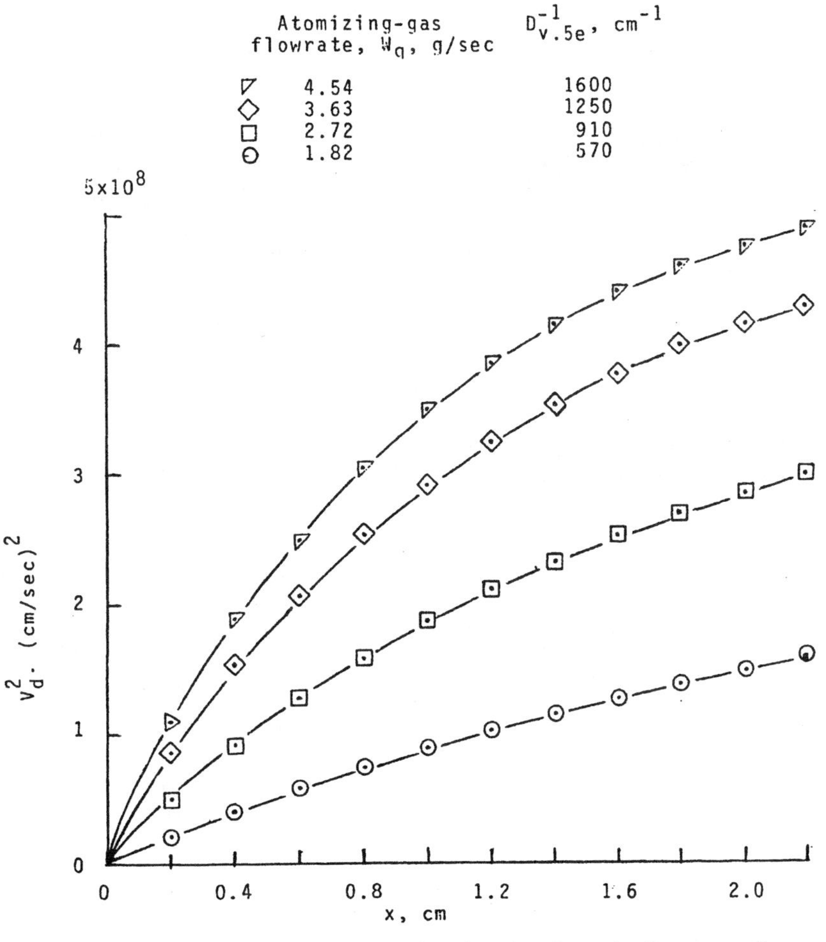


Figure 6.—Acceleration of volume-median dropsize, $D_{v.5e}$, in nitrogen gasflow, at $T_g$ = 293 K.

TABLE III.—ATOMIZING-GAS TRANSPORT PROPERTIES; Tg = 293 K and Wg = 4.54 g/sec

| Atomizing gas | $Vc\times10,^{-4}$ cm/sec | $\mu_r\times10,^{4}$ g/cm sec | $kf\times10,^{5}$ cal/sec sq cm, °C/cm |
|---|---|---|---|
| Helium | 9.10 | 1.98 | 26.3 |
| Nitrogen | 3.43 | 1.25 | 4.2 |
| Argon | 2.87 | 1.10 | 2.85 |

## Computation of Initial Characteristic Drop Size, $D_{v.5c}$

Vaporization rates of characteristic drop sizes of the cryogenic sprays were calculated according to the heat balance expression: $dm/dt = hA\Delta T/Ht$, where h is the heat transfer coefficient and A is spray surface-area based on $D_{v.5e}$. Also, $\Delta T = Tg - Td$ and $Ht = Hv + Cp\Delta T$, where Hv is latent heat of vaporization of $LN_2$ and Cp is the specific heat of nitrogen vapor. The droplet vaporization rate, i.e., the rate at which spray surface-area changes with time, was calculated from the following heat-balance expression:

$$\frac{-D^2_{v.5}}{\Delta t} = \frac{4kg\ \Delta t\ Nu}{\rho_g Ht} \tag{4}$$

where kg and $\rho_g$ are gas thermal conductivity and liquid density, respectively. Previous studies of fuel droplet vaporization, as reported in Ref. 2, used a high-speed droplet tracking camera to determine vaporization rates of jet-A fuel, n-octane, water and several other liquids. As a result, the following expression was obtained: Nu + 2 + 0.303 Re, where Nu is Nusselt number and $Re = D_{v.5e}\ \rho_g V_r/\mu_g$. $V_r$ is gas velocity relative to drop velocity, which is averaged over the incremental distance $\Delta x$. Atomizing-gas viscosity and thermal conductivity were evaluated at the average gas-film temperature. $LN_2$ temperature was kept at near the boiling point 77 K, i.e. sufficiently low to avoid flash boiling of the liquid jet. Latent heat of vaporization of $LN_2$ and the specific heat of the nitrogen vapor were evaluated at 77 K and the average gas-film temperature, respectively.

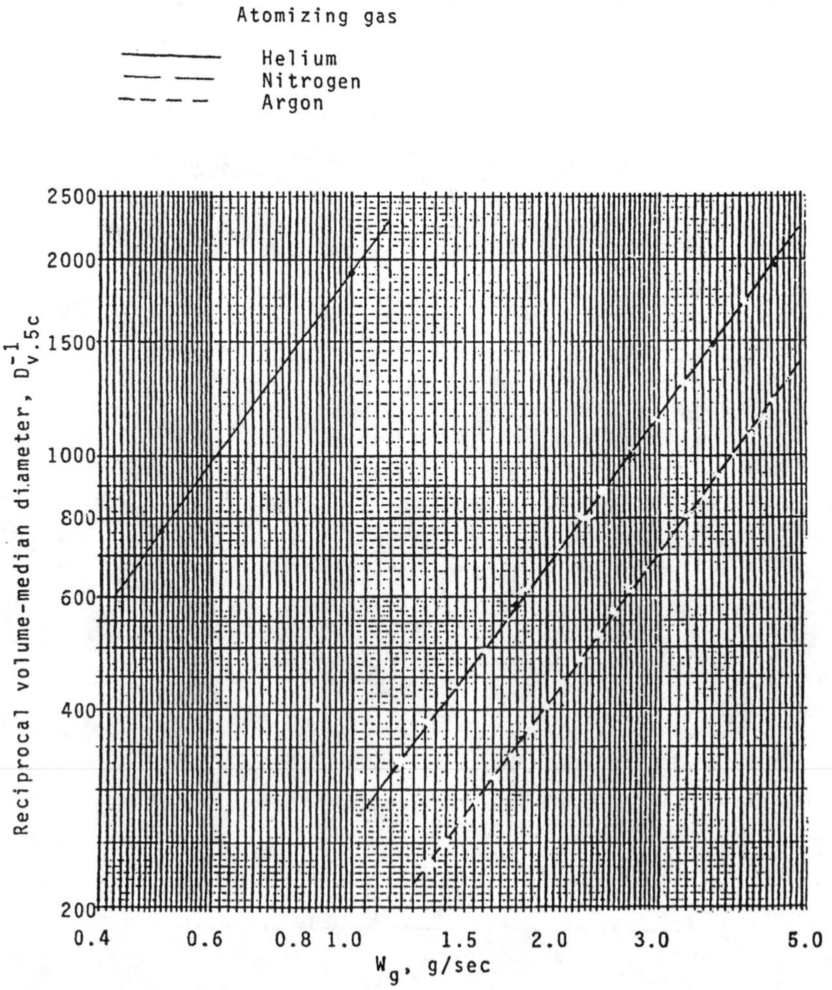


Figure 7.—Effect of $W_g$ on initially unvaporized $D_{v.5c}$, at fuel-nozzle orifice, x = 0, and $T_g$ = 293 K.

Values of the initially unvaporized volume median diameter squared, $D^2_{v.5c}$, were calculated from experimental measurements of $D^2_{v.5e}$ and values of $-\Delta D_{v.5}$ were obtained from Eq. (4), as follows:

$$-\Delta D^2_{v.5} = D^2_{v.5c} - D^2_{v.5e} \tag{5}$$

Values of $D_{v.5c}$ were correlated with atomizing-gas flowrate, $W_g$, as shown in Fig. 7 and the following general expression was obtained:

$$D^{-1}_{v.5c} = kc\ W_g^{1.33} \tag{6}$$

where $k_c$ = 264 for $GN_2$ at 293 K. For the same conditions, the experimental value of ke obtained from Eq. (1) was 275. Thus, values of $k_c$ and $k_e$ were nearly the same. However, the value of exponent n for partially vaporized sprays is considerably less, i.e., 1.11 as compared with 1.33 given in Eq. (6). This shows a marked effect of droplet vaporization on drop size measurements. Also, Eq. (6) shows good agreement of calculated values of n with atomization theory, i.e., Ref. 13 predicts n = 1.33 for liquid jet breakup in high-velocity gasflow. As a result of using heat-transfer and drag coefficient expressions reported in Ref. 2, it was possible to compute values of characteristic drop size $D_{v.5c}$ for unvaporized $LN_2$ sprays formed at the fuel injector orifice.

## Correlation of $D_{v.5c}$ with Dimensionless Groups

$D_{v.5c}$ was normalized with respect to $LN_2$ jet diameter, Do, and plotted against the product of We, Re and ρg/ρl, i.e., the Weber number, Reynolds number and fluid-density ratio, respectively, as shown in Fig. 8. This plot yields the following expression:

$$Do/D_{v.5c} = kc(WeRe\ \rho g/\rho 1)^{0.44} \tag{7}$$

where WeRe is the ratio of aerodynamic to liquid surface forces, i.e., $LN_2$ viscosity and surface tension. The three lines in Fig. 8 show, that the effect of physical properties of the three atomizing gases on $k_c$, needs to be determined to obtain a single correlating expression based on Eq. (7).

## Atomizing-Gas Property Effects on $LN_2$ Jet Breakup

In Ref. 14, a series of tests were made using He, $GN_2$ and Ar as atomizing gases to study the effect of atomizing-gas properties on $LN_2$ jet breakup at a gas temperature of 293 K. As a result of this study, a new dimensionless group was derived that gave the effect of rms gas molecular-velocity, Vm, and gas viscosity, μg, on $D_{v.5c}$, as shown by the following relationship:

$$Do/D_{v.5c} \sim (\rho 1 Vm^3/g\mu g)^{0.75} \tag{8}$$

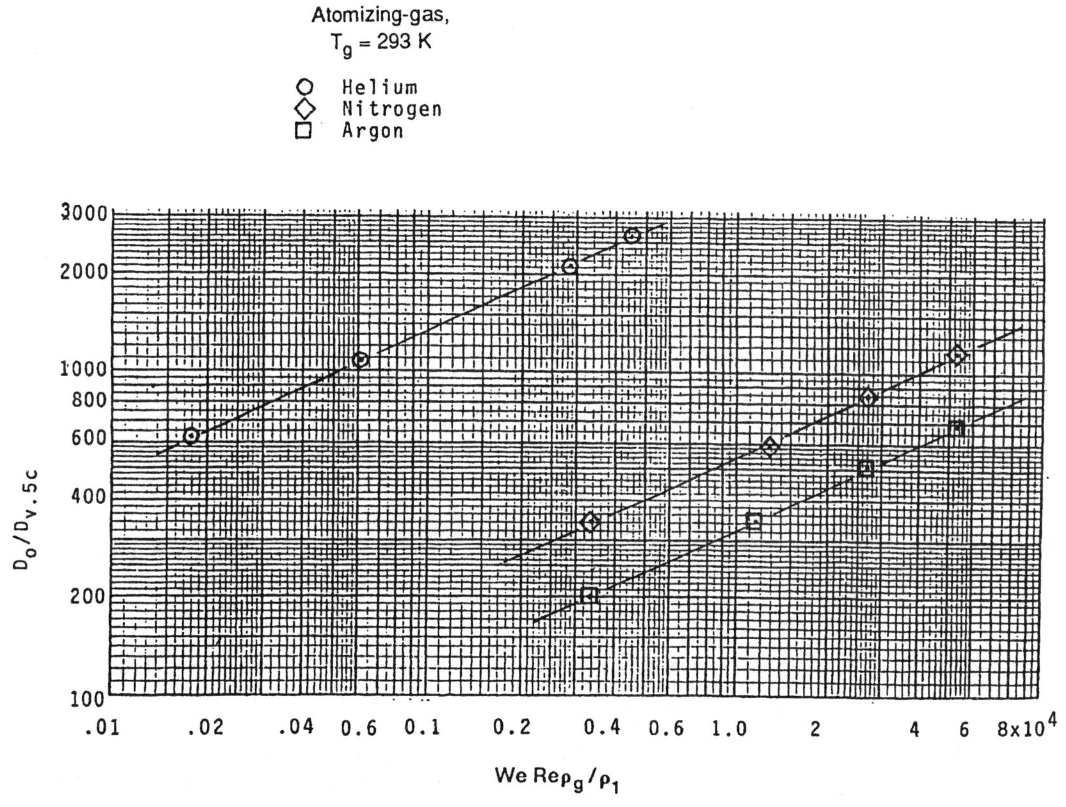


Figure 8.—Correlation of $D_o/D_{v.5c}$ with dimensionless groups.

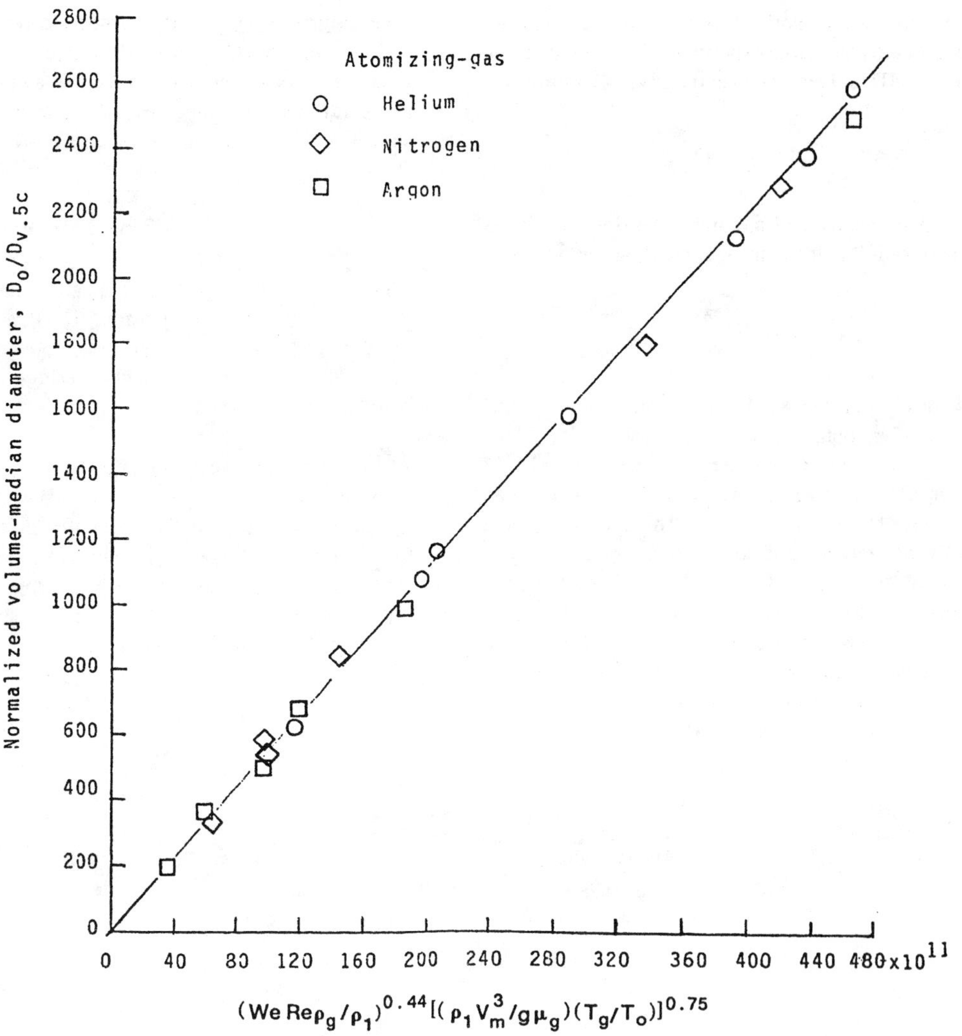


**Figure 9.—Correlation of volume-median diameter, $D_{v.5c}$ with dimensionless groups.**

where Vm is the RMS molecular gas-velocity and where the gas molecular-acceleration group is normalized with respect to the force due to gravitational acceleration, g.

To obtain a single correlating coefficient, k'c, for the three atomizing gases, $Do/D_{v.5c}$ is plotted against the three dimensionless ratios; WeRe ρl/ρg, $\rho l/V^3 m/g\ \mu g$ and Tg/To, as shown in Fig. 9. Thus, the following expression is obtained:

$$Do/D_{v.5c} = k'c(WeRe\ \rho l/\rho g^{0.44}[(\mu l\ V^3 m/g\mu g)(Tg\backslash To)]^{0.75} \qquad (9)$$

where $k'c = 5.7\times10^{-11}$, for $LN_2$ jet breakup in high-velocity gasflows of He, $GN_2$ and Ar over a gas temperature range of 111 to 293 K.

## CONCLUDING REMARKS

In the present study, an expression for cryogenic liquid jet breakup in two-fluid fuel nozzles was obtained that was valid over a wide range of atomizing-gas properties. Gas molecular weight and temperature were varied over ranges of 4 to 40 and 111 to 293 K, respectively. Also it should be noted that the two-fluid fuel nozzle used in this study is the same type of atomizer as that used in the $RL^{-10}$ main rocket engines of the space shuttle.

## REFERENCES


1. Ingebo, R.D., “Experimental and Theoretical Effects of Nitrogen Gas Flowrate on Liquid Jet Atomization,” Journal of Propulsion and Power, Vol. 4, No. 6, Nov.-Dec. 1988, pp. 406–411.
2. Ingebo, R.D., Atomization, Acceleration and Vaporization of Liquid Fuels,” Sixth Symposium (International) on Combustion, Reinhold Publishing Corporation, New York, 1957, pp. 684–687.
3. Kim, K.Y., and Marshall, W.R., Jr., “Drop Distributions from Pneumatic Atomizers,” AIChE Journal, Vol. 17, No. 3, May 1971, pp. 575–584.

4. Lorenzetto, G.E., and Lefebvre, A.H., "Measurements of Drop Size on a Plain-Jet Airblast Atomizer," AIAA Journal, Vol. 15, No. 7, July 1977, pp. 1006–1010.
5. Nukiyama, S., and Tanasawa, J., "Experiments on the Atomization of Liquids by Means of a Air Stream, Parts III-IV," Transactions of the Society of Mechanical Engineers, Japan, Vol. 5, No. 18, Feb. 1939, pp. 63–75.
6. Weiss, M.A., and Worsham, C.H., "Atomization in High Velocity Airstreams," American Rocket Society Journal, Vol. 29, No. 4, Apr. 1959, pp. 252–259.
7. Wolf, H.E., and Anderson, W.H., "Aerodynamic Breakup of Liquid Drops," Proceedings of the 5th International Shock Tube Symposium, edited by Z.I. Slawasky, J.F. Moulton, Jr., and W.S. Filler, Naval Ordnance Lab., White Oak, MD, 1966, pp. 1145–1169. (Avail. NTIS, AD–638011).
8. Buchele, D.R., "Particle Sizing by Weighted Measurements of Scattered Light," NASA TM–100968, 1988.
9. Swithenbank, J., Beer, J.M., Taylor, D.S., Abbot, D., and McCreath, G.C., "A Laser Diagnostic Technique for the Measurement of Droplet and Particle Size Distribution," Experimental Diagnostics in Gas Phase Combustion Systems, edited by B.T. Zinn and C.T. Bowman, Progress in Astronautics and Aeronautics, Vol. 53, AIAA, New York, 1977, pp. 421–447.
10. Felton, P.G., Hamidi, A.A., and Aigal, A.K., "Measurement of Drop-Size Distribution in Dense Sprays by Laser Diffraction," from No. 12: ICLASS–86; Proceedings of the Third International Conference on Liquid Atomization and Spray Systems, Vol. 2, edited by P. Eisenklam and A. Yule, Institute of Energy, London, UK, 1985, pp. IVA/4/1–IVA/4/11.
11. Ingebo, R.D., and Buchele, D.R., "Scattered-Light Scanner Measurements of Cryogenic Liquid-Jet Breakup," NASA TM–102432, 1990.
12. Bulzan, D.L., Levy, Y., Aggarwal, S.K., and Chitre, S., "Measurements and Predictions of a Liquid Spray From an Air-Assist Nozzle," Atomization and Sprays, Vol. 2, 1992, pp. 445–462.
13. Adelberg, M., "Mean Drop Size Resulting From The Injection of a Liquid Jet Into a High Speed Gas Stream," AIAA Journal, Vol. 6, No. 6, June 1986, pp. 1143–1147.
14. Ingebo, R.D., "Atomizing-Gas Temperature Effects on Cryogenic Spray Dropsize," NASA TM–106106, 1993.

FACT-Vol. 18, Combustion Modeling, Scaling and Air Toxins
ASME 1994

# COMBUSTION SCHEMES FOR TURBULENT FLAMES

Mingchun Dong and David G. Lilley
School of Mechanical and Aerospace Engineering
Oklahoma State University
Stillwater, Oklahoma

## ABSTRACT

Improved simulation of turbulent chemically-reacting flow is possible by appropriate choice of combustion model. In the present case, axisymmetric flow in a typical combustor is considered. Fuel enters centrally through a round cross-sectioned inlet while air enters through a concentric annular injet. Combustion flowfield predictions are given to illustrate the effects of various reaction schemes on the resulting flowfield (streamline patterns, velocities, species concentrations, temperature, etc.). It is deduced that higher-order chemical reaction schemes performed better than lower-order schemes. The two-step and four-step schemes both show promise for application in gas turbine combustors, because more kinetic information can be qualitatively calculated with these two schemes. The mechanism of the four-step chemical reaction scheme is appealing, and with further parameter evaluation and testing, it is likely to permit more accurate prediction of major species, such as unburned fuel, carbon monoxide and hydrogen.

## BACKGROUND

In the design of combustion equipment, lengthy and costly experimentation on full-scale hardware is required. These experiments can be diminished in number and economical design practices greatly facilitated by the availability and understanding of prior predictions of the combustor flowfield obtained via the use of mathematical models incorporating numerical prediction procedures. Such work combines the rapidly developing fields of theoretical combustion aerodynamics and computational fluid dynamics and their continued improvement and use will significantly reduce the time and cost of combustor development programs [1, 2]. Consequently, a phenomenological modeling approach to combustor design supplemented by key experiments which depict the fundamental physical phenomena present within combustors now is feasible and a necessity in minimizing combustor development time and cost requirements. This paper is concerned with a two-dimensional axisymmetric modeling approach with emphasis placed on combustion models and their application to reacting flows.

A primitive, pressure-velocity-variable, finite-difference code is now being applied to predict two-dimensional axisymmetric turbulent reacting flows. The method and program involve a staggered grid system for axial and radial velocities and a line relaxation technique for efficient solution of the equations. Turbulence simulation is by way of a two-equation k-e model and combustion via three chemical reaction schemes based on Arrhenius and eddy-breakup concepts for diffusion and premixed situations. The present solution method and computer code have been developed from previous work [3-10] and combustion flowfield predictions are now given to illustrate the effects of various reaction schemes on the flowfield (velocities, species concentrations, temperature and turbulence levels), and dramatic effects are illustrated via radial profiles and shaded contour plots.

The present contribution focuses on combustion simulation schemes for turbulent flames. In combustion simulation, the overall one-step reaction scheme fails to predict the formation of intermediates, such as carbon monoxide. The two-step scheme proposed by Westbrook and Dryer [11] can provide the prediction of carbon monoxide, which will improve the combustion simulation dramatically, but the formation of intermediate hydrocarbons is ignored. The four-step scheme proposed by Hautmanm et al. [12] describes a transformation mechanism of the hydrocarbon fuel into intermediate hydrocarbons and hydrogen, and the oxidation of intermediates to carbon monoxide, carbon dioxide, and water vapor. These three overall reaction schemes are investigated in this study for combustion simulation.

## COMBUSTION SCHEMES

### One-Step Scheme

The simplest global mechanism is the one-step reaction scheme. The advantage of this mechanism is its simplicity; it involves the solution of the conservation equations for stagnation enthalpy, unburned fuel and the mixture fraction (diffusion and premixed flames can be treated in the same way for one-step scheme). The heat release and the concentrations of other species are then obtained from linear functions of the amount of fuel consumed. But this model fails to predict the important characteristics of hydrocarbon oxidation, that is, the formation of intermediates and carbon monoxide, which influence the process considerately. As a result, it is inadequate to obtain quantitative predictions. The scheme is described as following:

$$C_xH_y + \left(x + \frac{y}{4}\right)\left(O_2 + nN_2\right) \rightarrow xCO_2 + \frac{y}{2}H_2O + \left(x + \frac{y}{4}\right)nN_2$$

where $C_xH_y$ is a typical hydrocarbon fuel, $O_2$ is oxygen, $N_2$ is nitrogen, $CO_2$ is carbon dioxide, and $H_2O$ is water vapour. It assumes that the fuel and oxidant react chemically in a unique proportion, combining with a stoichiometric oxidant/fuel ratio of i to form product plus release of energy by fuel burning (finite rate chemistry is assumed). The reaction rate expression to be used for fuel consumption is usually taken as the minimum of the Arrhenius and Eddy-Breakup models, which have been widely discussed, see Refs. 11 through 13, for example.

### Two-Step Scheme

A slightly more complex scheme is the two-step mechanism [11]:

$$C_xH_y + \left(\frac{x}{2} + \frac{y}{4}\right)\left(O_2 + nN_2\right) \rightarrow xCO + \frac{y}{2}H_2O + \left(\frac{x}{2} + \frac{y}{4}\right)nN_2$$

$$xCO + \frac{x}{2}\left(O_2 + nN_2\right) \rightarrow xCO_2 + \frac{y}{2}nN_2$$

Here $C_xH_y$ is a typical hydrocarbon fuel, and the additional species (which did not occur in the one-step scheme) is carbon monoxide CO. This scheme involves the solution of one additional equation for the concentration of carbon monoxide. The reaction rates to be used with these two equations are given in Refs. 11 and 13. Although the two-step scheme has been widely used, it is deficient in that the formation of intermediates is ignored.

### Four-Step Scheme

The complete oxidation of hydrocarbon fuel can be described by the following steps:

(a) Transformation of the hydrocarbon fuel into intermediate hydrocarbons and hydrogen with little release of energy
(b) Oxidation of intermediates to carbon monoxide and hydrogen
(c) Oxidation of carbon monoxide to carbon dioxide
(d) Oxidation of hydrogen to water vapour

Steps (b) through (d) are exothermic and are responsible for the release of energy and associated temperature rise. A reaction scheme, which is designed to model correctly the oxidation process, must include a description of these steps. The newly introduced species are intermediate hydrocarbons and hydrogen $H_2$, in addition to those of the one-step and two-step schemes.

The simplest mechanism that accounts for the essential features of the hydrocarbon oxidation is the following four-step scheme proposed by Hautman, et al. [12]:

$$C_NH_{2N+2} \rightarrow \frac{N}{2}C_2H_4 + H_2$$
$$C_2H_4 + O_2 \rightarrow 2CO + 2H_2$$
$$CO + \frac{1}{2}O_2 \rightarrow CO_2$$
$$H_2 + \frac{1}{2}O_2 \rightarrow H_2O$$

which is valid only for alphatic hydrocarbons (of the type $C_NH_{2N+2}$). To accommodate a general hydrocarbon $C_xH_y$, the first two steps have been modified by Srinivasan et al. [13] as following:

$$C_xH_y \rightarrow C_xH_{y-2} + H_2$$
$$C_xH_{y-2} + \frac{x}{2}O_2 \rightarrow xCO + \frac{y-2}{2}H_2$$

This scheme involves the solution of two more transport equations for the mass fractions of intermediate hydrocarbons ($C_xH_{y-2}$) and hydrogen, comparing with transport operations for unburned hydrocarbon fuel, carbon monoxide, and "total fuel" as outlined in the two-step scheme. References 12 and 13 may be relied upon to supply the reaction rate expressions to be used with these equations.

## COMPUTATIONAL SIMULATION

In the computer simulation of chemically reacting flowfields, the problem is simulated by simultaneous nonlinear partial differential equations. These are the full elliptic equations -- the turbulent Reynolds equations of conservation of mass, momentum (with x, r, θ velocity components u, v, w), stagnation enthalpy h, chemical species mass fractions $m_j$, turbulence energy k, and turbulence dissipation rate e, which govern the two-dimensional steady flow of turbulent chemically reacting multicomponent mixtures. The governing partial differential equations are reduced to a set of finite-difference equations for values at points of the grid system covering the solution domain. These, together with appropriate boundary conditions, constitute a system of strongly coupled simultaneous algebraic equations. The solution proceeds as described in earlier work [3-8]. The task is to obtain predictions for velocities, pressure, temperature, species concentrations, and turbulence quantities throughout the region of interest. The technique involves a staggered grid system for axial and radial velocities, a line relaxation procedure for efficient solution of the equations and a two-equation k-e turbulence model, together with an appropriate chemical reaction scheme as just described.

Consideration is given to recent work in the development of a primitive-variable finite difference solution procedure. The simulation and solution techniques focus attention directly on the primitive pressure and velocity variables. An Eulerian finite difference formulation is used with pressure and velocity as the main dependent variables. In addition, the velocity components are positioned between the nodes where pressure and other variables are stored. At each iterative step, the time-advanced expressions for u and v are substituted into the finite difference form of the continuity equation for each cell, and the pressure correction equation so-derived enables p, u and v to be further corrected to obey continuity. Iteration continues several hundred iteration steps until all the equations are sufficiently well satisfied and values have stabilized throughout the flow domain.

The prediction procedure is obtained by modifying the already-existing and freely-available STARPIC computer codes [7]. Production runs obtain predicted results showing the effect of the flow parameters on the flowfield. The code is a swirling flow version of the Imperial College, London, TEACH computer code [3]. The computer code methodology is described briefly in Ref. 1, pp. 161-165, and swirl flow predictions with codes of this type are documented in Ref. 2, pp. 253-267 (isothermal) and pp. 272-288 (with chemical reaction). This well-tried theoretical technique, based directly on the governing partial differential equations, complements well alternative methods of experimentation and deductions from empirically-based equations.

The mathematical problem, computer code and solution procedure are described well in Ref. 7. The finite difference equations and boundary conditions constitute a system of strongly-coupled simultaneous algebraic equations, with no linearities since the coefficients and source terms in the basic equations are themselves functions of some of the variables, and the velocity equations are strongly linked through the pressure. The TDMA (tri-diagonal matrix algorithm) is used in columns for iterative sweeps of the field, with underrelaxation being used at each step. Iteration monitoring and final convergence is decided by way of a residual-source criterion, which measures the departure from exactness (summed over all points) for each of the variables being solved. Since the ideas of the code have been described well in Refs. 5-8, it is not necessary to repeat the elaborate detail here.

## RESULTS AND DISCUSSION

In this application, the geometry of Lewis and Smoot [14], simulating an industrial furnace, is selected as the test case. In the experiment, coaxial streams of fuel (town gas) and air are injected into a suddenly-expanded combustion chamber, see Figure 1. The flame is stabilized at the dividing lip between the two streams. Measurements have been made of the temperature and time-mean species concentrations. The parameters and test conditions of the combustor are summarized as follows:

Air velocity, $U_{air}$ = 34.3 m/s
Air temperature, $T_{air}$ = 589°K
Fuel velocity, $U_{fuel}$ = 21.3 m/s
Fuel temperature, $T_{fuel}$ = 300°K
Inlet pressure, P = 94 kPa
Overall equivalence ratio ($CH_4/O_2$), OER = 1.18.

In this predictions, a non-uniform grid of 40 x 15 is used (NI = 40, NJ = 15). Uniform axial velocity and temperature profiles are prescribed for the fuel and air stream as given above, respectively. The inlet turbulence intensities for air and fuel are given in the measurements as 6%, and the length scales are assumed to be 0.0057 m for the air jet and 0.0016 m for fuel jet, respectively [15]. The walls are taken to be adiabatic, and the wall function treatment is employed in the momentum equations for wall drag calculations. The fuel mixture fraction is set equal to one in the fuel stream and zero in the air stream.

Predictions of temperature, unburned fuel, oxygen, water vapour, carbon dioxide, carbon monoxide and hydrogen, using three kinetic schemes respectively, are presented in detail in a recent study [16] with available experimental measurements. Figure 2 shows the comparison between the data and the predictions for the temperature. Near the inlet (at x/D = 0.06), the predicted temperature profiles are in good agreement with the data. However, in the developing and middle regions (up to x/D = 2.34), all three models over-predict the temperature level because of the over-estimated reaction rates given by the recommended rate constants of Srinivasan et al. [13]. Generally, the higher-order scheme shows better agreement with the data than the lower-order results due to the relatively slow reaction rate. The unburned fuel mole fractions are presented in Figure 3. These profiles are in good agreement with the measurements. However, the one-step and two-step schemes over-predict the unburned fuel, and the four-step scheme under-predicts the unburned fuel due to the overly-fast extimate of the pyrolysis rate transforming into the intermediate hydrocarbons and hydrogen with little release of energy. This conclusion is substantiated by the predicted temperature levels and oxygen concentrations.

Streamline patterns and contour plots of temperature and unburned fuel mass fractions are presented in Figures 4 through 9, which further illustrate and complement the results discussed above. Results are shown for the one-step and four-step combustion schemes, only. All these figures are stretched in the radial direction by 200 percent to aid observation. First, a large corner recirculation (due to the sudden expansion) is observed in the streamline pattern plots shown in Figures 4 and 5 with the one-step and four step combustion schemes, and approximately the same reattachment length of x/D = 2 is predicted. Second, the predicted flame zones are clearly delineated by the temperature contours as shown in Figures 6 and 7. Although they all look similar, the predicted maximum temperatures (given in the plots), which are used in scaling for each plot, are dramatically different. A sharp temperature gradient seen near the edge of the flame is a direct result of an insufficient diffusion along the radial direction. Additionally, contour maps for unburned fuel mass fractions (shown in Figures 8 and 9) and other species concentrations (see Ref. [16]) further illustrate the other kinetic facets of the entire reacting flowfield.

## CLOSURE

The development of a primitive-variable finite difference computer code has been discussed, with application to the turbulent combustion flowfield of an axisymmetric flow in a typical combustor. Improved simulation of turbulent chemically-reacting flow is possible by appropriate choice of combustion model. It is deduced that higher-order chemical reaction schemes performed better than lower-order schemes. The two-step and four-step schemes both show promise for application in gas turbine combustors, because more kinetic information can be qualitatively calculated with these two schemes. The mechanism of the four-step chemical reaction scheme is appealing, and with further parameter evaluation and testing, it is likely to permit more accurate prediction of major species, such as unburned fuel, carbon monoxide and hydrogen.

## REFERENCES


1. Gupta, A. K. and Lilley, D. G. "Flowfield Modeling and Diagnostics." Abacus Press, Tunbridge Wells, England, 1985.

2. Gupta, A. K., Lilley, D. G. and Syred, N. "Swirl Flows." Abacus Press, Tunbridge Wells, England, 1984.

3. Gosman, A. D. and Pun, W. M. "Calculation of Recirculating Flows." Report No. HTS/74/12, Dept. of Mech. Engng., Imperial College, London, England, 1974.

4. Lilley, D. G. "Primitive Pressure-Velocity Code for the Computation of Strongly Swirling Flows." AIAA Journal, Vol. 14, June 1976, pp. 749-756.

5. Novick, A. S., Miles, G. A., and Lilley, D. G. "Numerical Simulation of Combustor Flow Fields" Journal of Energy, Vol. 3, March-April 1979, pp. 95-105.

6. Novick, A. S., Miles, G. A. and Lilley, D. G. "Modeling Parameter Influences in Gas Turbine Combustor Design." Journal of Energy, Vol. 3, Sept.-Oct., 1979, pp. 257-262.

7. Lilley, D. G. and Rhode, D. L. "A Computer Code for Swirling Turbulent Axisymmetric Recirculating Flows in Practical Isothermal Combustor Geometries." NASA CR-3442, Feb. 1982.

8. Samples, J. W. and Lilley, D. G. "Numerical Simulation of Reacting Combustor Flowfields." Proc. ASME Comp. in Engng. Conf., Chicago, IL, July 20-24, 1986.

9. Dong, M. and Lilley, D.G. "A PC-Based Computer Code for Axisymmetric Turbulent Swirling Reacting Flows." Int. Joint Power Generation Conf., Atlanta, GA, Oct. 18-22, 1992.

10. Dong, M. and Lilley, D.G. "Parameter Effects on Flow Patterns in Confined Turbulent Swirling Flows." Int. Joint Power Generation Conf., Kansas City, MO, Oct. 17-21, 1993.

11. Westbrook, C.K. and Dryer, F.L. "Simplified Reaction Mechanisms for the Oxidation of Hydrocarbon Fuels in Flames." Combustion

Science and Technology, Vol. 27, 1981, pp. 31-43.

12. Hautmanm D.J., Dryer, F.L., Schug, K.P. and Glassman, I. "A Multi-step Overall Kinetic Mechanism for the Oxidation of Hydrocarbons." Combustion Science and technology, Vol. 25, 1980, pp. 219-225.

13. Srinivasan, R., Reynolds, R., Ball, I., Berry, R., Johnson, K., and Mongia, H. "Aerothermal Modeling Program, Phase I - Final Report." NASA CR-168243, Vol. 1, August 1983.

14. Lewis, M.H. and Smoot, L.D., "Turbulent Gaseous Combustion, Part I: Local Species Concentration Measurements." Combustion and Flame, Vol. 42, 1981, pp. 183-196.

15. Nikjooy, M. and So, Ronald M.C. "On the Modeling of Non-Reactive and Reactive Turbulent Combustor Flows." NASA CR-4041, Grants NAG3-167 and NAG3-260, April 1987.

16. Dong, M. "Prediction of Turbulent Swirling Reacting Flows" Ph.D. Thesis, Mechanical and Aerospace Engng., Oklahoma State University, Stillwater, OK, 1994.

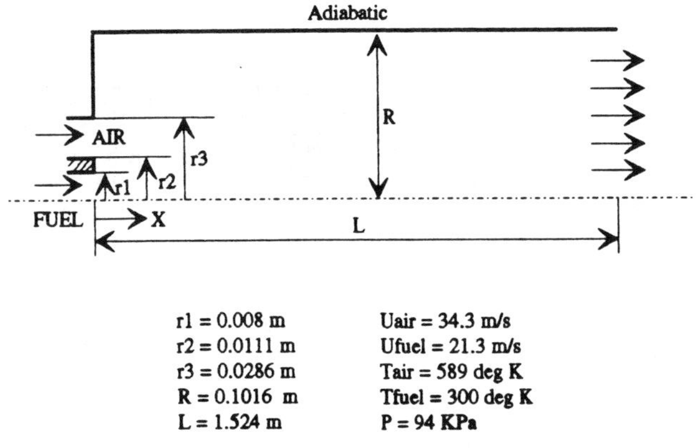


Figure 1. Geometry of Axisymmetric Combustor with Coaxial Fuel and Air Jets [14]

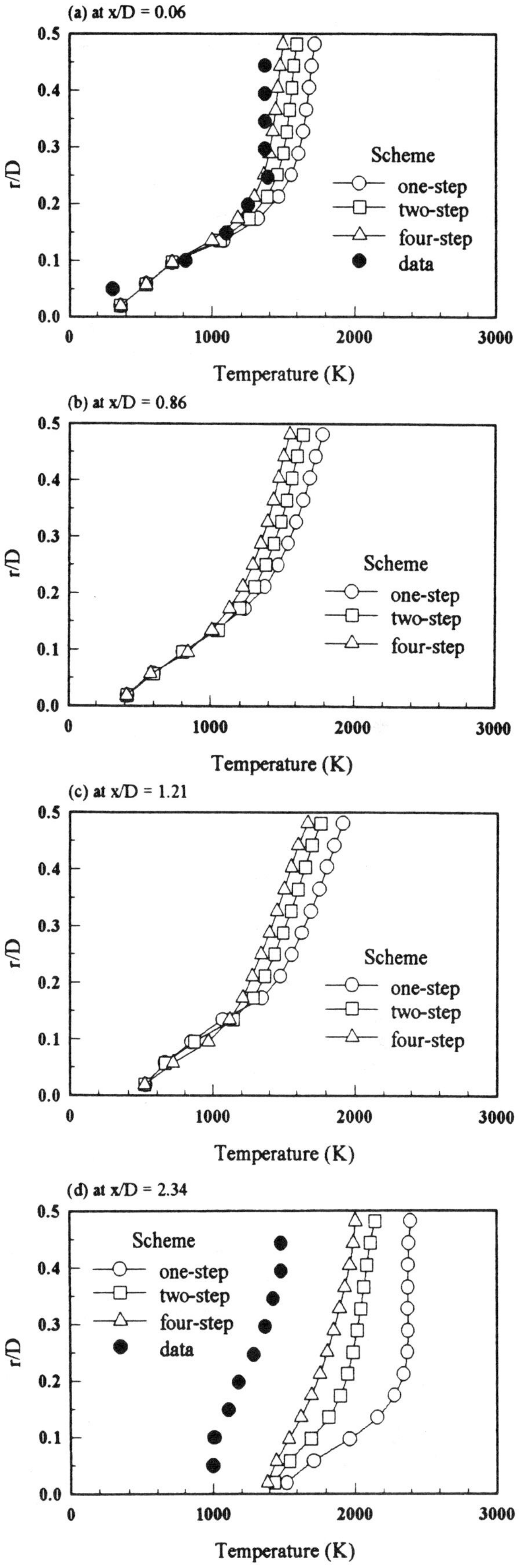


Figure 2. Temperature Profiles with Data of Lewis & Smoot [14]

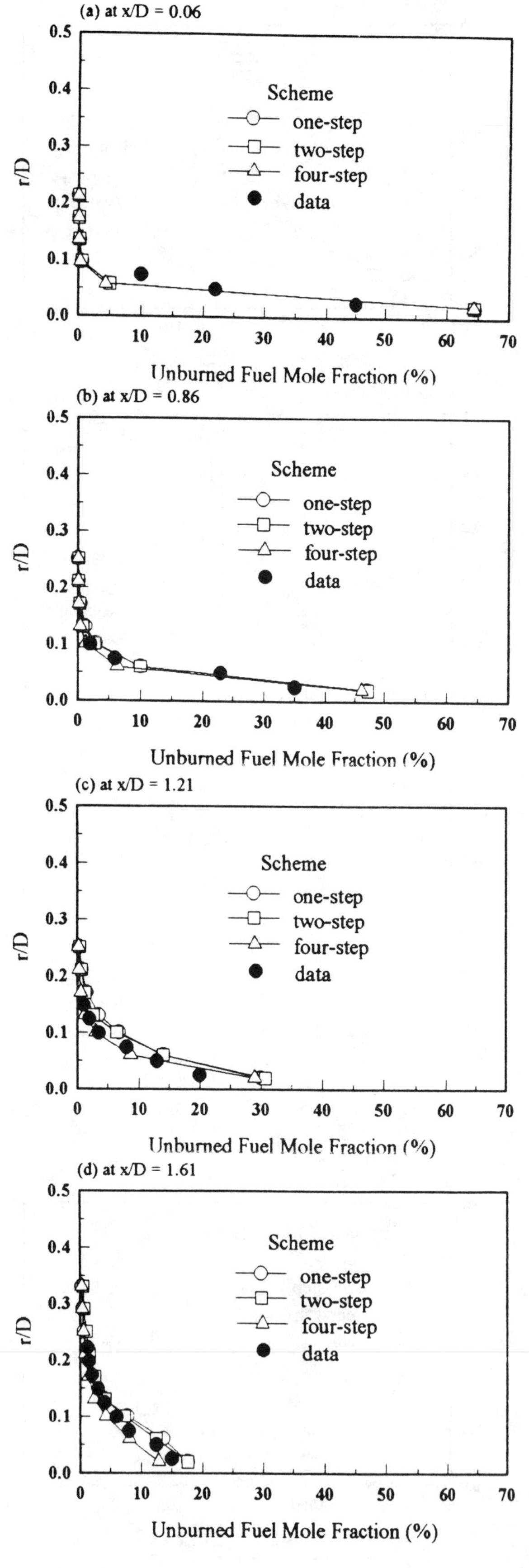


Figure 3. Unburned Fuel Profiles with Data of Lewis & Smoot [14]

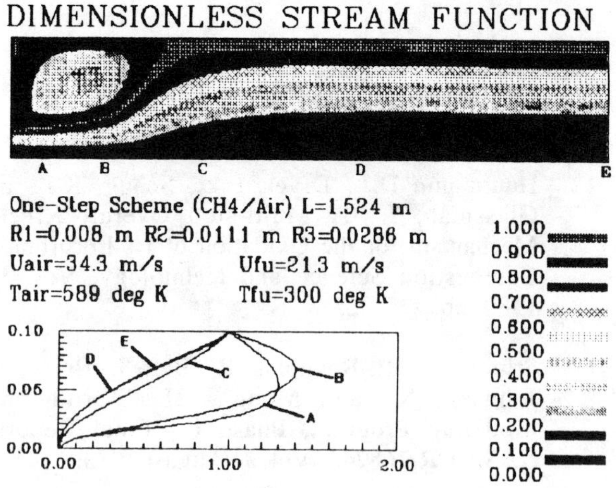


Figure 4. Streamline Patterns with One-Step Scheme

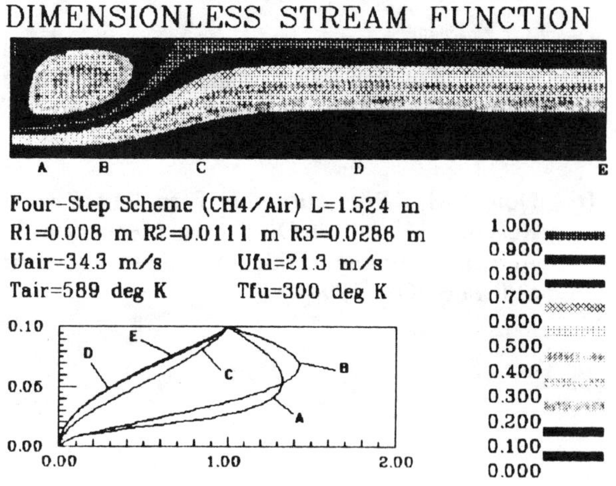


Figure 5. Streamline Patterns with Four-Step Scheme

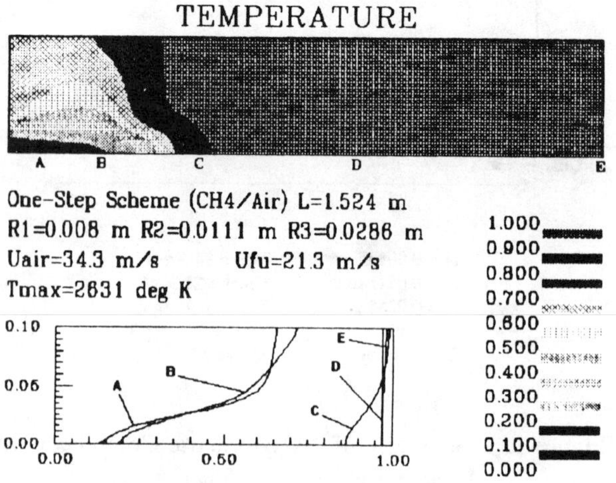


Figure 6. Temperature Contour Map with One-Step Scheme

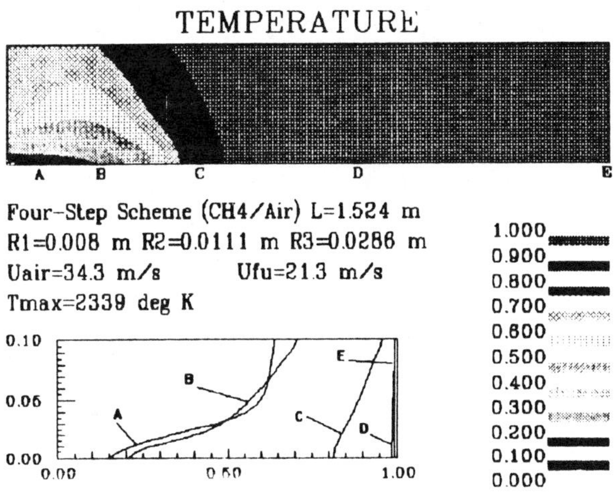


Figure 7. Temperature Contour Map with Four-Step Scheme

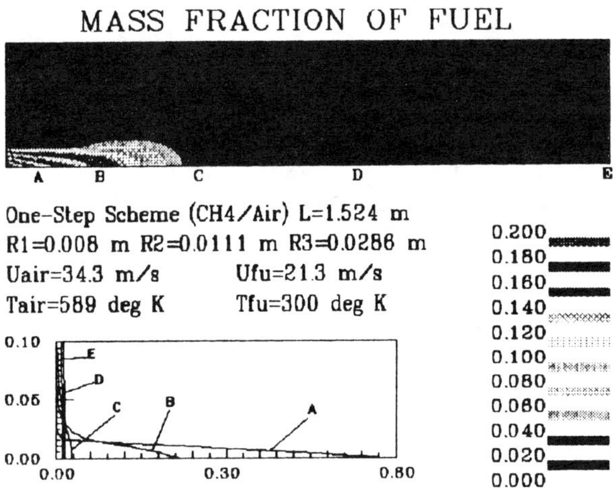


Figure 8. Unburned Fuel Mass Fraction Contour Map with One-Step Scheme

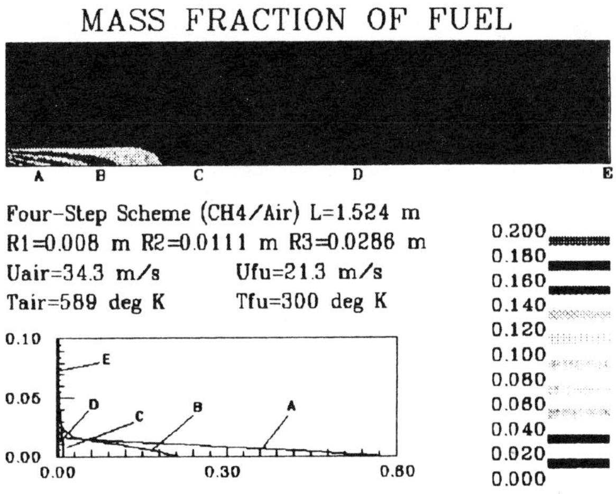


Figure 9. Unburned Fuel Mass Fraction Contour Map with Four-Step Scheme

FACT-Vol. 18, Combustion Modeling, Scaling and Air Toxins
ASME 1994

# COLOR GRAPHIC INTERPRETATION OF FLOWFIELD PREDICTIONS

T.-K. Lin, Mingchun Dong, and David G. Lilley
Oklahoma State University
Stillwater, Oklahoma

## ABSTRACT

Three-dimensional jet mixing is investigated via a three-dimensional fluid flow computer code. The code runs on microcomputers with effective 3-D perspective color graphic displays of the results. The present version of the code retains several simplifications, uses a uniform rectangular grid system, and runs on microcomputers, thus providing a powerful and economical software capability. Equations are solved for the fully 3-D problem, including pressure, three velocity components and species mass fractions. The simulation is a finite difference time-marching procedure using an explicit formulation of the conservation equations, followed at each step by an iteration updating pressures and velocities so as to impose the continuity requirement. The present paper documents the problem, describes briefly its simulation, theory and assumptions, and gives results showing an application of the code to co-flowing round jet mixing, illustrating velocity and density effects.

## INTRODUCTION

Our study is concerned with a method which, in a short computer program, solves the fully three-dimensional time-dependent flow equations in cartesian coordinates. As a sample computation, the solution domain is the mixing region downstream of a jet being injected under co-flow conditions with secondary flow, consistent with the experimental work described in Beer and Chigier [1]. The domain is chosen via inspection of the symmetry planes associated with this situation of a round jet injection into a co-flowing stream. The objective is to simulate the mixing of the jet fluid with the surrounding co-flowing stream, and illustrate velocity and density ratio effects.

## THE PREDICTION PROCEDURE

The 3-D solution domain is shown in Fig. 1. The top, bottom, front and back boundaries are symmetry planes, the left boundary inlet has specified values and the right boundary is given the parallel exit flow condition. The cross-section is in the yz-plane. The mathematical modeling techniques focus attention directly on the primitive pressure and velocity variables, following the Marker and Cell Los Alamos technique. This is one of the most well-known methods to solve time-dependent fluid flow problems; its conceptual simplicity is one of the main attributes. A 2-D version is described with FORTRAN computer code in Ref. 2, upon which the present work is based. The ideas of 3-D flow prediction are given for incompressible flows in cartesian geometries with sample predictions in Ref. 3, while Ref. 4 pp. 175-178 speculates on an application to reacting flow in cylindrical coordinates. An Eulerian finite difference formulation is used with pressure and velocity as the main dependent variables. In addition the velocity components are positioned between the nodes where pressure and other variables are stored.

In the computational sequence, at each time-step, the time-advanced expressions for u, v, and w are substituted into the finite difference form of the continuity equation for each cell, and an iterative process on pressure and velocity corrections is applied until the continuity equation is sufficiently well satisfied. The technique parallels that for a 2-D flowfield; its extension to nonreacting 3-D situations is described in Ref. 5. On the question of turbulence modeling, standard texts indicate that the time-mean behavior of certain turbulent jet flows may be simulated via a constant large viscosity, see Ref. 5. This is the approach taken here.

The justification for taking a constant value turbulence viscosity is well-known in jet mixing. It is asserted that the turbulent viscosity in a round free jet in stagnant surroundings is approximately constant and given by

$$\upsilon_t = 0.00196\,(x + a)U_m$$

in terms of station maximum axial velocity $U_m$. In terms of jet initial velocity $U_{jet}$ and, correcting for the co-flowing velocity, this can be written for the present study as

$$\upsilon_t = 0.00196\,A(U_{jet} - U_{sec})d$$

where A is an empirical constant (equal to 6.8 found in experiments, see Ref. 5) and d is the jet diameter. Other experimenters quote values of A from 5.9 to 6.4 depending upon jet exit conditions. An indication of how susceptible computed results are to this choice of value given to the constant turbulent viscosity may be found via computer experimentation. The results show that, even with this limitation, a useful simulation of the actual turbulent jet mixing problem is available.

The description about the technique, the governing equations, the grid system, the boundary conditions, the solution procedure, and accuracy and stability are well described previously [4,5]. The solution domain is a three-dimensional rectangular region to be considered is divided into rectangular cell divisions, with uniform $\Delta x$, $\Delta y$ and $\Delta z$ spacings. This solution domain is complemented by a layer of cells on all sides, so as to allow easy simulation of the required boundary conditions. These fictitious cells increase the total number of cells in each direction.

The total mesh arrangement is shown in Fig. 2 with a coarse grid, showing 30, 12 and 12 internal cell divisions in directions x, y and z, respectively. Appropriate specification of the spacings and I, J, K limits can assign the solution domain size. Also, the inlet velocities and species mass fractions can be specified on the cells with I = 1 via observation of the yz-section and inlet plane information required form Fig. 2.

The particular sample problem being solved has a solution domain of size:

| | |
|---|---|
| 0.372 | m in x-direction |
| 0.06 | m in y-direction |
| 0.06 | m in z-direction |

defined in Fig. 1. The secondary flow is air (or another gas of specified density) with velocity 0, 25 and 50 m/s in the 143 cells in the inlet cross-section. The primary jet flow of air with velocity 100 m/s enters through the one cell at the lower left of Figs. 1 and 2. Because of symmetry planes, this is equivalent to the jet being idealized with a square cross-section of size 0.01 x 0.01 $m^2$ with equivalent jet diameter of 0.011 m. Both streams enter at 298 K and atmospheric pressure. The primary and secondary jets are positioned as shown in Fig. 2. Figure 2 shows a coarse 34 x 12 x 12 cell mesh system placed over the quarter-section solution domain (a 32 x 14 x 14 mesh system including fictitious cells). Thus the cell sizes $\Delta x$, $\Delta y$, $\Delta z$ are 0.012, 0.005, and 0.005 m, respectively. Approximately 1500 time-steps are required with the coarse grid in order to reach satisfactory steady state results. Grid refinement tests were applied but general results for this strong advection flow were not very susceptible to the degree of refinement.

## RESULTS

To provide an example of an application of the computer code, a jet in co-flowing surroundings is considered. Several studies [6-14] have documented experimental evidence about jets in co-flowing streams. The effect of the velocity ratio $R = U_{sec}/U_{jet}$ [secondary to jet velocity ratio] and density ratio $\rho_{sec}/\rho_{jet}$ are particularly important, since their values are known to affect the downstream development and lateral spread. Our study is basically that of an unconfined jet emerging into an infinite co-flowing stream; in our study 'free slip' boundary conditions are applied at the sides of the domain which are located approximately 6 jet diameters to the side of the jet. From the entrainment point of view the jet is essentially unconfined, see Ref. 4.

A free jet emerging into co-flowing surroundings has been investigated previously, see Refs. 6-13. Ha and Lilley [14] provide a lengthy review of the situation and present predictions using a boundary layer jet mixing computer code. The decay of the axial velocity on the centerline has been plotted versus the downstream distance for this free jet.

Ha and Lilley [14] computed these flows for the axisymmetric flowfields of jets in co-flowing streams, by solving the boundary layer equations using the standard Prandtl mixing length model of turbulence. They investigated jet spread and decay parameter effects by making various production runs; their results confirmed previous experimental findings about the jet development in co-flowing streams, and extended knowledge to more practical situations. When the jet and surroundings have the same density, flowfields differ according to the ratio of secondary to jet velocities. Figure 3 illustrates the predicted effect of surrounding co-flowing velocities ($R = U_{sec}/U_{jet} = 0$, 0.25 and 0.5) on the axial velocity maximum decay. Notice that the one with higher surrounding co-flowing velocity produces less decay of the axial velocity along the centerline in the downstream direction. The calculated results show good agreement with the experimental data. The data shown also compares favorably with previous experimental data. This decay of axial velocity with downstream distance is well-known to combustion aerodynamicists involved with jet mixing.

The predicted effect of the ratio of secondary to jet density ($\rho_{sec}/\rho_{jet} = 0.25$, 0.5 and 1.0) on the axial velocity maximum decay is shown in Fig. 4. The lower density surroundings produce less decay of the axial velocity along the centerline in the downstream direction. This is expected, since there is less drag on the main jet flow from the lower density surroundings. Inspection of Figs. 3 and 4 reveals that the calculated results, are in quite good agreement with the previous numerical investigation by Ha and Lilley [14], which themselves are consistent with previous experimental data.

Flowfield illustrations now given for different velocity ratios R = 0, 0.25, and 0.5 are for the case of the same density in the jet and its surroundings. Figure

5 Parts (a) and (b) illustrates the mixing when the co-flowing stream is stagnant [that is, R = 0 and $U_{sec} = 0$]. The dimensionless axial velocity u is shown at three longitudinal slices and four lateral cross-sections via gray-scaled contour plots in yx-planes and yz-planes, respectively. Each cross-section plane is shown in the 3-D view at the top of the figure with 2-D contour plots shown below. In Part (a) of the figure, longitudinal A, B and C are located at 0.5, 1.8 and 3 jet diameters (approximately) to the side of the jet centerline. In Part (b) of the figure, lateral cross-sections A, B, C and D are located approximately at 4, 10, 16 and 21 jet diameters downstream of the jet injection location. The gray scale indicator is ranged from 0.0 to 1.0 and re-used for values above 1.0 and below 0.0 with the same indicated interval.

Figures 6 and 7 show calculated values of axial velocity when the secondary-to-jet velocity ratio is R = 0.25 and 0.5, respectively. These may be compared with and contrasted to corresponding data given in Fig. 5 which was for the stagnant surroundings case. The predictions illustrate the reduced rate of entrainment and a narrower jet mixing region when the secondary flow has a nonzero velocity. A longer thinner jet mixing region results, and progressively so as the secondary velocity increases. However, with the normalization used [division by $U_{jet}$] higher velocities are seen in the figures in the surrounding region. These and other predictions are extremely informative in understanding the mixing characteristics of any particular 3-D situation.

## CLOSURE

Computer predictions about the effect of flow and geometric parameters on the 3-D flowfield permit an easy quick judgment to be made about the merits of a particular flow mixing and reacting situation. This in turn provides a useful theoretical contribution to industrial thermofluid problems. Current studies are involved with extensions of the code, and application to fires, flames and combustion.

## REFERENCES


1. Beer, J.M. and Chigier, N.A. "Combustion Aerodynamics", Wiley, New York, 1972.

2. Hirt, C. W., Nichols, B. D., and Romero, N. C., "SOLA - A Numerical Solution Algorithm for Transient Fluid Flows", Report LA-5852, Los Alamos Scientific Laboratory, Los Alamos, NM., 1975.

3. Hirt, C. W. and Cook, J. L. "Calculating Three-Dimensional Flows Around Structures and Over Rough Terrain", J. Comp. Physics, Vol. 10, 1972, pp. 324-340.

4. Gupta, A. K. and Lilley, D. G., "Flowfield Modeling and Diagnostics", Abacus Press, Tunbridge Wells, England, 1985.

5. Lilley, D. G. "Three-Dimensional Flow Prediction for Industrial Mixing", Computers in Engng. Conf., San Francisco, CA, July 31 - Aug. 3, 1988.

6. Pai, S.I. "Fluid Dynamics of Jets", Van Nostrand, New York, 1954.

7. Abramovitch, G.N. "The Theory of Turbulent Jets", MIT Press, Cambridge, MA, 1963.

8. Rajaratnam, N. "Turbulent Jets", Elsevier, Amsterdam, Netherlands, 1976.

9. Schetz, J.A. "Injection and Mixing in Turbulent Flows", AIAA/Progress Series, New York, 1980.

10. Wygnanski, I. and Fielder, H. "Some Measurements in the Self-Preserving Jet", J. Fluid Mechanics, Vol. 38 Part 3, 1969, pp. 577-612.

11. Forstall, W. and Shapiro, A.H. "Momentum and Mass Transfer in Coaxial Gas Jets", Trans. ASME J. of Appl. Mech., Vol. 72, 1950, pp. 399-408.

12. Landis, F. and Shapiro, A.H. "The Turbulent Mixing of Co-axial Gas Jets", Proc. HT + FM Inst., Berkeley, CA, June 1951, pp. 133-146.

13. Gupta, A.K., Lilley, D.G. and Syred, N.A. "Swirl Flows", Abacus Press, Tunbridge Wells, England, 1985.

14. Ha, S. and Lilley, D.G. "Jets in Co-Flowing Streams", Proc. of ASME Computers in Engng. Conf., New York, NY, August 9-13, 1987.

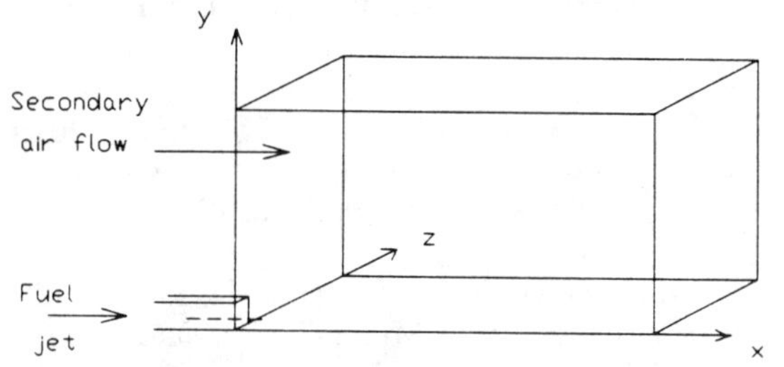


Fig. 1. 3-D Solution Domain.

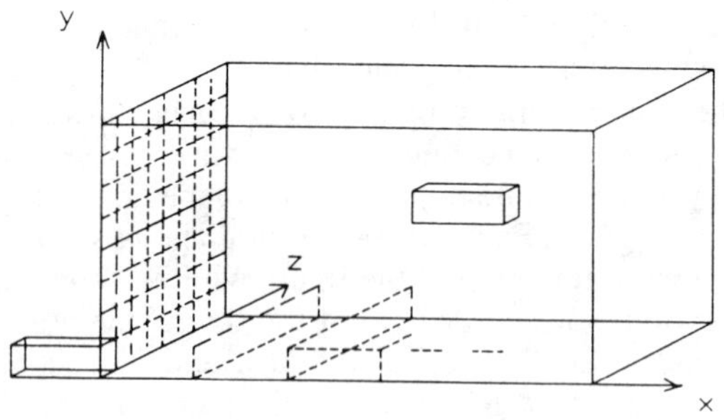


Fig. 2. Grid System Covering the Flow Domain.

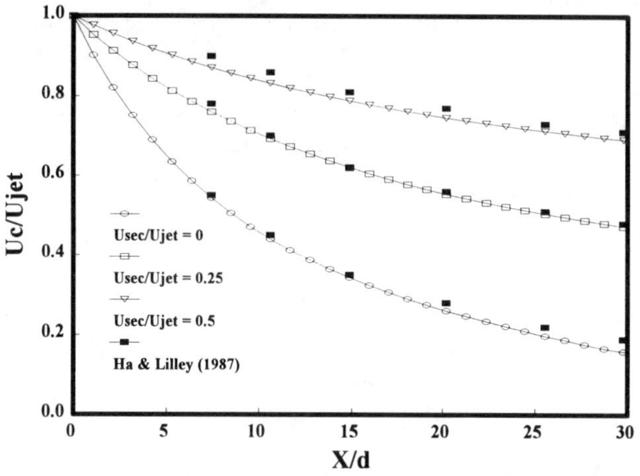


Fig. 3. Decay of Centerline Velocity Along the Jet Axis - Effect of Surroundings to Jet Velocity Ratio.

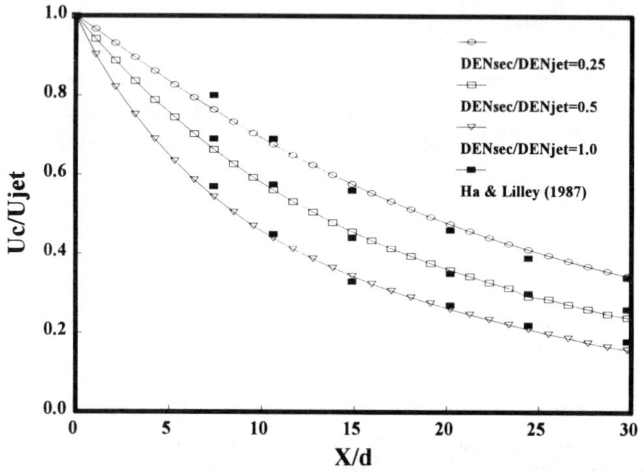


Fig. 4. Decay of Centerline Velocity Along the Jet Axis - Effect of Surroundings to Jet Density Ratio.

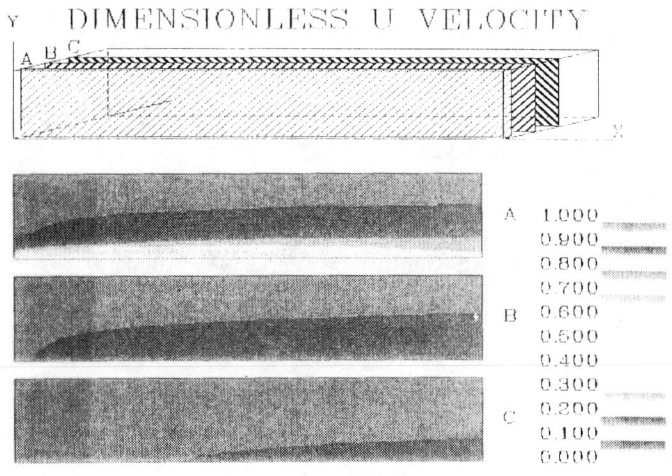


a) Longitudinal Slices

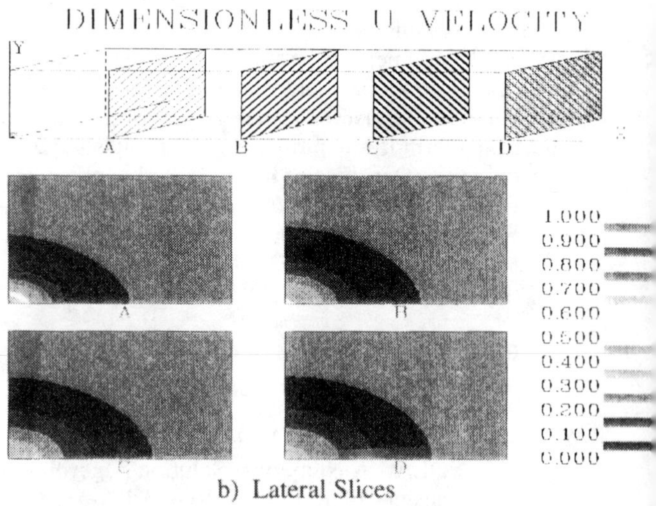


b) Lateral Slices

Fig. 5. Dimensionless Axial Velocity for the Stagnant Surroundings Case with R = 0

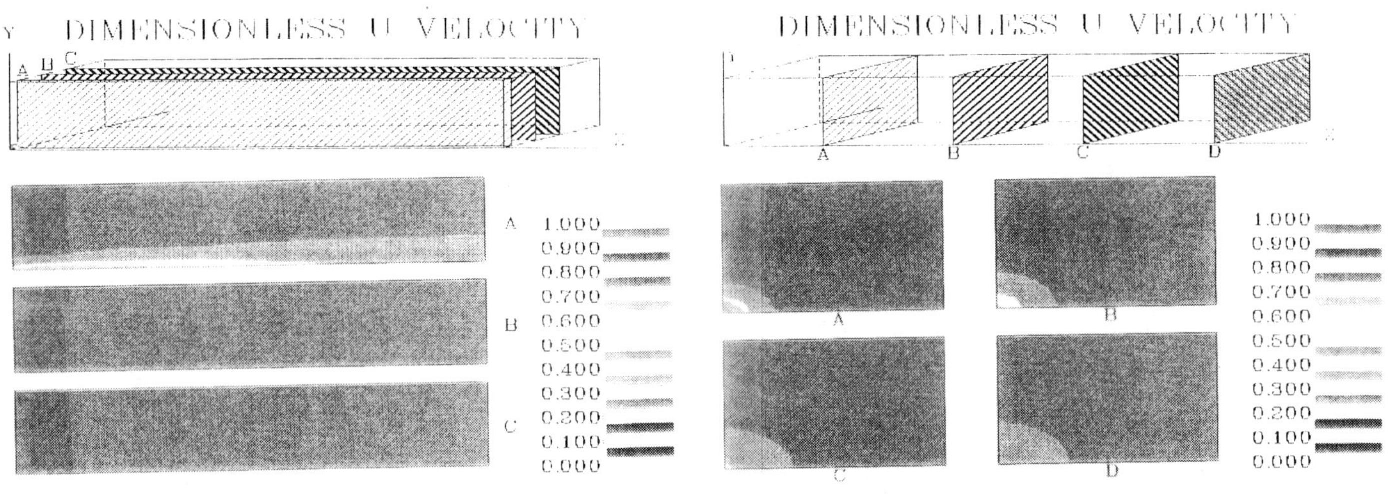


a) Longitudinal Slices

b) Lateral Slices

Fig. 6. Dimensionless Axial Velocity for the Co-Flowing Surroundings Case with R = 0.25

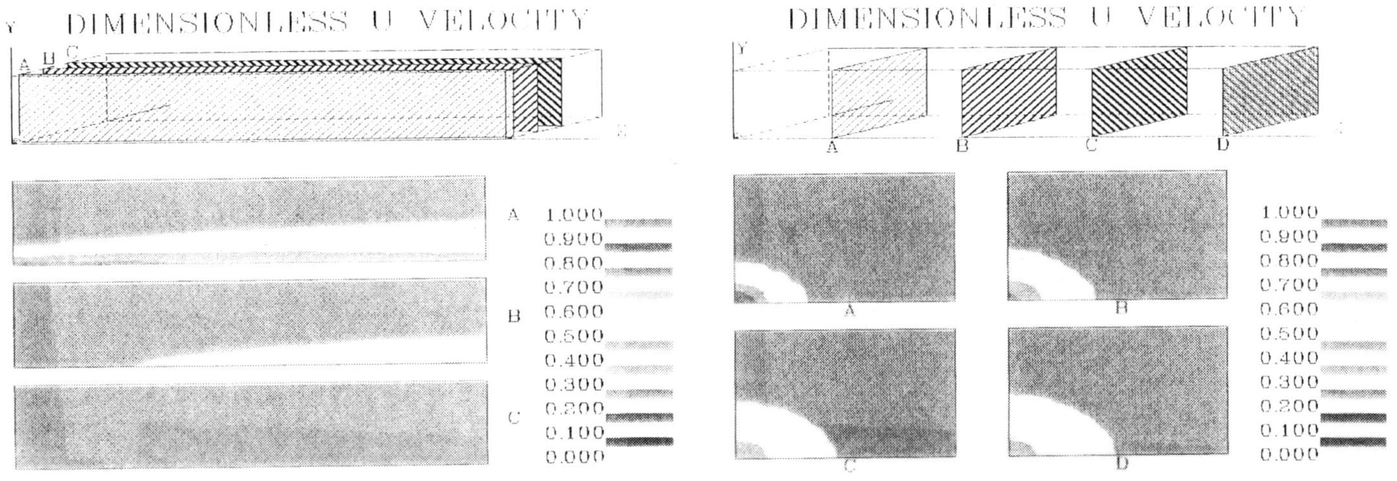


a) Longitudinal Slices

b) Lateral Slices

Fig. 7. Dimensionless Axial Velocity for the Co-Flowing Surroundings Case with R = 0.5

FACT-Vol. 18, Combustion Modeling, Scaling and Air Toxins
ASME 1994

# COMPARISON OF MODEL PREDICTIONS WITH FULL-SCALE UTILITY BOILER DATA

**C. E. Latham and C. F. Eckhart**
Research and Development Division
Babcock & Wilcox
Alliance, Ohio

**C. P. Bellanca and H. V. Duong**
J. M. Stuart Station
Dayton Power and Light Company
Aberdeen, Ohio

## ABSTRACT

Many aspects of boiler operation may be affected by a low-$NO_x$ combustion system including combustion efficiency, heat transfer and corrosion potential. Combustion efficiency benefits from rapid, complete mixing of fuel and air. Limiting the rate at which the fuel and air mix, particularly during early stages of combustion, can effectively control $NO_x$ formation, but these measures for coal-fired units have a tendency to reduce combustion efficiency. The delayed combustion can alter the heat release rate which affects the heat absorption patterns and may change the furnace exit gas temperature (FEGT). Air staging cause some or even all of the lower furnace volume to contain a substoichiometric mixture of combustion products. When fuels containing sulfur are fired, corrosive chemical species, like $H_2S$, can exist in the oxygen-lean environment, leading to accelerated wall tube corrosion rates. Consequently, one aspect can be controlled fairly easily, but controlling all can prove difficult and trade-offs are necessary. Mathematical modeling provides the means to evaluate the impact of low-$NO_x$ combustion systems on boiler combustion efficiency, heat transfer performance and life. Babcock & Wilcox's mathematical flow, combustion and heat transfer models were used during a Clean Coal Technology Demonstration project to assist the application of a novel pulverized coal burner technology. The models were used extensively in the development of the burner and to evaluate the impact on furnace performance during the full-scale demonstration. The objective of the modeling task was to provide a validated tool that could be used to assist the commercialization of the new technology. To validate the models, the predictions with the original cell burners and the new Low-$NO_x$ Cell™ burners were compared with measurements taken during the baseline and post-retrofit tests. Probing was conducted through observation ports at the furnace exit and at a plane 12 m (40 ft) above the burners. Velocity, temperature, and gas species measurements were made at two boiler loads. The comparison between the predictions and the data is presented and discussed.

## INTRODUCTION AND BACKGROUND

### Technology Demonstration

One of the Clean Coal Technology Demonstration projects is the "Full-Scale Demonstration of Low-$NO_x$ Cell Burner Retrofit" (DE-FC22-P0P90545). Project participants include the U.S. Department of Energy (DOE), the Electric Power Research Institute (EPRI), the State of Ohio Coal Development Office, Dayton Power & Light Company (DP&L), Babcock & Wilcox (B&W), Allegheny Power System, Centerior Energy, Duke Power Company, New England Power Company, and the Tennessee Valley Authority. The objective of the Low-$NO_x$ Cell™ burners (LNCB™) demonstration was to evaluate the applicability of this technology for reducing $NO_x$ emissions in full-scale, cell-burner-equipped boilers. The program objectives are:

1. Achieve at least a 50% reduction in $NO_x$ emissions
2. Reduce $NO_x$ with no degradation to boiler performance or life
3. Demonstrate a technically and economically feasible retrofit technology

The LNCB™ burner was developed as an economical, plug-in replacement for the cell burner. DP&L agreed to be the host utility for the full-scale demonstration of the LNCB™ burner, offering the use of its J.M. Stuart Station Unit No. 4. Unit No. 4 is a 605-MWe universal pressure (UP) boiler originally equipped with 24, two-nozzle cell burners arranged in an opposed-wall configuration.

### Role of Modeling

The objective of the modeling task of this project was to provide a validated tool that could be used to assist the commercialization of the new burner technology. To validate the models, the predictions with the original cell burners and the LNCB™s were compared with measurements taken during the baseline and post-retrofit testing at two boiler loads.

Initially, flow and combustion modeling were performed before the retrofit to determine the impact of the LNCB™s on furnace performance. Lower combustion temperatures were found implying lower thermal NO; insignificant increases in unburned carbon were predicted; and a more uniform distribution of furnace exit gas flow and temperature was predicted. Although the predicted average FEGT increased by 4K (7°F), the predictions showed less side-to-side variation and had lower peak-to-minimum values. The model predictions did indicate that carbon monoxide (CO) concentrations below the burners exceeded 100,000 ppm. However, the predicted concentrations above the burners and at the furnace exit were similar to the pre-retrofit predictions and measurements. Preliminary post-retrofit testing confirmed the unexpectedly high concentrations below the burners. Since CO measurements taken in the hopper and the predictions agreed well for three different sets of operating conditions, the models were used to investigate schemes to mitigate the CO. The models were used to identify an arrangement for the burners and air ports that eliminated the high CO concentrations below the burners that maintained the combustion efficiency and furnace heat transfer characteristics (Latham, et al., 1992). Following installation of the new arrangement, the post-retrofit testing resumed. The testing confirmed that the CO concentrations in the furnace hopper had been reduced to pre-retrofit levels of about 1000 ppm.

### Demonstration Summary

The overall objectives of the demonstration program were achieved. The average $NO_x$ emission rate for full load with all mills in service was reduced by 55%. $NO_x$ reductions exceeded 50% over the entire load range. The overall operation of Unit No. 4 was unchanged. The FEGT at full load was about 6K (10°F) lower than the baseline value. The reheat steam temperature was controllable to 814K (1005°F) at full and intermediate loads. The unburned carbon (UBC) loss increased slightly but averaged 1.2% (flyash catch) at full load with all mills in service. However, overall boiler thermal efficiency was unchanged. The effect of the increased UBC was offset by a decrease in dry gas loss due to a lower economizer gas outlet temperature and operation at lower excess air levels. Localized corrosion rates measured on bare SA-213T2 boiler tubes taken from a corrosion test panel on the furnace side wall in the burner zone show corrosion rates that are not significantly higher than those experienced with the original cell burners.

## BURNER DEVELOPMENT

### Cell Burners

Economic considerations which dominated boiler design during the 1960s led to the development of the cell burner for very compact, low-initial-cost boiler designs. Each cell burner consists of two or three coal feed nozzles that act as a single unit. A conventional two-nozzle cell burner is shown in Figure 1. Cell burners were designed for rapid mixing of the fuel and oxidant. The tight burner spacing and rapid mixing minimize the flame size while maximizing the heat release rate and unit efficiency. Consequently, the combustion efficiency is good, but the rapid heat release produces large quantities of $NO_x$. Typically, $NO_x$ levels associated with cell burners will range from 1.0 to 1.8 lb $NO_x$ as $NO_2$ per million Btu input (430-770 ng/J).

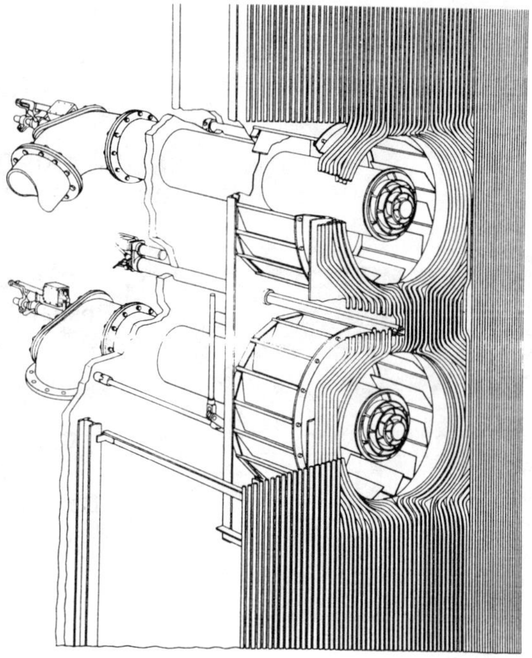

FIGURE 1. CONVENTIONAL CELL BURNER.

### LNCB™ Burner

The two-nozzle LNCB™ burner, shown in Figure 2, was developed by B&W in association with the Electric Power Research Institute (EPRI). The features of the LNCB™ burner were designed to restrict the formation of thermal and fuel $NO_x$. The LNCB™ burner was designed to be directly installed into the existing cell burner furnace wall openings (no pressure part changes), without affecting requirements for coal storage, handling or preparation. The original two coal nozzles in a cell burner are replaced with a single coal injection nozzle and a special secondary air injection port (or dedicated overfire air port). The flame shape is controlled using an impeller at the exit of the fuel nozzle and adjustable spin vanes in the secondary air zone. Unlike the conventional cell burner, secondary airflow can be balanced burner-to-burner using sliding dampers in the upper and lower throats. The air port louver dampers provide additional control over the mixing between the fuel and air streams. During operation, the lower fuel nozzle operates at a low stoichiometry, typically 0.6, with the balance of air entering through the upper port. The controlled mixing of the fuel and air delays the

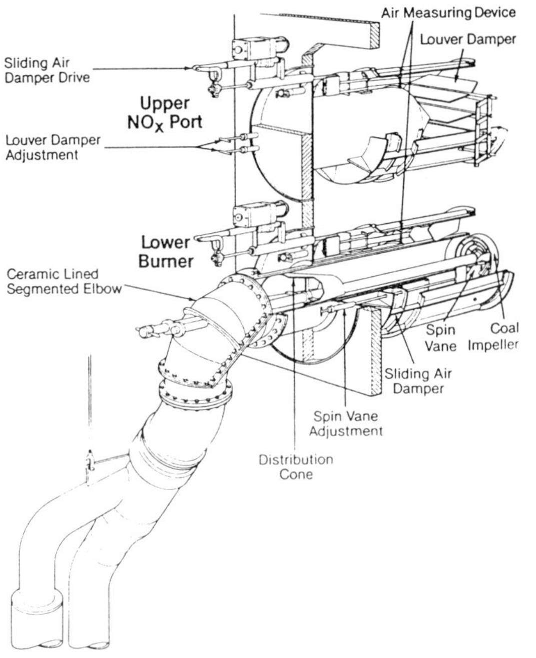


FIGURE 2. LOW-$NO_x$ CELL™ BURNER.

combustion, producing a longer flame that limits the production of $NO_x$.

The novel design of the burner necessitated characterizing the burner at several scales showing feasibility at each scale to settle concerns about maintaining combustion performance. An integrated numerical and experimental program was designed (LaRue and Rogers, 1985) to fully characterize the burner at several scales: 1.75 MW, 30 MW, and utility scale. Several aspects of the LNCB™ burner performance were investigated in the pilot-scale studies. $NO_x$ reductions greater than 50% were achieved. CO emissions were very low (less than 50 ppm) and comparable to cell burner field performance. Unburned carbon losses were low (less than 0.2%) for the standard cell and the LNCB™ burner. No changes in FEGT were measured. The LNCB™ burner flames were stable at lower loads, but were about twice as long as the standard cell flames at full load.

Numerical modeling was done before the pilot-scale testing to project burner performance and locate instrumentation. After the tests, predictions were compared to data, models were refined when required, and performance was scaled to the next level. In general, the pilot-scale numerical modeling agreed well with the data (Fiveland and Wessel, 1991). Consequently, the models were used as a tool to assess the performance of the LNCB™ burner in a full-scale utility boiler. The role of modeling throughout the development of the burner is described by Fiveland and Latham (1993).

## BOILER DESCRIPTION

DP&L's J.M. Stuart Station Unit No. 4 is a once-through supercritical pressure boiler with a single reheat manufactured by Babcock & Wilcox. A schematic of Unit No. 4 is shown in Figure 3. The 605-MWe boiler is fired by 12, two-nozzle cell burners on each of the front and rear walls, arranged two rows high and six columns wide. Six MPS pulverizers supply pulverized coal to the 24 LNCB™ nozzles. The burner throat diameter is 0.9652 m (38 inches). Unit No. 4 burns Kentucky, Ohio, and West Virginia high-volatile bituminous coals. At full load, the boiler produces 554 kg/s (4.4 x $10^6$ lbm/hr) of main steam at 814K (1005°F) and 26.34 MPa (3805 psia). The heat input per LNCB™ burner at full load is 64.3 MW (219.4 x $10^6$ Btu/hr). Since the original flue gas tempering system has been removed, the FEGT at full load is approximately 111K (200°F) above the original design temperature of about 1645K (2500°F).

## DESCRIPTION OF MODELS

Computational techniques have been developed to solve the governing, partial-differential equations for flow, combustion and heat transfer in fossil-fuel fired furnaces. In general, the models solve the fully-elliptic, three-dimensional, finite-difference approximations of the conservation equations for mass, momentum, turbulence, combustion, and heat transfer. The details of the models are documented in Fiveland and Wessel (1988a, 1988b, 1991). The governing equations are integrated over each control volume, discretized using finite-difference techniques and solved for each control volume. Detailed information for each control volume is produced for the flow characteristics, major chemical species ($CH_4$, C, $CO_2$, $O_2$, $SO_2$, $H_2O$), temperature, and heat flux distributions throughout the boiler -- information which is difficult or impractical to obtain experimentally. The model must be sufficiently complex to describe flow, combustion, and heat transfer, and yet simple enough for parametric application to practical systems.

Coal devolatilization and subsequent char oxidation are characterized by a total fuel evolution rate. The first stage of fuel evolution represents devolatilization and char formation processes due to pyrolysis. A second stage of fuel evolution represents the rate of char conversion to fuel before it forms products. Ash remains after the fuel evolves from the char. Combustion of the fuel is characterized by a two-step process: In the first step, fuel is assumed to react with oxygen to form carbon monoxide and products; in the second step, there is a mixing limited, kinetic reduction of carbon monoxide to carbon dioxide. This two-step reaction sequence is important because of its effect on heat transfer and nitrogen oxide pollutant processes. In many practical combustion devices operating with fuel-rich conditions like the LNCB™ burner, CO concentrations can reach $10^5$ ppm. Neglecting this intermediate combustion product could result in large inaccuracies in the distribution of gas temperature and oxygen concentration.

## MODELING APPROACH

The furnace is modeled from the burner throats through the horizontal section of the convection pass. To utilize the flow and combustion models, the furnace volume is divided into control

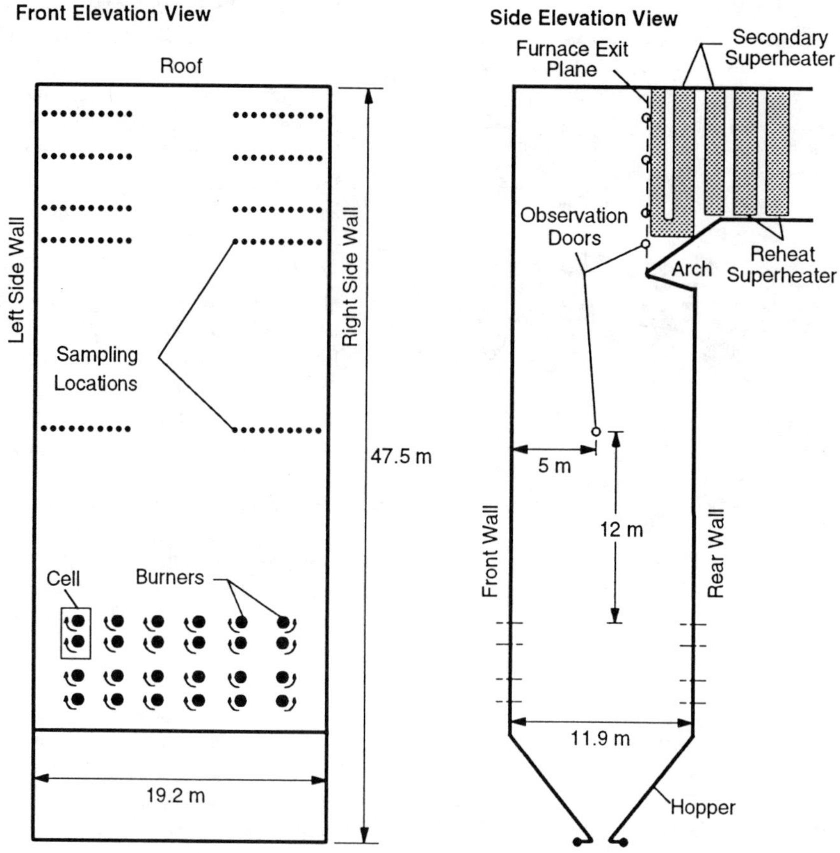


FIGURE 3. J.M. STUART STATION UNIT NO. 4.

volumes (52,000 active) as shown in Figure 4. All features of the furnace are simulated such as the burners, overfire air ports, hopper, arch, and pendant surfaces. The computational grid is finer in the burner zone and in the upper furnace where high gradients of density and species concentrations exist. Although considered relatively coarse when compared to current models for similar applications, the grid was refined as much as possible to allow timely solutions on the workstation computers available at the time the modeling task began in 1990. The same grid for the furnace model was used throughout this project to maintain consistency.

Convection pass heat transfer surfaces are modeled as porous media with convective surface heat transfer and modified radiation absorption and scattering properties. The furnace enclosure is modeled with convective surface heat transfer and constant emissivity. The net heat absorbed through the furnace walls and convection pass heat transfer surfaces can be adjusted by modifying the overall thermal boundary condition to allow for uncertainties in the thickness and thermal conductivity of furnace ash or slag deposits or tube bank fouling.

The combustion and heat transfer are calculated using the flow field along with the coal composition, fineness, temperature and the high heating value (HHV); air composition and temperature; and thermal wall boundary conditions. Furnace heat transfer performance data from the full-load tests were used to adjust the thermal boundary conditions for the furnace wall and convection pass heat transfer surfaces to compensate for uncertainty in furnace slagging and fouling conditions. The heat transfer boundary conditions were then assumed to remain constant for the intermediate-load modeling. The resulting thermal conductance of the furnace was slightly higher for the post-retrofit model than for the baseline model, indicating less overall slagging and fouling in the furnace.

The numerical modeling simulated the operation of Unit 4 at full and intermediate loads using the actual operating conditions during tests in which the in-furnace probing was conducted. The operating conditions and fuel analyses used for the modeling are listed in Table 1.

The furnace flow field is calculated based on the inlet flow rate and momentum of the burners and air ports. The mass flow, the axial momentum, and the swirl intensity of the burners and air ports are calculated using the geometry and flow control settings. The baseline arrangement and swirl pattern of the cell burners on Unit 4 are shown in Figure 3, and the post-retrofit arrangement and swirl of the LNCB™s are shown in Figure 5. For the baseline modeling, the air and coal distribution to the cell burners was assumed to be uniform. A uniform air distribution is unlikely with the wrap-around windbox and the relatively low pressure

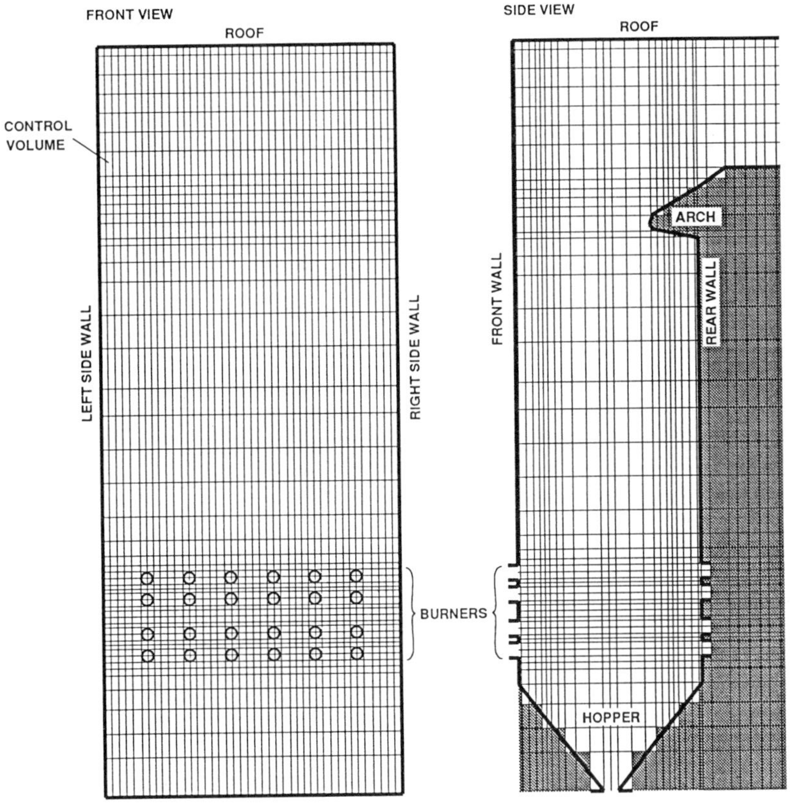


FIGURE 4. COMPUTATIONAL GRID.

drop of the cell burners. Less air enters through burners near the side walls than through those near the center of the firing walls. The lower excess air levels lead to reducing conditions near the side walls, as evidenced by tube wastage experienced on units of this vintage. The models do not require that uniform distributions be used, but no reliable data were available that could be used to determine the actual burner-to-burner distribution of the air or coal. Consequently, the comparisons with the data will be influenced by this assumption. Flow modeling of the windbox could have been performed to predict the air flow distribution but was beyond the scope of this project.

During the post-retrofit testing, the coal and primary air flow rates were measured for each mill. The coal and primary air flow rates for individual burners were not measured, so any possible burner-to-burner variations could not be determined. The secondary air flow distribution was determined from impact-suction pitot tube arrays on each burner (see Figure 2). The air-fuel ratio for each burner was calculated using average coal and primary air flow rates calculated from the mill totals and the measured secondary air flow rate. Since more detailed burner-to-burner measurements do not exist and the existing measurements indicate that the distribution is relatively uniform, a uniform coal and air distribution to each burner was assumed that would give the calculated average air-fuel ratio. Although the assumption of uniform distributions is supported by the post-retrofit measurements, some localized effects will appear in the data but may be smoothed or not appear at all in the predictions. Additionally, the level of detail with which the boundary conditions are specified must be consistent with capacity of the grid resolution to predict meaningful differences.

Each air port was also equipped with an impact-suction pitot tube array. At full load, the measured flow rates for 21 of the 24 air ports were within plus-or-minus 5%. A uniform distribution between these 21 air ports was assumed. The flow rates for each of the remaining three air ports were about 50% lower than the average of the other 21. The air flow through these three ports, which are located on the rear wall in the lower row inverted cells (see Figure 5), was restricted to further reduce $NO_x$ emissions and direct more air to other ports for better carbon conversion. Uniform distributions were assumed between these three ports.

The locations of the four out-of-service burners and air ports at intermediate load are shown in Figure 5, also. For this test, the secondary air flow to the out-of-service burners was the same as the flow to the in-service burners. As with full load, a uniform coal and primary air distribution between in-service burners was assumed. The measured air flow rates for 14 of the 20 in-service air ports were within plus-or-minus 5%, so a uniform air distribution between these ports was assumed. The air flow rate to each of the six remaining in-service ports was approximately 30% less than the average of the other 14 in-service ports, and a uniform

TABLE 1
MODELED OPERATING CONDITIONS

| Parameter | Units | Baseline Load Full | Baseline Load Intermediate | Post-Retrofit Load Full | Post-Retrofit Load Intermediate |
|---|---|---|---|---|---|
| **Output** | MWe | 605 | 458 | 603 | 461 |
| **Temperatures** | | | | | |
| Windbox | K | 599 | 583 | 588 | 568 |
| Pulverizer Outlet | K | 338 | 338 | 339 | 339 |
| **Coal Flow Rate** | kg/sec | 57.2 | 44.2 | 56.7 | 45.6 |
| **Air Flow Rates** | | | | | |
| Burners | kg/sec | 618.2 | 519.2 | 278.2 | 254.1 |
| Air Ports | kg/sec | N/A | N/A | 318.1 | 270.9 |
| Total | kg/sec | 618.2 | 519.2 | 596.2 | 525.0 |
| **Excess Air** | % | 21 | 33 | 18 | 33 |
| **Burners Out of Service** | | None | 8 | None | 4 |
| **Ultimate Analysis** (As Received) | | | | | |
| Carbon | | | | | |
| Hydrogen | % | 66.52 | 65.95 | 65.75 | 63.49 |
| Oxygen | % | 4.49 | 4.48 | 4.64 | 4.50 |
| Nitrogen | % | 7.51 | 7.91 | 7.78 | 7.14 |
| Sulfur | % | 1.23 | 1.16 | 1.41 | 1.37 |
| Moisture | % | 1.00 | 1.12 | 0.78 | 0.74 |
| Ash | % | 5.55 | 5.39 | 4.76 | 6.10 |
| | % | 13.70 | 13.99 | 14.88 | 16.66 |
| | | 100.00 | 100.00 | 100.00 | 100.00 |
| **Proximate Analysis** (As Received) | | | | | |
| Moisture | | | | | |
| Fixed Carbon | % | 5.55 | 5.39 | 4.76 | 6.10 |
| Volatile Matter | % | 47.61 | 47.24 | 38.54 | 35.28 |
| Ash | % | 33.14 | 33.38 | 41.82 | 41.96 |
| | % | 13.70 | 13.99 | 14.88 | 16.66 |
| | | 100.00 | 100.00 | 100.00 | 100.00 |
| **Higher Heating Value (HHV)** | J/kg | 2.7631E+07 | 2.7536E+07 | 2.7550E+07 | 2.6445E+07 |

distribution between these six ports was assumed. The locations of the six ports with reduced flow are shown on Figure 5. The flow was restircted to these ports for the same reasons noted for full load. The air flow rates for the ports adjacent to the out-of-service burners were about 40% less than the flow rate of the 14 in-service ports. A uniform air flow distribution to the out-of-service air ports was assumed.

## MEASUREMENT METHODS AND UNCERTAINTIES

Full-scale utility boiler data sets that are reliable and detailed enough for comparisons with three-dimensional numerical model predictions are extremely rare. The data are difficult and costly to obtain, and boiler operating conditions are often difficult to define completely. Large boilers with limited wall openings prevent detailed mapping of entire planes, often restricting access to a few locations. Attempts to obtain data near the burner zone are hindered by even more restricted access and a more hostile environment.

During baseline testing, measurements of gas velocity, gas temperature and chemical species (CO and $O_2$) were made at full (605 MWe) and intermediate (458 MWe) loads. The probing was repeated during the post-retrofit testing at full (603 MWe) and intermediate (461 MWe) loads. An attempt was made to select post-retrofits tests from the matrix that repeated, as closely as possible, the conditions during baseline testing. The duration of each test ranged between 8 and 12 hours. The normal operation of the boiler was not interrupted for the probing, including sootblowing cycles. The operating conditions were well defined and remained relatively constant during the probing.

Observation ports at the furnace exit plane and a horizontal plane about 12 m (40 feet) above the burners were used for the probing. The data were taken by inserting a water-cooled, high-velocity thermocouple (HVT) Fechheimer probe 6.1 m (20 feet) into the furnace and recording measurements at 0.61-m (2-foot)

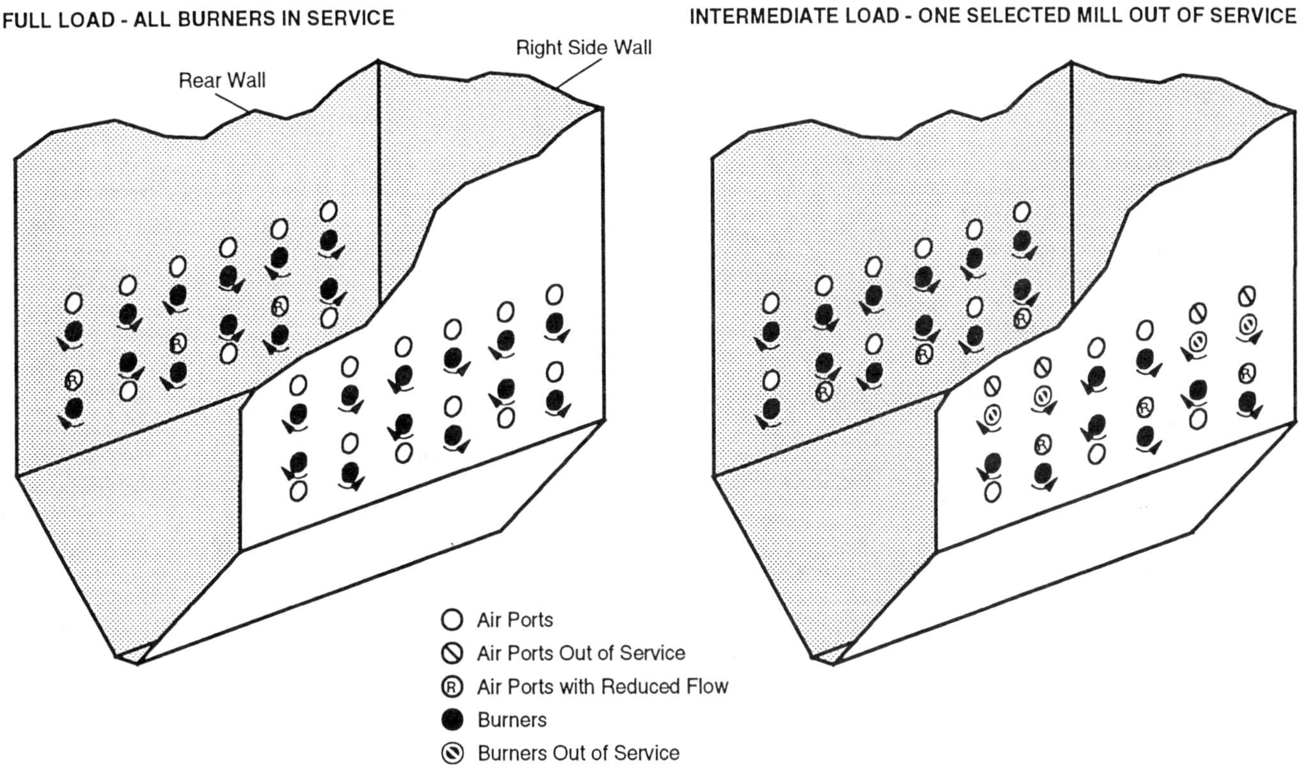


FIGURE 5. BURNER ARRANGEMENT AND OPERATION.

intervals as the probe was retracted. Since the probes were inserted through both side walls and extended only 6.1 m (20 feet) into the furnace, no data were obtained for the 7 m (23 feet) in the middle of the furnace.

The locations of the probe traverse planes at the furnace exit are illustrated in Figure 3. Eight ports (four on each side wall) were used at the exit plane. The ports were about 2.4 m (8 feet) apart. Only four ports would accommodate probing at the plane 12 m (40 feet) above the burners. The traverses for probing through the side walls were 0.9 m (3 feet) from the centerline of the furnace toward the front wall. Measurements were made along two additional traverses through the rear wall about 0.46 m (18 inches) from each side wall.

Temperatures were measured with a single-shield HVT probe. During the testing, the probing was hindered by slag and ash plugging of the HVT probe shield, especially at full load. Accuracy of the gas temperature, $O_2$ and CO and measurements depends on the cleanliness of the probe tip (Babcock & Wilcox, 1992). The pluggage formed gradually as the probe was retracted from the furnace at 0.61-m (2-foot) intervals. Temperatures measured with a partially plugged probe will be lower than the actual gas temperature. Consequently, the most reliable temperature measurements were those taken the farthest from the walls. The $O_2$ and CO are less sensitive to the pluggage as long as a sufficient volume of gas is provided to the analyzers. The $O_2$ and CO data reported are averages based on about two minutes of sampling at each point. If severe plugging occurred, the probe was retracted from the furnace and new ceramic shielding was installed. The probe was reinserted into the furnace and testing resumed. The velocity measurements were obtained using the same HVT probe. Total and static pressure taps were integrated into the design of the probe body, allowing the HVT probe to function as a Fechheimer (three-hole pitot) probe as well. An air purge was applied through the pressure taps when no measurements were being taken to prevent slag and ash from clogging the holes.

Even though the probing data do not provide a complete mapping of the measurement planes and may not be as reliable as that taken in a controlled laboratory environment, the full-scale data sets are among the most extensive and reliable in existence. The defined sets of operating conditions for baseline and post-retrofit operation at two loads and the extensive in-furnace probing data make the test data well suited for comparison with numerical model predictions.

## COMPARISON BETWEEN PREDICTIONS AND DATA

Predicted gas velocities, gas temperatures, and CO and $O_2$ concentrations were compared with the measurements for baseline and post-retrofit operation at full and intermediate loads. The comparisons of gas velocities and temperatures will be discussed but will not be represented. The emphasis of the modeling done on the project was related to CO. Consequently, the comparisons between predicted and measured CO and $O_2$ concentrations will be presented and discussed.

### Gas Velocity Magnitudes

The same trends exist in the predictions and the data for baseline and post-retrofit operation. Both the predictions and the data indicated that the gas velocities at the furnace exit were higher near the arch than near the roof. No significant side-to-side flow upsets are indicated by the predictions or the data at the exit or 12 m (40 feet) above the burners. The predicted distributions at intermediate load are not as uniform from side-to-side as at full load. The data indicate slightly more side-to-side variation at intermediate load than at full load. The agreement between the predictions and data was generally better at intermediate load than at full load.

The flow near the arch and in the upper furnace is very complex. The local effect of the arch and the higher velocity flue gas flowing up the center of the furnace and turning 90 degrees to enter the convection pass result in a high-velocity region near the tip of the arch. The agreement should be improved by increasing the fineness of the computational grid (adding more control volumes) to the model in the upper furnace.

### Gas Temperatures

For the baseline and post-retrofit tests at both loads, the predicted temperature distributions were less uniform than the measured distributions. The predicted temperatures at the plane 12 m (40 feet) above the burners were higher than measured. The overpredicted temperatures indicate that the lower furnace heat absorption in the model may have been slightly low. Consequently, the predicted peak temperatures near the arch in the furnace exit plane were higher than measured. Measured temperatures 12 m (40 feet) above the burners were typically between 1390K to 1667K (2500°F to 3000°F). The uncertainty in the temperature measurements at full load is relatively high since the probing was plagued by slagging in the HVT shield, causing the measured temperatures to be lower than the actual gas temperatures. The agreement between the predictions and data is slightly better at intermediate load than at full load. The gas temperatures at the exit plane were typically 170K (300°F) lower at intermediate load than at full load. Some side-to-side variation was predicted and measured at the furnace exit and at the plane 12 m (40 feet) above the burners. An inherent asymmetry in the furnace exists due to the arrangement of the bottom row of burners (see Figure 5).

### Carbon Monoxide and Oxygen Concentrations

The model predictions are compared graphically to the data in Figures 6, 8, 9, and 10. The predictions are presented on shaded contour plots with labeled contour lines. The measurement locations are shown as circles on the contour plots. The circles are filled with the color that corresponds to the same color table as the predictions. The numbers indicate the measured value at each location. Average values are listed for repeated measurements. The reader is viewing into the burner zone from above at the plane 12 m (40 feet) above the burners. At the furnace exit plane, the reader is viewing into the convection pass from the furnace.

The predicted CO concentrations for full and intermediate loads at the plane 12 m (40 feet) above the burners are shown in Figure 6. The baseline predictions and data agree very well. Both indicate that the CO concentrations are about 1000 ppm. The low CO concentrations are expected due to the rapid combustion characteristics of the cell burners and the high excess air levels. The side-to-side variation indicated by the data is not evident in the predictions, which is a result of the assumption of uniform air and coal to all burners. The post-retrofit predictions and data indicate that the CO levels are higher. At this elevation, higher CO is expected due to the delayed combustion characteristics of the LNCB™ burners. The agreement is reasonably good at both loads, except in the left-rear corner where the data indicate CO levels much higher than predicted. At full load, the predictions indicate zones with CO exceeding 10,000 ppm near the front and rear walls. The 3D surface plot in Figure 7 shows the volume enclosed by CO concentrations of 10,000 ppm. The high CO concentrations above the burner zone are not from products of incomplete combustion at the center of the furnace, but from products that passed through the spaces between burners in the combustion zone and drift upward along the front wall. The CO is oxidized and as the flow patterns mix these products with oxygen-rich flow. The predicted locations of the high CO regions may be slightly different than the actual locations in the furnace, which may account for the differences in the left-rear corner.

At intermediate load, the baseline predictions (not shown) do not exceed about 100 ppm, while some of the data exceed 800 ppm. The lack of agreement is due to the high excess air level associated with reduced-load operation. The model tends to underpredict CO concentrations in high-temperature regions with very high $O_2$ levels. The predicted post-retrofit distributions agree better with the data, although the peak CO levels are about 1000 ppm too low. The CO concentrations are very low along the front wall due to the operation of the unit at intermediate load. The four outer cells in the upper row on the front wall were out of service (see Figure 5). Since the secondary air flow to the out-of-service cells was essentially the same as the in-service cells, high $O_2$ and low CO concentrations exist near the front wall.

The $O_2$ concentrations for the same three planes are shown in Figure 8. The baseline predictions and data indicate that the distributions are relatively flat. The data indicate that the $O_2$ is slightly higher on the left than on the right, which is consistent with the lower CO measured on the left than on the right. The predictions and data are considerably less uniform for the post-retrofit cases, which is a result of the staged combustion. The high $O_2$ levels in the center of the furnace indicate that sufficient penetration of the overfire air is achieved with the upper row of air ports. More oxygen exists near the side walls than the front and rear walls. The asymmetry in the burner arrangement causes the corner-to-corner variation at full load. At intermediate load, the $O_2$ levels are much higher near the front wall at intermediate load due to the cooling air from out-of-service cells. Most of the $O_2$ remains near the front wall as it passes up the furnace, causing the higher concentrations in the upper section of the furnace exit plane.

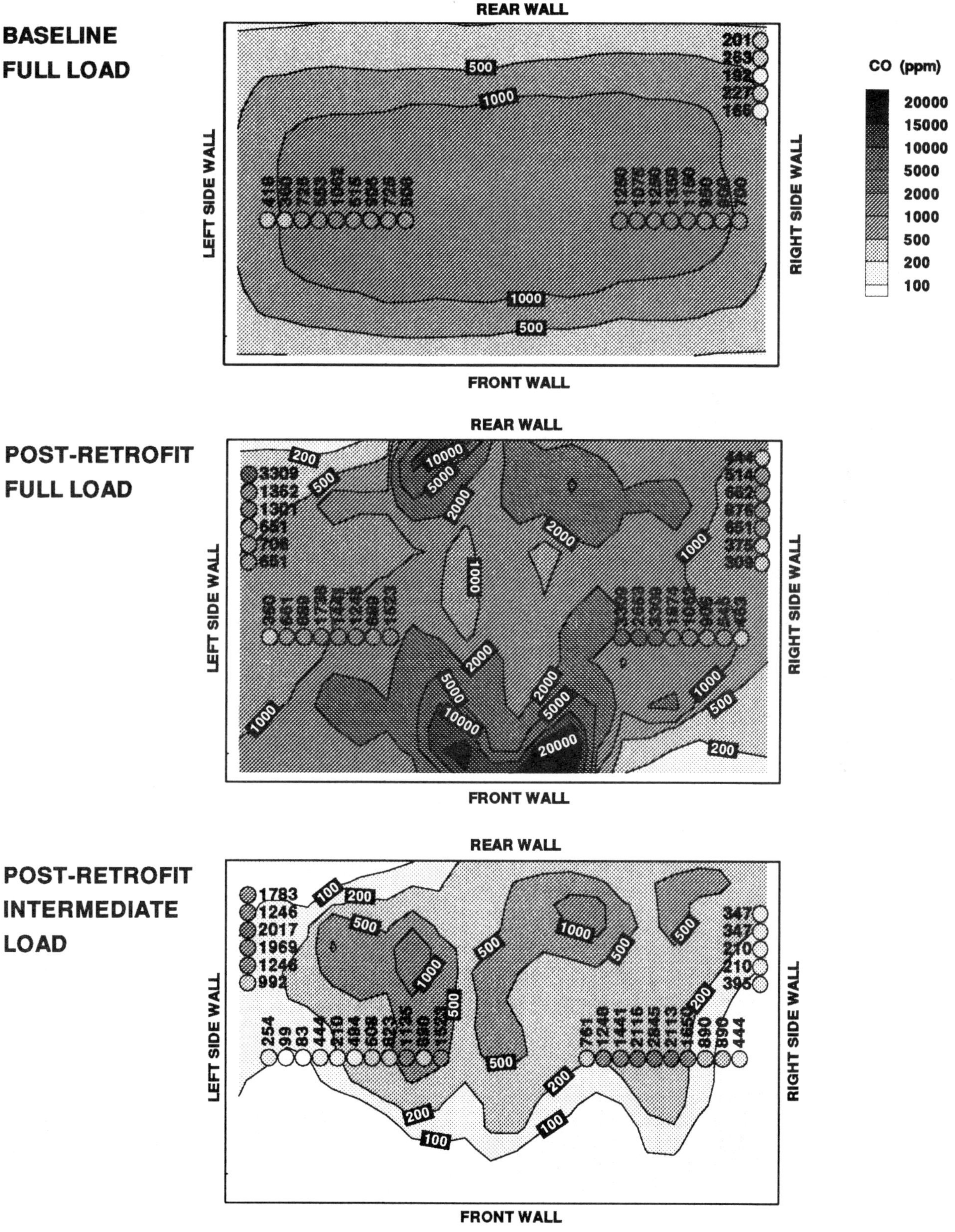


FIGURE 6. CO CONCENTRATIONS 12 M (40 FEET) ABOVE THE BURNERS.

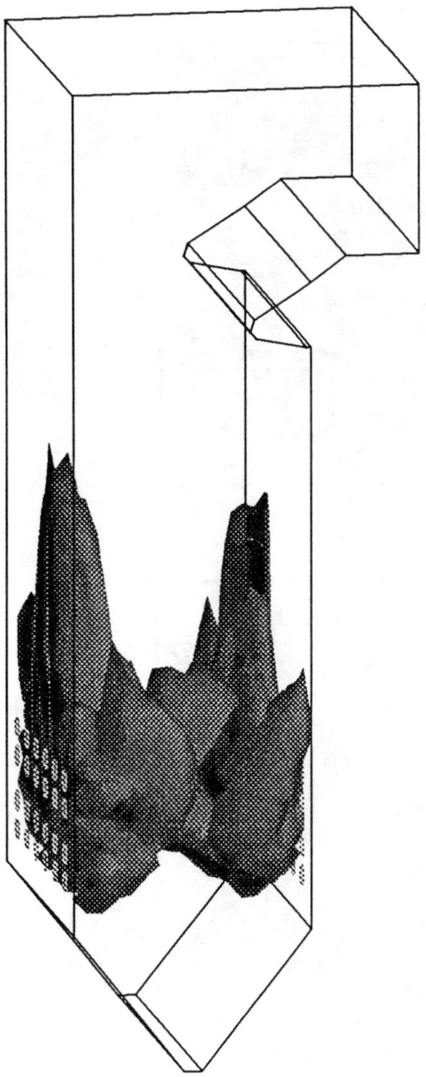

FIGURE 7. SURFACE CONTOURS OF 10,000 PPM CO.

The CO concentrations at the furnace exit are shown in Figure 9. The predictions and limited amount of reliable data agree well for the baseline except near the arch where the measured CO is much lower than the predictions. The post-retrofit predictions and data agree reasonably well, except for two regions where the CO data exceeds 1000 ppm (lower left side and upper right side). The source of the high CO in the exit plane may be localized regions in the furnace flow field that contain high CO concentrations but limited oxygen, like those predicted along the front and rear walls. As the flow passes the arch and turns to exit the furnace, these localized regions mix with the surrounding flow that contains excess oxygen. The post-retrofit data indicate that the high CO concentrations on the left side of the exit plane exist for both full and intermediate loads. Although not possible to obtain, a more detailed mapping of the plane 12 m (40 feet) above the burners could help to explain the origin of the high CO concentrations at the furnace exit.

The post-retrofit $O_2$ concentrations at the furnace exit are shown in Figure 10. The measured $O_2$ ranges from 1.2% to 2.6% at the locations in the furnace exit plane where CO concentrations greater than 1000 ppm were measured. The CO converts rapidly to carbon dioxide ($CO_2$) with the high temperatures and the availability of excess oxygen. The measured CO concentrations at the economizer outlet were between 28 and 55 ppm at full load with all mills in service. The predictions and data at the furnace exit exhibit similar characteristics to those observed 12 m (40 feet) above the burners. At full load, the predictions and data exhibit corner-to-corner variations but to opposite corners. At intermediate load, the high $O_2$ concentrations observed along the front wall appear near the roof in the exit plane.

## CONCLUSIONS

The operating conditions during testing were well defined and extensive measurements were made at the furnace exit and at a plane 12 m (40 feet) above the burners. Even with the noted limitations, the data set was very valuable for comparing with predictions for a large utility furnace. The modeling predictions and data gave the same indications about the baseline and post-retrofit operation of the 605 MWe utility boiler. Comparisons not only provided insight into the operation of the boiler, but increased confidence in using the models as evaluation tools. Furthermore, the comparisons were used to identify needed model enhancements. The disparities between predictions and data include the flow field near the furnace arch, the lower furnace heat absorption and the furnace exit CO concentrations. The overall agreement between the predictions and the data are expected to be better with a more refined computational grid. However, grid refinement is not a substitute for constantly improving the models of the fundamental processes. B&W continues to support its internal combustion model development program and is actively improving its models to achieve better agreement with data from this boiler and other sources, as well. noted in this paper. With this successful validation effort, the models have been and are being used confidently to characterize the performance of other full-scale utility boilers equipped with Low-NOx Cell™ burners.

## ACKNOWLEDGEMENTS

Project participants in the CCT Demonstration Program include the U.S. Department of Energy (DOE), the Electric Power Research Institute (EPRI), the State of Ohio Coal Development Office (OCDO), Dayton Power & Light Company (DP&L), Babcock & Wilcox, Allegheny Power System, Centerior Energy, Duke Power Company, New England Power Company, and the Tennessee Valley Authority (TVA).

The data acquisition was made possible by the concerted effort of individuals from B&W Combustion Systems, Results Engineering, Field Service, Engineering Technology and the Research and Development Division.

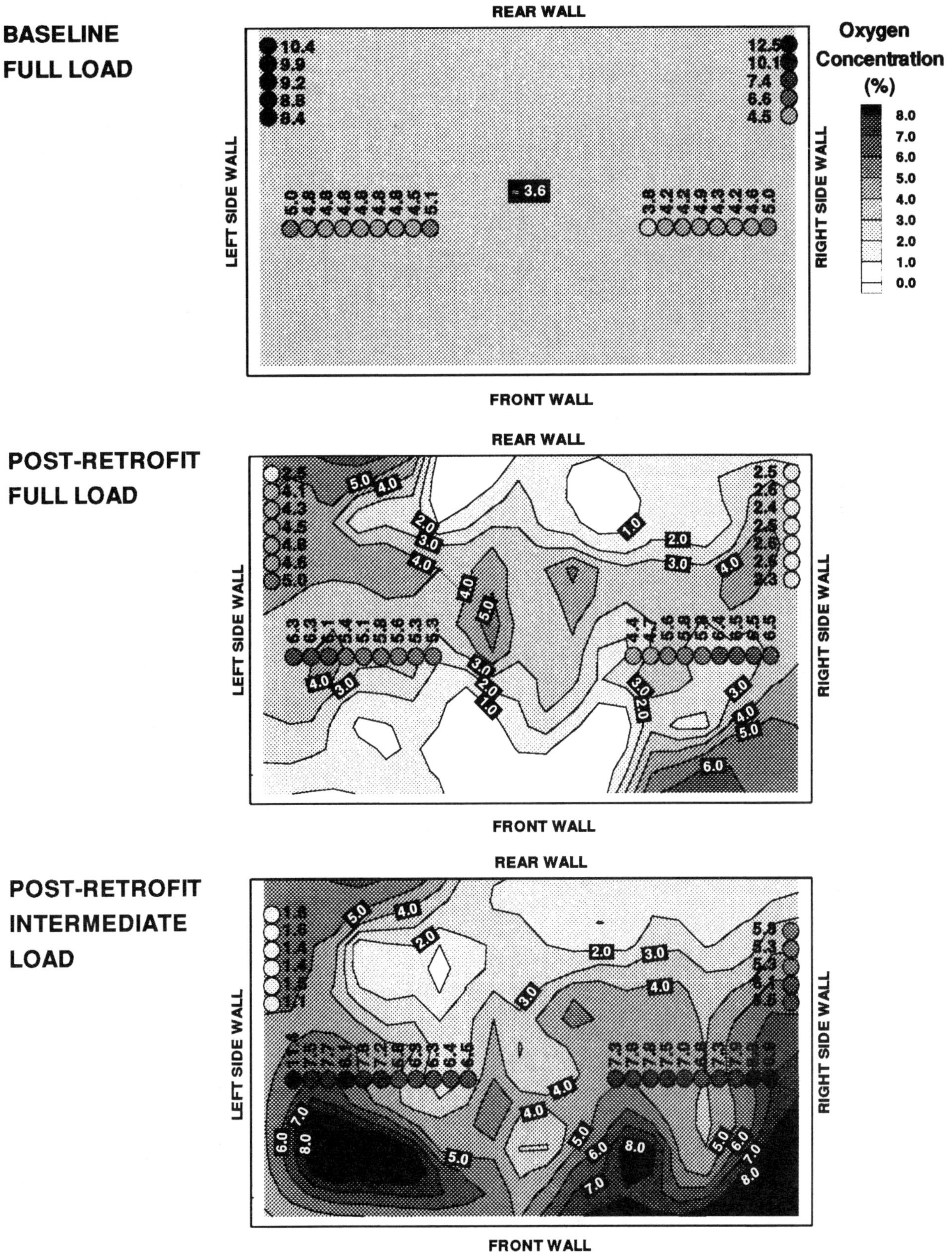


FIGURE 8. $O_2$ CONCENTRATIONS 12 M (40 FEET) ABOVE THE BURNERS.

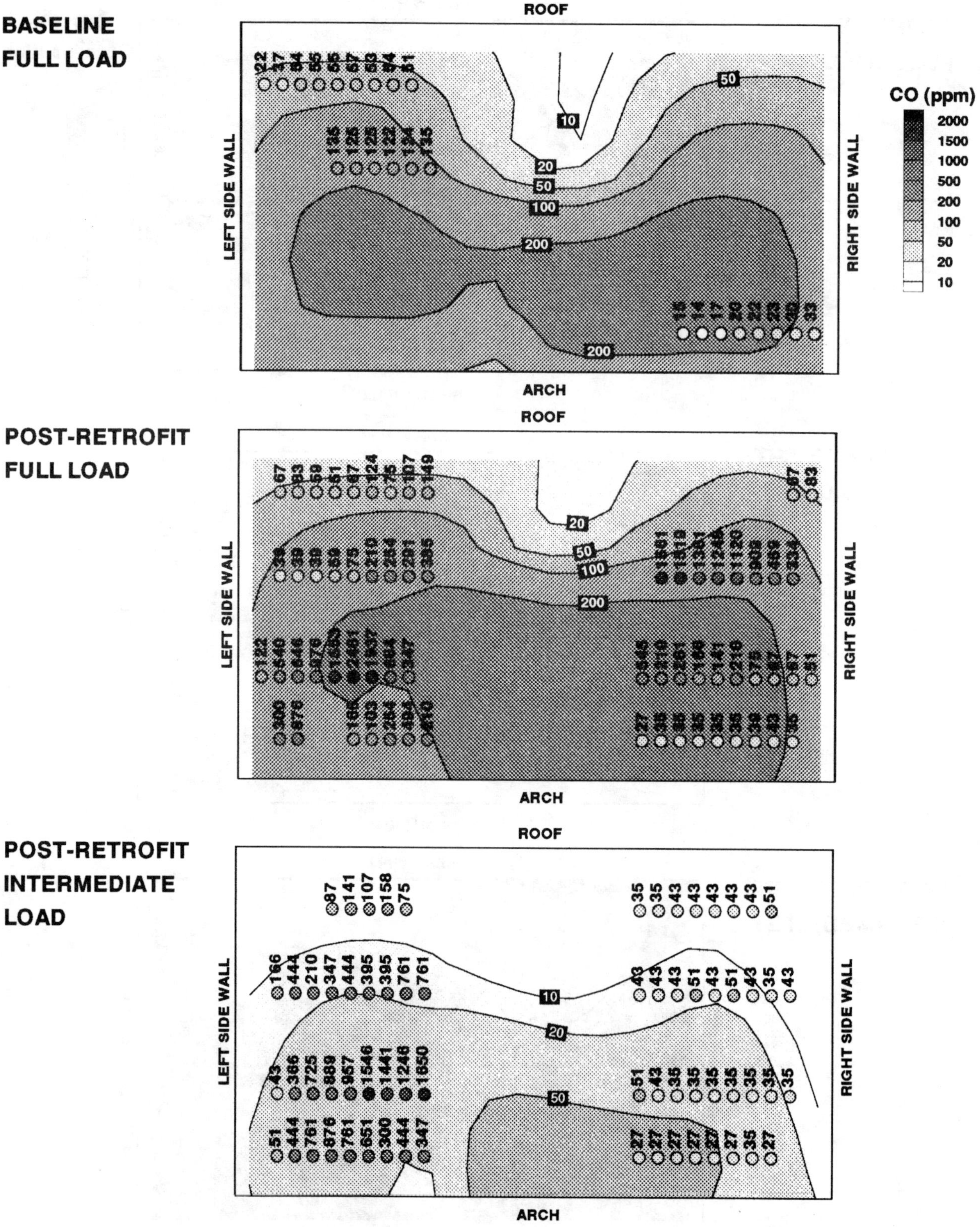


FIGURE 9. CO CONCENTRATIONS AT THE FURNACE EXIT PLANE.

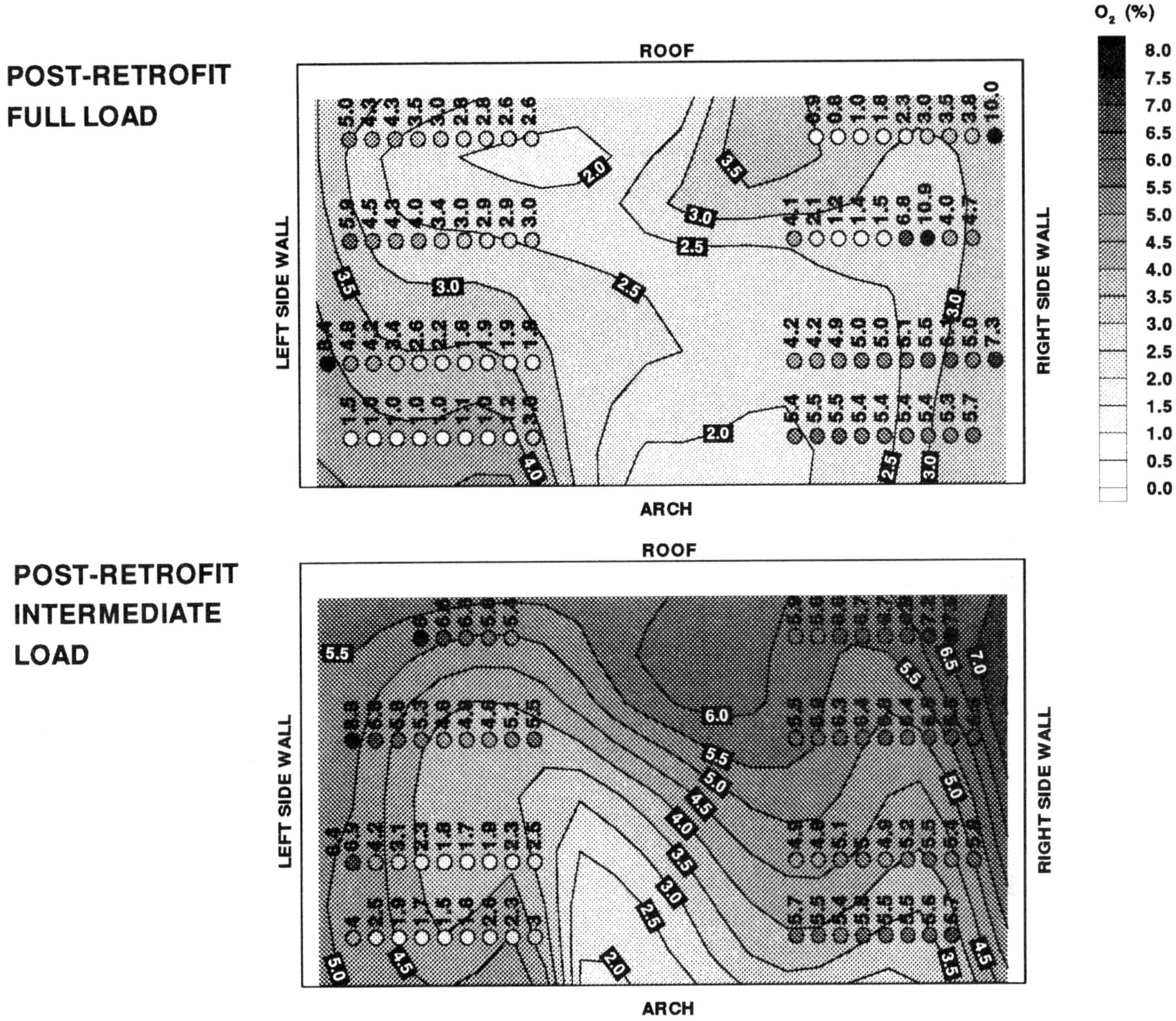


FIGURE 10. O2 CONCENTRATIONS AT THE FURNACE EXIT.

## REFERENCES


Babcock & Wilcox, *Steam/Its Generation and Use*, 40th edition, 1992, pp. 40-13.

Fiveland, W.A. and Wessel, R.A., "Numerical Model for Predicting the Performance of Three-Dimensional Pulverized-Fuel Fired Furnaces," *Engineering for Gas Turbines and Power*, Vol. 110, No. 1, p. 117, 1988a.

Fiveland, W.A. and Wessel, R.A., "Pulverized Coal Combustion Model With Finite-Rate Carbon Monoxide Oxidation," Presented to the American Flame Research Committee, Pittsburgh, Pennsylvania, October, 1988b.

Fiveland, W.A. and Wessel, R.A., "Model for Predicting the Formation and Reduction in Nitric Oxide Pollutants in Three-Dimensional Furnaces Burning Pulverized Fuel," *Journal of the Institute of Energy*, Vol. 64, No. 458, 1991.

Fiveland, W.A. and Latham, C.E., "Use of Numerical Modeling in the Design of a Low-NOx Burner for Utility Boilers," *Combustion Science and Technology*, Vol. 93, pp. 53-72, 1993.

LaRue, A.D. and Rodgers, L.W., "Development of Low-NOx Cell Burners for Retrofit Applications," presented at the 1985 EPA/EPRI Joint Symposium on Stationary Combustion NOx Control, Boston, Massachusetts, May 6-9, 1985.

Latham, C.E., Laursen, T.A., Duong, H., "Application of Numerical Modeling in a Clean Coal Demonstration Project," presented at the 1992 International Joint Power Conference, Atlanta, GA, October 18-22, 1992.

FACT-Vol. 18, Combustion Modeling, Scaling and Air Toxins
ASME 1994

# TRANSPORT AND DEPOSITION OF PARTICLES IN GAS TURBINES: EFFECTS OF CONVECTION, DIFFUSION, THERMOPHORESIS, INERTIAL IMPACTION AND COAGULATION

**David P. Brown[2] and Pratim Biswas***
Department of Civil and Environmental Engineering
Aerosol and Air Quality Research Laboratory

**Stanley G. Rubin**
Department of Aerospace Engineering and Engineering Mechanics
[2]Digital Simulation Laboratory

University of Cincinnati
Cincinnati, Ohio

## ABSTRACT

Aerosols are produced in a large number of industrial processes over a wide range of sizes. Of particular importance is deposition of coal and oil combustion aerosols in turbines. A model coupling the transport and the dynamics of aerosols to flow characteristics in gas turbines is presented. An order of magnitude analysis is carried out based on typical operational conditions for coal and oil combustion (neglecting coagulation) to determine the relative importance of various mechanisms on particle behavior. A scheme is then developed to incorporate a moment model of a lognormally distributed aerosol to predict aerosol transport and dynamics in turbine flows. The proposed moment model reflects the contributions from convection, intertia, diffusion and thermophoresis. Aerosol behavior in various laminar 2-D and axisymmetric flows is considered in this study. Results are compared to published work in 1-D and 2-D planar and axisymmetric.

## NOMENCLATURE

| | |
|---|---|
| H | Total enthalpy |
| g | Jacobian of the coordinate transformation |
| Kt | Thermophoretic constant |
| $M_\infty$ | Inflow Mach number |
| $M_k$ | kth volume moment of the distribution |
| n(v) | Normalized particle number concentration |
| Pe | Peclet number |
| Re | Reynolds number |
| St | Stokes number |
| Sc | Schmidt number |
| T | Fluid temperature |
| $T_w$ | Wall temperature |
| U, V, W | Contravariant components of fluid velocity |
| $U_p$ | Particle velocity |
| $U^*_\infty$ | Characteristic fluid velocity |
| x, y, z | Cartesian coordinates in physical space |
| $\beta$ | Collision frequency function |
| $\delta$ | Boundary layer thickness |
| $\gamma_k$ | kth coagulation coefficient |
| $\eta$ | Collection efficiency |
| $\mu$ | Fluid viscosity |
| $\rho$ | Fluid density |
| $\sigma$ | Standard deviation of aerosol size distribution |
| $\xi,\eta,\zeta$ | Transformed coordinated in computational space |

Subscripts

| | |
|---|---|
| $\infty$ | initial, free stream or characteristic value |
| D | diffusional |
| p | particle |
| g | geometric mean |
| k | related to the zeroth, first and second moment |
| T | thermal |
| v | viscous |
| $\xi,\eta,\zeta$ | differentiated with respect to $\xi,\eta,\zeta$ |

Superscripts

| | |
|---|---|
| * | dimensional quantity |

## INTRODUCTION

The use of coal based fuels in gas turbines is being studied by a number of research groups (Logan et. al. 1990; Parsons et. al., 1990; Wenglarz and Fox, 1990). One of the major challenges in the successful use of coal based fuels is minimizing the impact of

* To whom correspondence should be addressed.

fuel contaminants on engine components. With the passage of the 1990 Clean Air Act Amendments, there is also a renewed interest in controlling emissions (Quarles and Lewis, 1990). Several studies have been conducted on examining erosion and corrosion problems, and gas clean-up technologies in coal-fired gas turbines (Logan et. al., 1990; Parsons et. al., 1990, Ahluwalia et. al.; 1989 Beacher and Mansour, 1986).

Particles in the hot gas stream are a potential cause of concern. Firstly, by impinging on the turbine surfaces (primarily due to inertial effects), they can cause erosion. Erosion has been widely studied by several authors, and particle transport and deposition due to inertial effects have been successfully modeled (Tabakoff, 1988). These large particles can be readily removed in low pressure hot gas clean-up devices such as cyclones (Wheeldon et. al., 1990). On the other hand, a large number of submicrometer sized particles are also formed in the coal combustion process (Kauppinen et. al., 1989; Mamane et. al., 1986; Smith et. al. 1979). Few studies have been performed with regards to transport behavior of these submicrometer sized particles in gas turbines. For instance, alkali metals nucleate to form particles in this submicrometer size range (Quann et. al., 1982) and could also deposit onto surfaces due to thermophoresis, diffusion (Bai and Biswas, 1990) as well as differential impaction (Biswas, 1988). The deposition of submicrometer particles (alkali metal salts and molten flyash) do not pose problems of erosion; however, they could potentially cause fouling and corrosion of surfaces. Most studies on corrosion assume deposition of vapor phase species, and neglect the deposition of submicrometer aerosols. However, a number of measurements in coal combustion systems indicate a significant fraction of alkali metals and other corrosive species in the submicrometer size range (Campbell et. al., 1978: Davison et. al., 1974 ). Moreover, these submicrometer particles are not readily removed in hot gas clean up devices such as cyclones .

The objective of this study is to examine the transport of sumicrometer sized particles in a gas turbine. The fluid flow and temperature fields are first established by solving the Reduced Navier Stokes equations (Rubin and Tannehill, 1992). The particle transport equations are then solved to account for inertia, thermophoresis, diffusion, and particle-particle interaction due to coagulation. An order of magnitude analysis is carried out to establish the dominant mechanisms for conditions typically encountered in coal combustors. The numerical code is verified by solving for limiting conditions and comparing to solutions available in the literature. The equations are then solved to determine the deposition characteristics in gas turbines under various conditions. The implications of the simulation results are discussed.

## GOVERNING EQUATIONS

There are essentially two sets of governing equations which describe aerosol problems of this type: those related to the fluid flow, and those which determine the behavior of particles within a given flow. Under typical combustion conditions, particles in the submicron size ranges do not to appreciably alter the flow field.

The fluid flow is described by the compressible RNS (Reduced Navier-Stokes) (Rubin and Tannehill., 1992) equations. For general 2-D and axisymmetric curvilinear non-orthogonal $(\xi,\eta)$ coordinates they are given by (Pordal et. al., 1991):

Continuity (1)

$$(\rho grU)_\xi + (\rho grV)_\eta = 0$$

$\xi$-Momentum

$$(\rho grU^2x_\xi)_\xi + (\rho grUVx_\eta)_\xi + (\rho grUVx_\xi)_\eta + (\rho grV^2x_\eta)_\eta + r(Py_\eta)_\xi - r(Py_\xi)_\eta -$$
$$1/Re\left(\mu rx_\xi\left(U\left(x_\xi y_\xi - x_\eta y_\xi\right) + V\left(x_\eta x_\xi - y_\eta y_\xi\right)\right)_\eta / g\right)_\eta -$$
$$2/3\,Re\left(\mu ry_\xi\left(2\left(Ux_\xi y_\xi - Vx_\eta y_\xi\right)_\eta + \left(Vry_\eta x_\xi - Ury_\xi x_\xi\right)_\eta / r\right)/g\right)_\eta = 0$$

$\eta$-Momentum

$$(\rho grU^2y_\xi)_\xi + (\rho grUVy_\eta)_\xi + (\rho grUVy_\xi)_\eta + (\rho grV^2y_\eta)_\eta + r(Px_\eta)_\xi - r(Px_\xi)_\eta -$$
$$1/3\,Re\left(\mu ry_\xi\left(U\left(y_\xi y_\xi - x_\xi x_\xi\right)_\eta + V\left(y_\eta x_\xi - y_\xi x_\xi\right)_\eta\right)/g\right)_\eta -$$
$$2/3\,Re\left(\mu rx_\xi\left(3\left(Uy_\xi x_\xi + Vy_\eta x_\xi\right)_\eta - \left(Vry_\eta x_\xi + Ury_\xi x_\xi\right)_\eta / r + \left(Uy_\xi x_\xi + Vx_\eta y_\xi\right)\right)/g\right)$$
$$+2/3\,Re\left(\mu ry_\xi\left(U\left(y_\xi y_\xi - x_\xi x_\xi\right) + V\left(y_\xi y_\eta - x_\eta x_\xi\right)\right)_\eta / g\right)_\eta = 0$$

where U, V, r, P, x and y are all non-dimensionalized quantities; g is the Jacobian of the coordinate transformation; $\rho$ is the fluid density; U and V are contravariant velocities; Re is the Reynolds number and M is the Mach number. Subscripts connote the derivative of the subscripted quantity with respect to the subscripted variable. As thermophoresis is important in many circumstances, the full energy equation is required for calculation of fluid temperature gradients in cooled passages.

(2)

$$(\rho grHU)_\xi + (\rho grHV)_\eta = -\left(\left(\mu ry_\xi (Hy_\xi)_\eta / g\right)_\eta + \left(\mu rx_\xi (Hx_\xi)_\eta / g\right)_\eta\right)/(Re_\infty Pr)$$

where H is the total enthalpy, $\gamma$ is the ratio of specific heats and Pr is the Prandtl number. Closure of the equations is achieved with the equation of state and the power law for viscosity/temperature dependence.

$$P + (\gamma - 1)\rho V^2 / 2\gamma = (\gamma - 1)[1/(\gamma - 1)M_\infty^2 + .5](\rho H/\gamma) \quad (3)$$

$$\mu = T^{0.72}$$

In general, there are six major mechanisms with which particles can be transported to the blade surface: convection, diffusion, thermophoresis, inertia, gravitation and electrical potential. Transport of particles due to electrical forces are neglected. Additionally, since turbomachinery typically rotate at high speeds, the direction of gravitational force changes rapidly with respect to the blade local coordinate system and can be ignored. With these assumptions, the remaining dominant mechanisms for particle transport are convection, inertia, diffusion and thermophoresis.

When concentrations are high or temperature gradients are severe, particle/particle and particle/gas dynamics must be considered. In general nucleation, condensation and coagulation (Brownian and shear) can be important mechanisms for particle dynamics. Brownian coagulation is the most important consideration in many post combustion environments where aerosols concentrations are high. The particle conservation equation for an aerosol in a flow stream with coagulation effects is given as (Bai and Biswas, 1990)

$$\nabla \cdot \left(n(v)U_p\right) = \frac{1}{2}\int_0^v \beta(v', v-v')n(v)n(v-v')dv' - n(v)\int_0^\infty \beta(v,v')n(v')dv' \qquad (4)$$

where b is the collision frequency function and v and v' are single particle volumes. It is assumed that the particle size distribution conforms to a lognormal pattern, $n(v) = 1/\left(3\sqrt{2\pi}v\ln\sigma\right)\exp\left\{-\ln^2(v/v_g)/18\ln^2\sigma\right\}$, where v and $v_g$ are normalized particle volume and geometric mean particle volumes, and σ is the geometric standard deviation of the size distribution. In general, three parameters are required to define the lognormal distribution. These can be obtained by multiplying the conservation equation by $v^k$ and integrating over all aerosol volumes to give the moment form of the particle conservation equation and then setting k=0,1,2 to find the zeroth, first and second moment conservation equations. The moment equation can be written in condensed form as

$$\nabla \cdot \left[M_k\left(U - \frac{1}{S_{c_k}\mathrm{Re}}\nabla(\ln M_k) - St_k(U\cdot\nabla)\bar{U} - \frac{K_t}{\mathrm{Re}}\frac{\nabla T}{T+\tilde{T}}\right)\right] = \gamma_k M_k^2 \qquad (5)$$

where $\gamma_k$ represents the zeroth, first and second coagulation coefficients calculated from the harmonic average of the limiting cases for small (free molecular) and large (continuum) aerosols. $St_k$, and $Sc_k$ are functions of the local fluid and particle properties and the local aerosol distribution function.

## SOLUTION PROCEDURE FOR FLOW / MOMENT MODEL

The details of the solution technique of the RNS equations have been described by Pordal et. al. (1991). In this study, the boundary conditions are: no-slip conditions at the blade surface; $H_w$ specified at inflow and walls ($H_w$=0.75 is the condition of a cooled passage with a temperature of 0.75 times the inflow value); inflow density and temperature and outflow pressure of unity . $\partial^2P/\partial\xi^2=0$ at the inflow. For the Mach numbers considered here (less than 0.25), H and T vary approximately in the same manner and are used interchangeably.

The coupled, second-order partial differential equations for the aerosol moments were solved using a finite difference scheme with the flow and temperature fields obtained from the RNS procedure described above. Parabolic convective and inertial terms were upwind differenenced with respect to the dominant flow direction (ξ), and central differenced in the cross flow direction (η). Diffusive and thermophoretic terms are elliptic in nature; thus they are central differenced in both directions. Stability considerations have required the use of artificial time terms, in certain circumstances, when the diffusion term is small. At the inflow, boundary values of $M_0$, $M_1$ and $M_2$ are set to non-dimensional inflow conditions of unity. Boundary conditions at the walls can be specified as "slip" ($\partial M_k/\partial\eta$) or "non-slip" in which $M_k$ can be set to any constant value to reflect the nature of the boundary. $M_k$ = 0 reflects total capture of all particles that reach the wall. At the outflow, diffusion and thermophoretic terms are dropped so that no outflow boundary condition on $M_k$ is required. Since it was desired to be able to determine the moments of the distribution from any flow field, and since it is necessary to resolve the solution in zones of high gradients, the finite difference representations of the moment equations are written in general two-dimensional non-orthogonal curvilinear coordinates. This facilitates the necessary stretching at the boundaries and for arbitrary geometries. This resulting system of equations is coupled (through $St_k$ and $Sc_k$) and is highly nonlinear. It is solved in a semi-coupled fashion, in which the three equations are considered sequentially and nonlinear terms are lagged one iteration from the current solution, except for the coagulation term, which is quasi-linearized. Since the resulting coefficient matrix is block-tridiagonal, an efficient block tridiagonal solver was used to invert them at each iteration. Individual terms in the moment equations can be dropped to determine their influence on the solution, however some form of diffusion was required for numerical stability.

## SIMULATION CONDITIONS

Kang et al. (1990) reported that a bimodal ash particle size distribution with peaks at around 8 and 0.25 μm was produced in pulverized coal combustion. Kauppinen et. al. (1989) reported bimodal mass distributions with peaks at 0.2 μm and 10 μm. Similar results have been published by Davison et. al. (1974). For coal combustion in a fluidized bed combustor with 1% coal and 99% coal ash and limestone, Carpenter et. al. (1984) found particles to be clustered around the 3 μm and the 0.25 μm ranges. Temperatures ranged from 700 °C to 1480 °C. The total fly ash concentration was approximately 1.5 g/m$^3$. The mass fraction of submicrometer particles was 4.3%. Similar results were reported by Murthy et. al. (1979). Oil-fired burner results from Markowski et al. (1984) show that for No. 4 fuel oil (the normal fuel for oil-fired burners) operating at low NOx firing modes, there were also two modes in the particle size distribution. These were centered around the 0.0105 micrometer and 0.20 micrometer range. Both modes were reported to be lognormal with peak volume concentrations of about $7x10^{-5}$ $cm^3/m^3$ and a standard deviation of 1.4. Results were essentially the same for low NOx and normal firing modes. Mamane et. al. (1986) found size distributions similar to Markowski et. al. (1984) with two modes in similar size ranges. The operational parameters used in the turbine calculations are summarized in Table I.

### TABLE I.
### SUMMARY OF SIMULATION PARAMETERS REPRESENTATIVE OF COAL/OIL COMBUSTORS

**Inlet Conditions**

| | |
|---|---|
| $T_{\infty}$* Inlet fluid temperature | 1500 °K |
| $p_{\infty}$* Inlet fluid pressure | 5 atm |
| $U_{\infty}$* Inlet fluid velocity | $100\ m\ s^{-1}$ |
| $m_{\infty}$* Inlet fluid viscosity | $56\times10^{-5}\ kg\ m^{-1}s^{-1}$ |
| $\rho_{\infty}$* Inlet fluid density | $1.2\ kg\ m^{-3}$ |
| $\rho_p$* Particle density (Constant) | $1600\ kg\ m^{-3}$ |
| $\sigma_{g\infty}$ Inlet standard deviation of size distribution | 1.4 |
| $dp_{g\infty}$ Inlet geometric mean particle diameter | 5, 0.2, 0.01 μm |
| $M_{0\infty}$ Inlet number concentration of particles | $1\times 10^{16}\ \#\ m^{-3}$ |

**Turbine Wall Conditions**

| | |
|---|---|
| $T_w$* Wall fluid temperature | $0.75\ T_{\infty}$* |
| $U_w$* Wall fluid velocity | $0\ m\ s^{-1}$ |
| $M_{0w}$ Wall number concentration of particles | 0 |

## RESULTS AND DISCUSSION

An order of magnitude analysis is carried out to establish the important transport mechanisms. Estimates are based on the peaks of the bimodal distributions for typical combustion aerosols. Assuming particle density, fluid density, fluid velocity, fluid viscosity and characteristic length are equivalent for all cases, the relative importance of each term in the governing equations is estimated. The contribution to particle transport from each of these mechanisms is based upon the non-dimensional parameters, Sc, Re (or Pe), St and Kt as well as the gradients in U, n and T. Thus to calculate the relative importance of each term in the governing equation, we must also have approximations for the gradients in **U**, T and n(v). From boundary layer theory (Schlichting, 1975), the thickness of the laminar viscous boundary layer is approximated by $\delta_v=\wp(1/Re^{1/2})$. The thermal boundary layer thickness can be related, via the Prandtl number (Pr), by $\delta_T=\delta_v/Pr^{.4}$. Bird et. al. (1960) gives an approximate relationship between the particle diffusivity, the diffusion boundary layer and the viscous boundary layer: $\delta_D=\delta_v/Sc^{1/3}$. For a non-dimensional equation, the gradient is of order $1/\delta$ (since all non-dimensional properties become order one outside the boundary layer and are zero at the wall). Thus, in the normal direction to the wall where gradients are expected to be the largest,

$$U_p \propto U - \frac{1}{Sc^{4/3}Re^{1/2}} - St\,Re^{1/2}\,U - \frac{K_t}{Re^{1/2}}Pr^{0.4} \qquad (6)$$

U and $K_t$ are both of order one. Prandtl number is generally of order 1. Thus the contributions from each term are:

Convective = $\vartheta(1)$; Diffusive = $\vartheta(\frac{1}{Sc^{4/3}Re^{1/2}})$;

Inertial = $\vartheta(St\,Re^{1/2})$; Thermophoretic = $\vartheta(\frac{1}{Re^{1/2}})$

The relative importance of each the these terms is now calculated from representative peaks in the distributions of pulverized coal and oil combustion (Table II). Thus, for the fine mode of coal combustion aerosols and the coarse mode of oil combustion aerosols, diffusion and inertia can be safely ignored as dominant transport mechanisms. For the coarse mode of coal combustion aerosols, diffusion can be safely ignored. For the fine mode of oil combustion aerosols, inertial terms can be ignored. In general, thermophoretic terms dominate.

### TABLE II.
### SUMMARY OF ORDER OF MAGNITUDE ANALYSIS FOR COAL AND OIL COMBUSTION

| Mechanism | Coal (1st) & Oil (2nd) dp = 0.20 μm | Coal (2nd) dp = 5.0 μm | Oil (1st) dp = 0.01 μm |
|---|---|---|---|
| Diffusion | $0.22x10^{-5}$ | $0.41x10^{-9}$ | $0.65x10^{-2}$ |
| Inertia | $0.29x10^{-3}$ | $0.72x10^{-2}$ | $0.14x10^{-4}$ |
| Thermophoresis | $0.22x10^{-2}$ | $0.22x10^{-2}$ | $0.22x10^{-2}$ |

A number of simulations were carried out as summarized in Table III. Initial runs were made under limiting conditions to validate the code by comparing to solutions in the liturature.

Comparisons with Pratsinis (1988) and Lee and Chen (1984) for coagulation rates show exact agreement for all cases in the free-molecular, slip and continuum regimes. Comparisons with Pratsinis (1988) and Lee et. al. (1984) for aerosol distributions as time approaches infinity were performed by running pseudo 1-D cases. This was achieved by setting a slip boundary condition at the wall, in a uniform flow, and then setting U=0.05 m/s for various mean particle radii. For uniform flow for 1m, a residence time of 20 sec is obtained. Initial distributions of $\sigma_{\infty}$=1, 2 and 3 were chosen. For all prescribed conditions, $\sigma$ approaches the theoretical value of 1.355 (Lee et. al., 1984) for free-molecular and 1.32 (Pratsinis, 1988) for continuum cases.

The effects of particle size and polydispersity on coagulation rates compare well to Freidlander (1977) and Lee and Chen (1984). Freidlander (1977) reports that twelve seconds are required for 0.1 micrometer particles at a concentration of $1\times10^{15}$ #/m$^3$ to reach a concentration of one tenth the original value. Results from this code give a coagulation time of 15 seconds for the same conditions. Results show that larger standard deviations cause more rapid particle coagulation and thus more rapid increases in mean particle diameter.

**TABLE III.**
**SUMMARY OF SIMULATIONS**

**Code Validation**

| Conditions | Comparison Source |
|---|---|
| 1) Coagulation in uniform flow. | Lee & Chen (1984), Pratsinis (1988), Lee et. al. (1984) |
| 2) Effects of initial particle size, standard deviation and number concentration on Mo evolution in uniform flow. | Lee & Chen (1984), Freidlander (1977) |
| 3) Depositional efficiency due to diffusion for devoping flow in a straight channel. | Hinds (1982) |
| 4) Depositional efficiency due to diffusion and thermophoresis for devoping flow in a straight channel. | Chang et. al. (1990) |

**Turbine Flows**

| |
|---|
| 1) Convection, Diffusion and Inertia Considered ($dp_g$=0.01, 0.2 μm) |
| 2) Convection, Diffusion, Inertia and Thermophoresis Considered ($dp_g$=0.01, 0.2, 5.0μm) |
| 3) Convection, Diffusion, Inertia and Coagulation Considered ($dp_g$=0.01, 0.2 μm) |

Diffusion and thermophoresis were validated using developing viscous flow in a 2-D channel. The equations of fluid motion where solved only once on a 25 x 50 grid with a stretching factor of 1.2 in the y-direction and 1.05 in the x-direction. A channel of length 1 m and height 0.1 m was used as the test geometry. At STP (1 atm and 298 K), this gives a flow Reynolds number of 654 based on the inlet dimension at a velocity of 0.1 m/s. A wall temperature of 223.5 K (0.75 $T_{\infty}$) was used for calculations with thermophoresis. The moment equations where then solved on the same grid in the straight channel using these flow conditions and various moment model conditions. Typical results are shown in Figure 1 for 0.01 μm particles using diffusion, thermophoresis and coagulation.

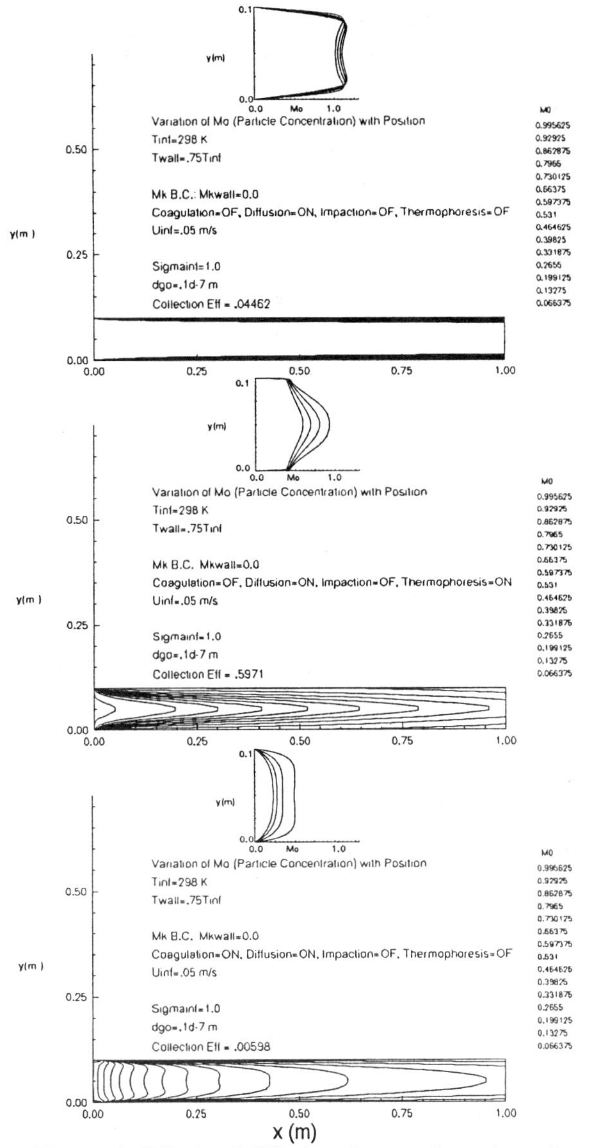


**Figure 1:** Effects of diffusion, thermophoresis and coagulation in a 1m x 0.1m straight channel: Contours & profiles of dimensionless number concentration at x=0.2, 0.4, 0.6, 0.8. ($M_{0\infty}=1\times10^{15}$ #/m$^3$)

To validate the diffusion terms in the code, the moment equations where solved in the straight channel for various initial aerosol sizes. The results are compared with Hinds' (1982) diffusional deposition equations. Cases were run for a range of inflow particle radii with $\sigma$=1.0 (monodisperse distribution) at the inflow. The flux of particles in and out were calculated from the resulting moment and velocity fields using trapezoidal integration. Excellent agreement was obtained with the results of Hinds (1982).

Additional computations were performed to validate the thermophoretic results. Experimental results for a two dimensional flow with thermophoresis were unavailable, however, there have been a number of studies for axisymmetric flow through a pipe with cooled walls (Chang et. al., 1990; Montassire et. al., 1990; Lipatov and Chermova, 1990). Results are compared qualitatively to Chang et al (1990). A typical profile is shown in Figure 1 for 0.01 micrometer particles and a wall temperature of 0.75 the inflow value. No published results have been found to validate the interesting nature of these profiles. Note the thickness of the diffusion boundary layer as opposed to the thermophoretic layer. Cases were run with wall temperatures of 0.75, 0.5, 0.25 and 0.1 times the inflow values. The trend, is the same for the results in this study and Chang et. al. (1990): deposition initially rises as the wall is cooled, but levels off when the wall is cooled below about 50% of the initial temperature. Beyond this point thermal gradients in the cross flow direction become negligible.

Calculations were considered with coagulation terms in the viscous channel flow field. Results are shown in Figure 1 for realistic conditions, with coagulation and diffusion terms retained in a highly concentrated particulate environment ($1\times10^{15}$ #/m$^3$). For the case shown ($dp_g$ = 0.01 μm), both Brownian diffusion and coagulation effects are significant. Note that the slower fluid velocity near the wall permits more time for Brownian coagulation and thus significantly reduces the number concentration as compared to the fast moving centerline of the channel. Also note that the thickness of the diffusion boundary layer is greatly altered due to the reduced mobility of larger coagulated particles. Collection efficiencies with the coagulation terms considered and not considered indicates this effect markedly. Collection efficiency drops by almost order of magnitude with coagulation present: 4.4% are collected without coagulation as compared to 0.6% with coagulation considered. This is due to coagulation leading to formation of larger particles which diffuse more slowly to the walls.

A qualitative example of the effects of inertial forces in particle distributions is shown in Figure 2. For 10 μm particles, all the number concentration ($M_0$) profiles are approximately equal across the channel width if coagulation is neglected. Under the same conditions, with inertial terms included, $M_0$ profiles evolve significantly as the flow negotiates the bend. Centrifugal forces cause particles to deviate from the streamlines and collect at the upper wall as indicated by the larger particle concentrations in the outside of the bend.

The flow in the turbine passages for the conditions listed is in the turbulent regime. However, for a first calculation, laminar conditions were assumed (eddy diffusivity and viscocity terms neglected) in this work. Current efforts are underway at performing turbulent calculations. The representative turbine flow was calculated on a 50 x 200 grid at one tenth the actual Reynolds number. Wall temperature was set to 0.75 $T^*_\infty$ to represent an internally cooled turbine blade. Flow solutions for parameters used in the moment calculation (velocity and temperature) are shown in Figure 3. A matrix of moment cases were run with this base flow. Example $M_0$ profiles at the suction (1, 2, 3, 4) and pressure (A, B, C, D) surface of the blade are shown in Figure 4 for 0.01 μm particles due to convection, diffusion and inertial terms, and due to convection diffusion, inertial and thermophoretic terms. Collection efficiencies for all the cases are summarized in Table IV.

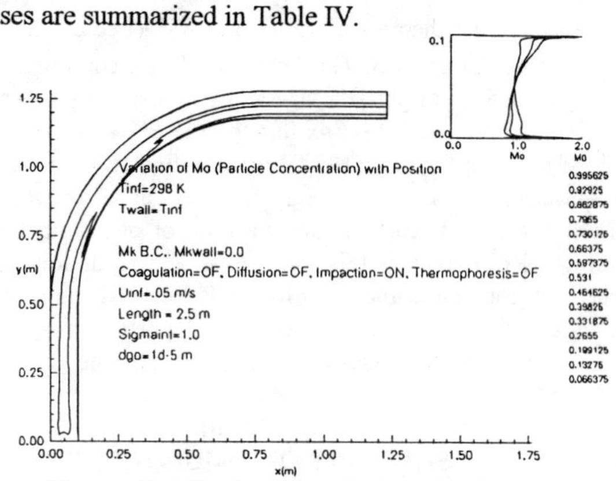


**Figure 2:** Effects of inertia in a 90° bend: Contours and profiles of number concentration at start of bend (y=0.5), middle of bend, and end of bend (x=0.75)

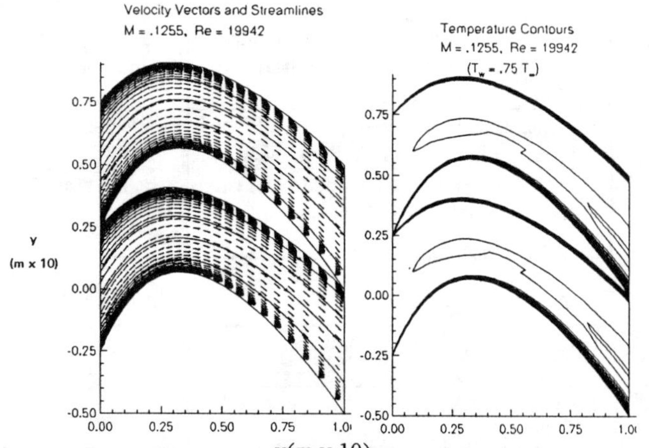


**Figure 3:** Relevant flow properties for moment calculations in turbine geometry

**TABLE IV.**
**SUMMARY OF COLLECTION EFFICIENCIES (%) IN TURBINE BLADE CALCULATIONS**

| Mechanisms | Geometric Mean Diameter (μm) | | |
|---|---|---|---|
| | 0.01 | 0.20 | 5.0 |
| Diffusion, Inertia | 0.19 | 0.02 | -- |
| Diffusion, Inertia, Thermophoresis | 1.70 | 1.59 | 1.59 |
| Diffusion, Inertia, Coagulation | 0.17 | 0.02 | -- |

It can be seen that thermophoretic forces dominate the deposition process for this wall temperature condition for all initial particle sizes as predicted in the order of magnitude analysis. In fact there is only a 6% change in total deposition from 0.01 μm particles and 5 μm particles, when thermophoresis is considered. At high particle concentrations, coagulation is found to reduce the collection efficiency, since the slow moving fluid at the walls allows sufficient time for aerosols to coagulate into larger non-diffusive conglomerates. The effect is greater for smaller aerosols where coagulation is more rapid. The net change in collection, due to the increase in the inertial force of larger coagulated particles, is found to be negligible. In areas where coagulation is significant, the fluid velocity is too small for appreciable centrifugal influences. Diffusion becomes negligible for initial particle diameters above 0.2 μm. Since these terms tend to condition the coefficient matrix, the equations become unstable for the 5 μm case and no solutions were obtained. Note that the diffusional boundary layer is of the order $1x10^{-3}$ m for the 0.01 μm particles (10,000 particle diameters) and $1x10^{-4}$ m for the 0.2 μm particles (500 particle diameters). This implies that the continuum assumption is still valid for intermediate particles. For 5 μm aerosols, the diffusion boundary layer becomes the same order of magnitude as the particle diameter and the continuum assumption becomes questionable near the wall.

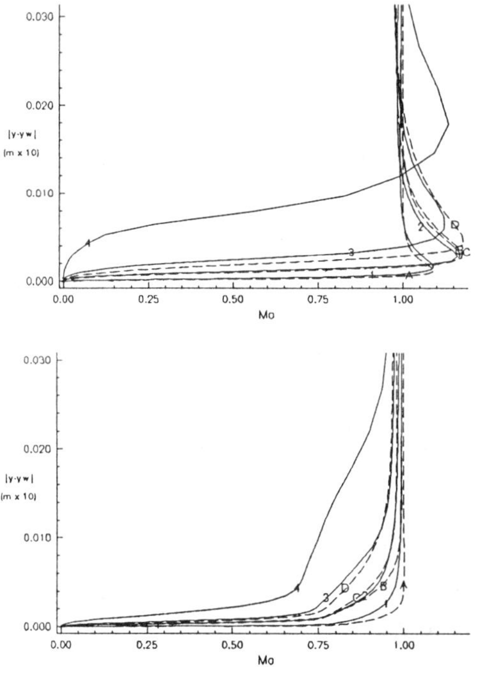


**Figure 4:** Moment results in turbine ($dp_g$ =0.01 μm)
Mo profiles with convection, diffusion and inertia (top)
Mo profiles with convection, diffusion, inertia, thermophoresis (bottom)
(A—B—C—D =Pressure Side, 1---2---3---4 =Suction Side)

It must be noted that the predicted deposition efficiencies are rather small (<5%). However, the calculations have been carried out under a specific set of conditions and work is underway to perform simulations under a variety of other conditions. Also, additional mechanisms that lead to submicrometer particle deposition such as turbulent diffusion (Freidlander, 1977) and differential impaction (Biswas, 1988) have not been considered.

## CONCLUSIONS

A 2-D moment model incorporating the effects of coagulation, diffusion, inertial impaction and thermophoresis has been developed in general non-orthogonal coordinates for general flows. Preliminary comparisons with 1-D published results and correlations for coagulation rates are excellent. Additional calculations for viscous flow in a 2-D straight duct and a 90° bend using the RNS equations are computed. Comparisons with published results for diffusion rates are excellent. It is found that diffusion and thermophoretic forces increase collection efficiency while collection efficiency drops when coagulation is considered. An approximate laminar calculations in a cooled turbine blade geometry indicates that thermophoretic forces are most important for deposition for the 1st and 2nd modes oil combustion particles and the 1st mode coal combustion particles. At high particle concentrations, coagulation is found to reduce the collection efficiency, since the slow moving fluid at the walls allows sufficient time for aerosols to coagulate into non-diffusive particles.

It has been demonstrated that the present moment model is effective and efficient in solving for deposition of polydisperse aerosols in general flow applications. Depositional efficiencies can be calculated directly without having to sum up results for monodisperse aerosols over the entire particle spectrum. Additionally, other important operations parameters, such as outflow distributions and number concentrations, can be readily calculated. Research is underway to perform simulations under turbulent flow and other conditions.

## REFERENCES

Ahluwalia, R. K., Im, K. H. & Wenglarz, R. A. (1989) "Flyash Adhesion in Simulated Coal-Fired Gas Turbine Environment", *Journal of Engineering for Gas Turbines and Power* v 111 pp. 672-678.

Bai, H., Biswas, P. (1990) "Deposition of Lognormally Distributed Aerosols Accounting for Simultaneous Diffusion, Thermophoresis and Coagulation", *Journal of Aerosol Science,* v 21 n 5 pp. 629-640.

Beacher, B.F., and Mansour, M. L. (1986) "Erosion of Coal-Burning Gas Turbine by Ash Particles", *Turbomachinery Performance Deterioration,* AIAA 4th Joint Fluid Dynamics, Plasma Dynamics and Laser Conference.

Bird, R. B., Stewert, W. E., and Lightfoot, E. N. (1960) *"Transport Phenomena",* John Wiley & Sons.

Biswas, P. (1988) "Differential Impaction of Aerosols", *Journal of Aerosol Science,* v 19 n 5 pp. 603-610.

Campbell, J. A., Laul, J., C., Nielson, K., K., and Smith, R. D. (1978) "Separation and Chemical Characterization of Finely-Sized Fly-Ash Particles", *Analytical Chemistry,* v 50 n 8.

Carpenter, R. L., Cheng, Y. S., and Barr, E., B. (1984) "Size Distribution of Fine Particle Emissions from a Fluidized Bed Coal Combustor at a Steam Plant", *Aerosols,* Liu, Pui and Fissan editors.

Chang, Y. C., Ranade, M. B., and Gentry, J. W. (1990) "Thermophoretic Deposition of Aerosol Particles", *Journal of Aerosol Science,* v 21 Suppl. 1 pp. s81-s85.

Davison, R. L., Natusch, D. F. S., and Wallace, J. R. (1974) "Dependence of Concentration on Particle Size", *Environmental Science and Technology* v 8 n 16.

Freidlander, S. K. (1977) *"Smoke, Dust and Haze"* John Wiley & Sons.

Hinds, W. C. (1982) *'Aerosol Technology: Properties, Behavior and Measurement of Airborne Particles"*, John Wiley & Sons.

Kang, S. G., Kersein, A. R., Helble, K. K. and Sarofim, A. F. (1990) "Simulation of Residual Ash Formation During Pulverized Coal Combustion: Bimodal Ash Particle Size Distribution", *Aerosol Science and Technology.*, v 13, pp. 401-412.

Kauppinen, E., Heiskanen, M., and Hahkala, M. (1989) "Real Time Measurement of Combustion Aerosol Particle Size Distributions", Technical Research Center of Finland, Report No. VTT-TUTK-619.

Lee, K. W., and Chen, H. (1983) "Coagulation of Polydisperse Particles", *Journal of Aerosol Science and Technology*, v 3 pp. 327-334.

Lee, K. W., Chen, H., and Geiseke, J. A. (1984) "Log-Normally Preserving Size Distribution for Brownian Coagulation in the Free-Molecular Regime", *Aerosol Science and Technology,* v 3 pp. 53-62.

Lipatov, G. N. and Chermova, E. A. (1990) "Several Features of Aerosol Behavior in Laminar Flow with Transverse Temperature Gradient", *Journal of Aerosol Science,* v 21, Suppl. 1 pp. s89-s92.

Logan, R. G., Richards, G. A., Meyer, C. T., Anderson, R. J. (1990) "Study of Techniques for Reducing Ash Deposition in Coal-Fired Gas Turbines", *Progress in Energy and Combustion Science* v 16 n 4 pp. 221-223.

Mamane, Y., Miller, J. L., and Dzubay, T. G. (1986) "Characterization of individual Fly Ash Particles emitted from Coal -and Oil-Fired Power Plants", *Atmospheric Environment* v 20 n 11.

Markowski, G. R., Downs, J. L., and Reese, J. L. (1984) "Submicrometer Size Distributions from an Oil-Fired Burner", *Aerosols,* Liu, Pui and Fissan editors.

Montassire, N., Bouldaud, D., and Stratmann, F. (1990) "Comparison Between Experimental Study and Theoretical Model of Thermophoresis Particle Deposition on Laminar Tube Flow", *Journal of Aerosol Science,* Vol. 21, Suppl. 1 pp. s85-s88.

Murthy, K. S., Howe, J. E, Nack, H., Hoke, R. C. (1979) "Emissions from Pressurized Fluidized-Bed Combustion Processes", *Environmental Science and Technology,* v 13 n2 pp. 197-204.

Parsons, E. L., Webb, H. A., and Zeh, C. M.(1990) "Assessment of Hot Gas Clean up Technologies in Coal-Fired Gas Turbines", International Gas Turbine and Aeroengine Congress and Exposition, American Society of Mechanical Engineers, GT293.

Pordal, H., Khosla, P. K. and Rubin, S. G., (1991) "Pressure Flux-Split Viscous Solutions for Subsonic Diffusers", *AIAA* 22nd Fluid Dynamics, Plasma Dynamics & Lasers Conference, June 1991.

Pratsinis, S. (1988) 'Simultaneous Nucleation Condensation and Coagulation in Aerosol Reactors' *Journal of Colloidal and Interface Science,* v 124 n 2.

Quann, R. J., Neville, M., Janghorbani, M., Mims, C. A., Sarofirm, A. F. (1982) "Mineral Matter and Trace-Element Vaporization in a Laboratory-Pulverized Coal Combustion System", *Environmental Science and Technology* v 16 n 11 pp. 776-781.

Quarles, J., and Lewis, W. H. (1990) *"The New Clean Air Act: a Guide to the Clean Air Program as Amended in 1990"*, Morgan, Lewis & Brockius.

Rubin, S. G., and Tannehill, J. C. (1992) "Parabolized/Reduced Navier-Stokes Computational Techniques", *Annual Review of Fluid Mechanics.* v 24, pp. 117-144.

Schlichting, H. (1975) "Boundary Layer Theory", 7th edition, McGraw Hill.

Smith R. D., Campbell, J. A., Nielson K., K. (1979) "Concentration Dependence upon Particle Size of Volatilized Elements in Fly Ash", *Environmental Science and Technology,* v 13 n 5.

Tabakoff, W. (1988) "Modeling Erosion Behavior at High Temperatures for Fluidized Beds", International Symposium on Materials Behavior in High Temperature and Energy Conversion Systems.

Wenglarz, R. A., and Fox, R. G. (1989) "Chemical Aspects of Deposition/Corrosion from Coal Water Fuels under Gas Turbine Conditions", *American Society of Mechanical Engineers,* 1989 International Conference on Fluid Bed Combustion.

Wheeldon, J. M., Burnard, G. K., Snow, G. C., and Svarovsky, L., (1990) "The Performance of Cyclones in the Off-Gas Path of a Pressurized Fluidized Bed Combustor", *European Federation of Chemical Engineers*, Symposium #99, Gas Cleaning at High Temperatures.

## ACKNOWLEDGEMENTS

David Brown acknowledges support from a NASA Graduate Research Fellowship.

FACT-Vol. 18, Combustion Modeling, Scaling and Air Toxins
ASME 1994

# THERMAL DESTRUCTION BEHAVIOR OF PLASTIC AND NON-PLASTIC WASTES IN A LABORATORY SCALE FACILITY

**A. K. Gupta, E. Ilanchezhian, and A. Missoum**
Department of Mechanical Engineering
College Park, Maryland

**E. L. Keating**
Consultant
NSWC
Annapolis, Maryland

## ABSTRACT

Experimental and theoretical studies are presented from a laboratory scale thermal destruction facility on the destruction behavior of surrogate plastic and non-plastic solid wastes. The non-plastic waste was cellulosic while the plastic waste contained compounds such as polyethylene, polyvinyl chloride, polystyrene, polypropylene, nylon, rubber and polyurethane or any of their desired mixtures. A series of combustion tests were performed with samples containing varying composition of plastic and non-plastic. Experimental results are presented on combustion parameters (CO, excess air, residence time) and toxic emissions (dioxin, furan, metals). Equilibrium thermochemical calculations are presented on the thermal destruction behavior of samples under conditions of pyrolysis, combustion, and pyrolysis followed by combustion. Special interest is on the effect of waste properties and input operational parameters on chemistry and product composition. STANJAN and SOLGASMIX computer codes were used in the chemical equilibrium study. Analysis and interpretation of the data reveal the effect of waste feed composition on combustion parameters and dioxin, furan and metals emission. Equilibrium calculation results are used to describe the experimentally observed trends for the thermal destruction behavior of the wastes. The results show significant influence of plastic on combustion characteristics, and dioxin, furan and metals emission.

## 1.0 INTRODUCTION

Numerous environmental catastrophes resulting from the improper disposal practices of the wastes of different kinds have caused increased public awareness on the growing problem of waste generated in all sectors of public, industrial and government. Waste minimization and recycling provides only a partial solution. Further stringent measures must be taken in order to solve the problem of waste disposal completely. The United States generates approximately $10^{12}$ lbs of waste every year and even with extensive waste minimization plans, this amount is projected to increase at a rate of 1% annually[1]. Thermal destruction in a broad scope includes the application of three fundamental reacting processes of pyrolysis, gasification and combustion. Thermal destruction technologies include: mass burn type incinerators, fluidized bed, rotary kiln, molten salt bed, low or high temperature oxygen/air enriched systems, low or high temperature starved air systems. More recently electric heating, microwave and plasma assisted systems have also appeared[2]. Thermal destruction offers distinct advantages over the other methods since it provides maximum volume reduction, permanent disposal, energy recovery and the by-products can be used in several ways in building material and road bed construction[3]. Of all the permanent treatment technologies, thermal destruction provides the highest overall degree of destruction. In addition it provides maximum volume and mass reduction and has the potential to provide energy recovery. However the disadvantage is that it must be environmentally clean and accepted by the public since the by-products can be health hazardous and detrimental to our environment.

Municipal solid wastes (MSW) presently use landfills (83%), incineration (6%) and recycling (11%) for waste disposal[4, 5]. The most common disposal of MSW is landfill. Landfill disposal creates problems of odor, methane, carbon-dioxide, toxic gases and leachate generated from landfill site to soil and ground water[6]. Special interest on air toxic organic pollutants and trace metals emission from incinerators came after the risk assessment findings towards human life[4]. In addition to the concern on pollutants such as $NO_x$, $SO_2$, HCl, CO, $CO_2$, unburnt hydrocarbons, particulates and the emission of dioxins, furans and volatile metals have received increased attention from many countries around the world[4]. The concerns of pollutants emission which are produced as by-products from the direct result of the combustion process[7], are common to all combustion sources. In the present paper an experimental and numerical investigation is described with the objective to explore thermal destruction behavior of different plastic and non-plastic mixtures in a laboratory scale facility and its effect on carbon monoxide, particulates, dioxin, furan and toxic metals emission. The results show that the composition of the waste has a significant influence on the emissions and that the blending approach for enhanced burning of waste has the benefit of reducing toxic emissions and solid residues.

## 2.0 EXPERIMENTAL

A schematic diagram of the experimental facility is shown in Figure 1. The laboratory scale Thermal Destruction Facility (TDF) consists of a destruction chamber, controls and instrumentation for TDF, EPA method 23 sampling train, continuous gas monitoring system for gas analysis, Scanning Electron Microscopy (SEM) for fly ash metal analysis and Neutron Activation Method (NAM) for bottom ash analysis. The chamber is 12" X 12" X 18" in size. A natural gas burner is used for start up and auxiliary burning. Inside of the chamber was lined with a 2 inch thick insulating refractory cement, type ASTM C 449, to minimize heat transfer to the surroundings. Continuous feeding is achieved by a rotary feeder driven by an electric motor. The fly ash and product gases leave through an exhaust pipe while the bottom ash is removed through the ash and slag pipe located at the bottom of the chamber. Gas sampling and temperature measuring ports are provided on the side and exit of the TDF for continuous monitoring and measurement of operating parameters. The TDF has all the necessary provisions to study the effect of operating parameters, such as fuel type and size, feed rate, excess air, residence time and temperature on thermal destruction and products formation.

The combustion air and fuel gas flow to the gas burner are measured using flowmeters. The waste feed rate is calculated from the speed of the feeder motor. Two digital temperature indicators are used for continuously monitoring the temperatures in the furnace and exhaust duct respectively. The furnace temperature is measured using a type R (Pt-Pt/13% Rh) thermocouple of wire size 0.01" while the exhaust gas temperature is measured using a type K (Cr-Al) thermocouple having 0.025" wires diameter. A microprocessor based data acquisition and analysis system was used to measure the instantaneous and time averaged temperature in the combustion chamber and exhaust duct.

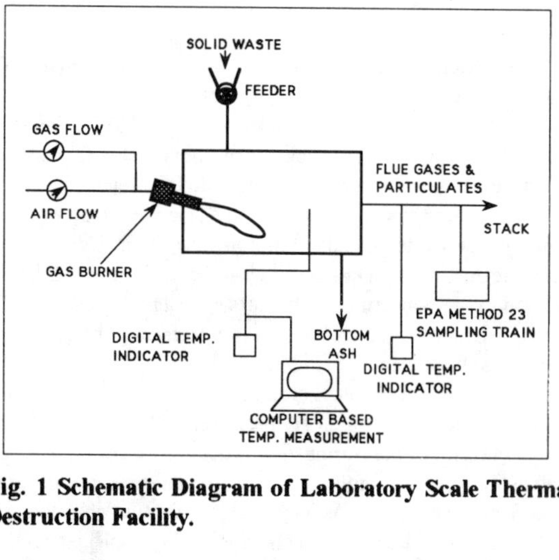


**Fig. 1 Schematic Diagram of Laboratory Scale Thermal Destruction Facility.**

Experiments were designed to study the effect of varying composition of plastic and non-plastic fraction of the fuel on emission characteristics. Seven different samples were examined (see Table 1). The chemical composition of the non-plastic and plastic materials and their heating values are given in Table 2.

The bulk density of all the samples was determined by weighing a known volume of the sample, to be between 5.4 to 8.5 lb/ft$^3$.

**Table 1 Composition of the Examined Non-plastic and Plastic Samples.**

| Sample No. | Non Plastic (% volume) | Plastic (% volume) |
|---|---|---|
| 1 | 100* | 0 |
| 2 | 85 | 15 |
| 3 | 50 | 50 |
| 4 | 50 | 50** |
| 5 | 50 | 50*** |
| 6 | 50 | 50**** |
| 7 | 0 | 100***** |

* =Paper and cardboard
** =Plastic mixture without polyurethane and latex
*** =Plastic mixture without polyvinyl chloride
**** =Plastic mixture without nylon, latex, polyurethane and acetate
*****=Plastic mixture of polyethylene, PVC, polystyrene, polypropylene, polyethylene, nylon, latex, acetate, polyurethane

The TDF chamber was heated up with a natural gas burner and kept in pilot running mode to achieve the desired temperature. The air flowrate was kept constant at 6.75 scfm and the mixture feed rate varied from 0.128 lb/min. to 0.441 lb/min. as a result of variation in the heating value of plastic and non-plastic in each sample. The equivalence ratio and gas residence time was varied from 0.58 to 0.93 and 0.67 to 1.00 sec. respectively. The combustion process is maintained to be in a pseudo-steady state with the TDF temperature maintained in the range of 1400-1600 °F. During each test, fly ash was collected while the bottom ash was collected at the end of the test. The online monitoring of CO during all the tests was carried out using an infrared CO analyzer. EPA method 23 (Modified method 5)[8] was used to sample PCDD and PCDF. The subsequent analysis was carried using a high resolution gas chromatograph combined with a mass spectrometer. Two different ions for dioxin (304.2, 306.2) and furan (320.2, 322.2) groups were monitored to obtain the peaks using selective ion monitoring technique. Fly ash samples were tested for unburnt carbon loss by evaluating Loss On Ignition (LOI) using a high temperature furnace. Fly ash samples were heated to a high temperature of 1700 °F and the weight loss measured provided an indication of unburnt carbon present in ash. Bottom ash analysis for toxic metals was carried out by Neutron Activation method. This was carried for short irradiation metals where the sample is made radioactive. A detector is used to count the isotopes coming out of the spectrum. However this method will not detect some long irradiating metals like cadmium, lead, and iron because of their longer waiting time. The semi-quantitative analysis of fly ash samples were performed using Scanning

Electron Microscopy (SEM). The spectrum of different metals in the fly ash and their weight percentage were determined by an X-ray micro analysis system.

## 3.0 RESULTS AND DISCUSSION

### 3.1 Numerical Results

Computer codes STANJAN[9] and SOLGASMIX[10] were used to calculate the equilibrium reaction conditions. A large thermodynamic data file compiled from JANAF tables is used in these codes[11]. Equilibrium thermochemical calculations of a mixture of non-plastic and plastic surrogate solids were carried out under conditions of pyrolysis and combustion. The non-plastic material is assumed to be cellulose while the plastic material may contain any or all of the following plastics: polyethylene, polyvinyl chloride, polystyrene, polypropylene, polyethylene teraphathalic, nylon, latex in the form of rubber, polyurethane and acetate. Cellulose represents the organic portion of the waste such as paper and cardboard.

**Table 2 Chemical Composition and Heating Value of the Examined Materials.**

| Material | Chemical formula | Heating Value (Btu/lb) |
|---|---|---|
| Polyethylene | $(CH_2H_2)n$ | 19932 |
| Polyvinyl chloride | $(C_2H_3CL)n$ | 7875 |
| Polystyrene | $(C_8H_8)n$ | 17838 |
| Polypropylene | $(C_3H_6)n$ | 19948 |
| Polyethylene | $(C_{10}H_8O_4)n$ | 12700 |
| Nylon | $(C_6H_{11}ON)n$ | 13640 |
| Latex (Rubber) | $(C_4H_8O_2S)$ | 19465 |
| Polyurethane | $(C_{12}H_{22}O_4N_4)n$ | 11203 |
| Acetate | $(C_4H_6O_2)n$ | 12050 |
| paper | $(C_6H_{10}O_5)n$ | 7200 |
| cardboard | $(C_6H_{10}O_5)n$ | 7000 |

The adiabatic flame temperature and product composition under conditions of direct combustion was predicted over a range of conditions extending from fuel-lean to fuel-rich modes (up to 10 moles of air). The analysis could provide up to 400 species and molecules but in the present analysis only the following major species were considered: C, $CH_4$, CO, $CO_2$, $C_3H_8$, H, HO, $H_2$, $H_2O$, N, NO, $NO_2$, O and $O_2$. Equilibrium predictions are affected by the species considered. Results shown in Figure 2 for the seven samples reveal the importance of plastic on the distribution of adiabatic flame temperature. Increased plastic content in the waste yields higher temperature than 100% non-plastic and the maximum temperature is obtained when burning 100% plastic. The adiabatic flame temperature shifts to lower number of moles of air with increase in plastic content. Inclusion of plastic to non-plastic materials has a significant influence on temperature whereas exclusion of certain plastic component within the mixture has negligible influence on the peak temperature. The higher predicted flame temperature is a result of higher heat content of plastic material.

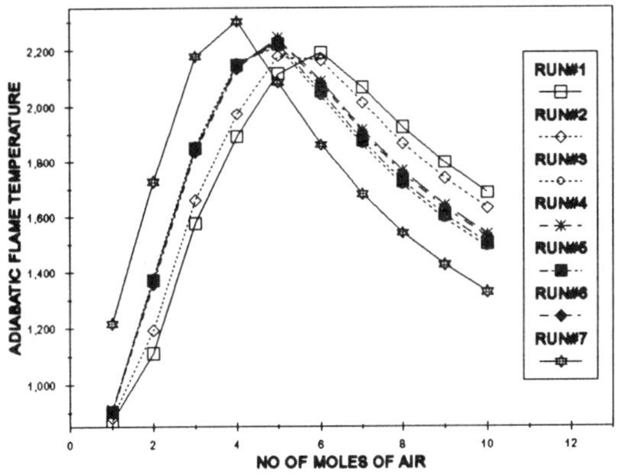


**Fig. 2 Combustion of Plastic and Non-plastic Materials in Air.**

The product mole distribution is calculated using SOLGASMIX[9] for different moles of air and the corresponding combustion temperature. Specifically the products formed during the combustion of seven different samples were examined according to:

$$plastic\,/\,non\text{-}plastic + a\,(O_2 + 3.76\,N_2) \rightarrow products$$

The results shown in figures 3(A and B) and 4 (A and B) are for 100% non-plastic and 100% plastic respectively.

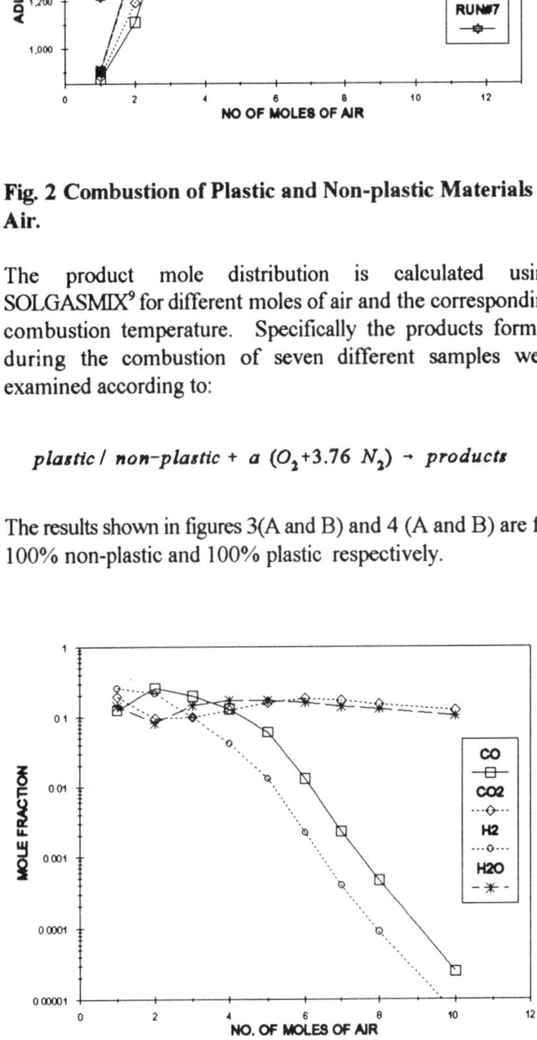


**Fig. 3A Evolution of CO, $CO_2$, $H_2$ and $H_2O$ from the Combustion of 100% non-plastic in air**

Yield of $CO_2$ and $H_2O$ stays almost constant for both the 100% non-plastic and 100% plastic. Concentrations of CO and $H_2$ decay more rapidly with 100% non-plastic than with plastic due to the variation in the reaction temperatures. Emission of NO and $NO_2$ from 100% non-plastic increases with increase in moles of air whereas their level decreases from 100% plastic after stoichiometric condition. Formation of compounds such as $H_2S$ and HCl from the combustion of plastic is due to the presence of sulfur and chlorine in different kinds of plastics (see Table 2).

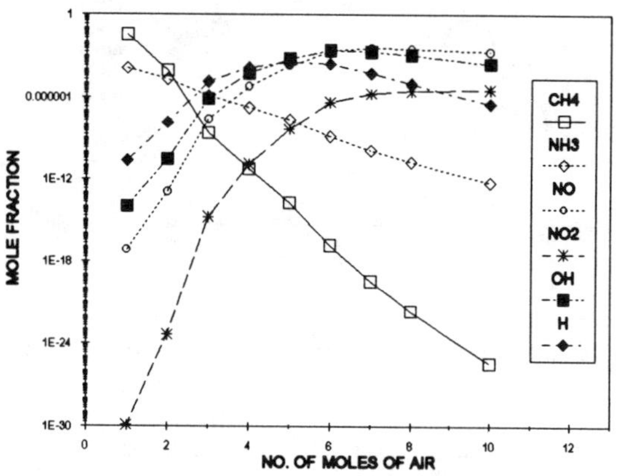


**Fig. 3B Evolution of $CH_4$, $NH_3$, NO, $NO_2$, OH and H from the Combustion of 100% Non-plastic.**

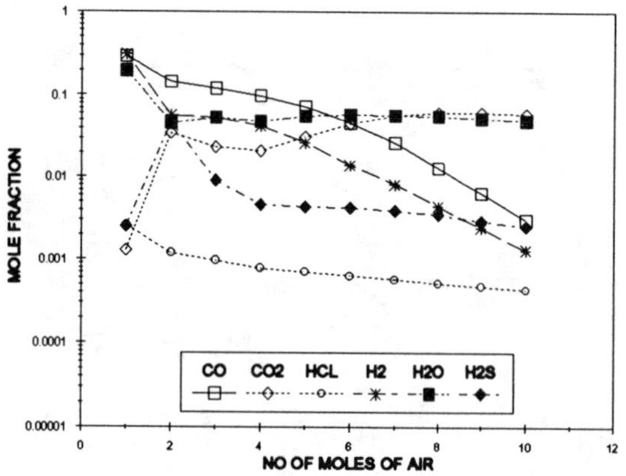


**Fig. 4A Evolution of CO, $CO_2$, HCl, $H_2$ and $H_2O$ from the Combustion of 100% Plastic in Air.**

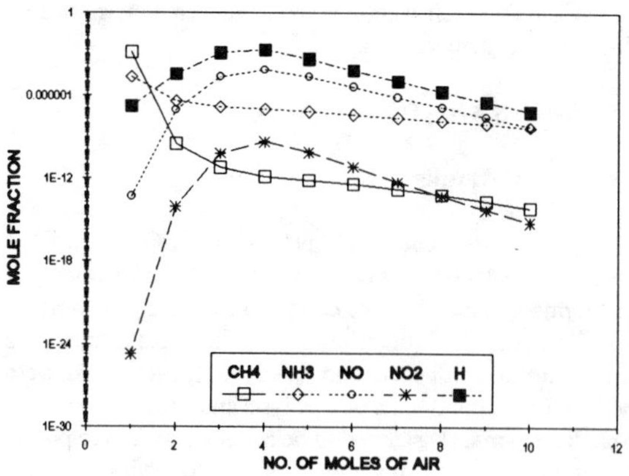


**Fig. 4B Evolution of $CH_4$, $NH_3$, NO, $NO_2$ and H from Combustion of 100% Plastic in Air.**

### 3.2 Experimental Results

The mean values of operating variables monitored during the tests include furnace temperature, exhaust temperature, waste feed rate, air flowrate and carbon monoxide concentration and are given in Table 3.

**Table 3 Experimental Conditions and Measured CO Levels.**

| Run No. | Mean Furn. Temp (°F) | Mean Exh. Temp (°F) | Waste Flow (lb / min) | Gas Flow rate (scfm) | CO at 1500 °F (Vol.%) | CO at 1350 °F (Vol.%) |
|---|---|---|---|---|---|---|
| 1 | 1500 | 1270 | 0.44 | 46.11 | 0.1 | 0.14 |
| 2 | 1400 | 1200 | 0.32 | 39.04 | 0.1 | 0.18 |
| 3 | 1600 | 1300 | 0.36 | 43.08 | 0 | 0.1 |
| 4 | 1400 | 1200 | 0.30 | 37.95 | 0.04 | 0.12 |
| 5 | 1500 | 1236 | 0.31 | 39.49 | 0.02 | 0.04 |
| 6 | 1400 | 1250 | 0.29 | 38.69 | 0.02 | 0.04 |
| 7 | 1600 | 1250 | 0.12 | 31.0 | 0 | 0.04 |

The above data shows that 100% non-plastic feed rate is less than 100% plastic for the same inlet air and furnace temperature (compare sample Nos. 1 and 7). Higher heating value of the plastic and air stoichiometry requirements are the reasons for lower feed rate and air requirement. The furnace temperature was observed to be fluctuating at the measurement location as a result of large temperature gradient in the flame zone. The unique burning characteristics of plastic and non-plastic as well as the local furnace conditions are possible sources for this contribution. However, in the present study, the exhaust temperature was maintained constant at 1250 within ±50°F.

From the test operating data, the fundamental combustion parameters such as excess air, equivalence ratio, gas residence time were calculated and these are presented in Table 4.

**Table 4 The Evaluated Flow Parameters.**

| Run No. | Gas vel. (fps) | Res. Time (sec.) | Fuel / air Ratio (Actual) | Fuel / air Ratio (Theo.) | Equi. Ratio | Exc. Air |
|---|---|---|---|---|---|---|
| 1 | 1.73 | 0.67 | 0.145 | 0.17 | 0.874 | 1.14 |
| 2 | 1.46 | 0.79 | 0.119 | 0.18 | 0.658 | 1.51 |
| 3 | 1.62 | 0.72 | 0.185 | 0.20 | 0.925 | 1.08 |
| 4 | 1.42 | 0.82 | 0.158 | 0.20 | 0.789 | 1.26 |
| 5 | 1.48 | 0.79 | 0.163 | 0.20 | 0.815 | 1.22 |
| 6 | 1.45 | 0.80 | 0.154 | 0.20 | 0.774 | 1.29 |
| 7 | 1.16 | 1.00 | 0.147 | 0.25 | 0.589 | 1.69 |

The residence time was computed by dividing the gas volumetric flow by the furnace volume. The calculated residence time for all types of waste was somewhat lower (<1sec) than that often used in many commercial incinerators (1-2 sec). The actual fuel/air ratio was evaluated from the waste and air flowrates. The theoretical fuel/air ratio was obtained from SOLGASMIX calculation for combustion and used in the calculation of equivalence ratio and excess air for the test operating conditions.

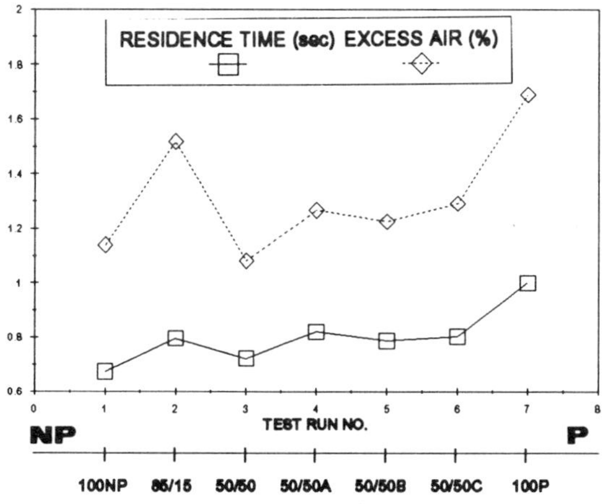


**Fig. 5 Variation of Residence Time and Excess Air with Sample Composition.**

Figure 5 shows the variation of residence time and excess air level inside the reactor for varying waste composition. The excess air level was very low with 100% non-plastic and very high with 100% plastic and in between for the mixtures. Consequently the residence time is highest with 100% plastic and lowest with 100% non-plastic and almost constant with mixtures. High excess air level was observed with 100% plastic case, since lower waste feed rate was adequate to maintain the constant furnace temperature at constant air flowrate. Although the test operating conditions were fuel lean for all the test samples, some runs were close to stoichiometric condition. No significant variation in excess air level and residence time for the 50/50 plastic and non-plastic mixtures was observed. This is due to the fact that the inclusion or exclusion of certain type of plastics within the plastic mix has little effect on the overall waste composition.

*Emission of Carbon Monoxide*

The concentrations of carbon monoxide (CO) in volume percentage, monitored for the lowest and highest furnace temperature condition are given in Table 4. The CO emission is an indicator of the combustion efficiency and performance of the combustion devices. The variation of CO concentration with sample composition is given in Figure 6. The CO emission indicates the temperature dependence of waste samples which is also a function of the sample composition. As expected the reactor operation at higher temperatures decreases CO. The results show high CO content with 100% and 85% non-plastic test runs (samples Nos. 1 and 2). This is an indication of incomplete combustion because of the lower heating value of paper as compared to plastic. The 50/50 mixture of plastic and non-plastic produces less CO than non-platsic alone. Absence of one or more components of plastics in the mixture changes somewhat the CO concentration. The CO concentration during the test runs varied between 1000 to 1800 ppm.

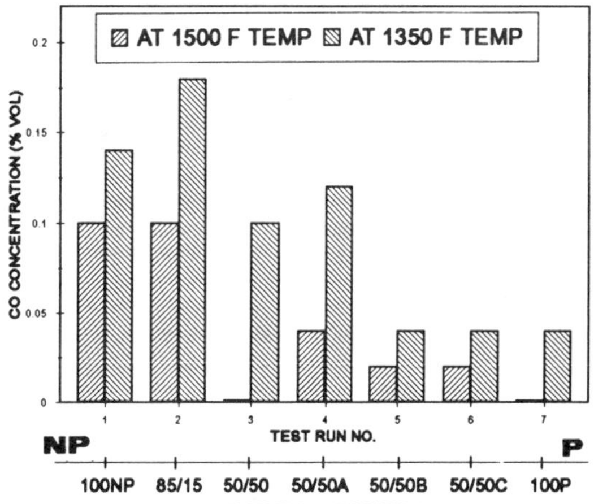


**Fig. 6 Variation of Carbon Monoxide Concentration with Sample Composition.**

A comparison of the mean value of CO with residence time shown in Figure 7 reveals higher CO concentration at lower residence time for all samples except in the case of sample # 4

(50% non-plastic/50% plastic A).

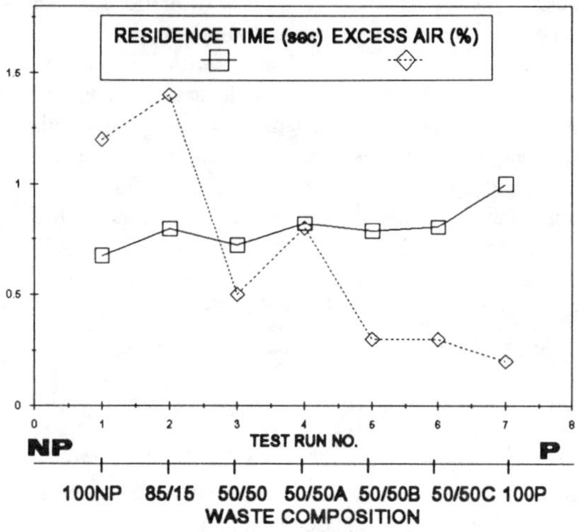


**Fig. 7 Variation of Mean Carbon Monoxide Concentration and Residence Time**

*Emission of Particulate Matter*

Particulate matter leaving the TDF is determined from the fly ash samples collected in the exhaust duct. From the gas flowrate exiting the chamber and gas temperature, the particulate loading was calculated in mg/m$^3$. The results given in Table 5 form a basis to make a qualitative comparison between tested samples.

**Table 5 Particulate Matter Emission for Different Samples.**

| Sample No. | 1 | 2 | 3 | 4 | 5 | 6 | 7 |
|---|---|---|---|---|---|---|---|
| Waste feed (kg/hr) | 12.0 | 8.88 | 9.84 | 8.22 | 8.64 | 7.98 | 3.5 |
| Firing density (kg/hr/m$^3$) | 801 | 592 | 656 | 548 | 576 | 532 | 233 |
| Total fly ash (gm) | 16.1 | 10.45 | 4.89 | 3.13 | 4.39 | 5.53 | 0.32 |
| Flue gas flow (scfm) | 46.1 | 39.05 | 43.0 | 37.9 | 39.4 | 38.6 | 31.0 |
| Parti. matter (mg/m$^3$) | 88.3 | 105.0 | 66.8 | 44.8 | 60.4 | 91.7 | 105 |

Maximum non-plastic containing samples (# 1 and 2) has the maximum amount of particulate emission and is attributed to higher ash content of paper in addition to higher flue gas flowrate. The lowest particulate emission was from 100% plastic (sample# 7). Other test samples provided results that bounded between samples 1 and 7. The particulate emission level decreases with increase in plastic content due to reduced ash content in the plastic fraction of the mixture. Higher particulate loading from sample# 2 (85% non-plastic/15% plastic) is due to higher firing density. Figure 8 shows a good similarity between the particulate matter emission and gas volumetric flowrate for all the samples. The particulate loading is very low in case of 100% plastic (sample# 7) and is due to low flue gas volumetric flowrate and ash content. A lack of good correlation for mixtures is probably due to the complex burning characteristics associated with inhomogeneity of the complex fuel mixture.

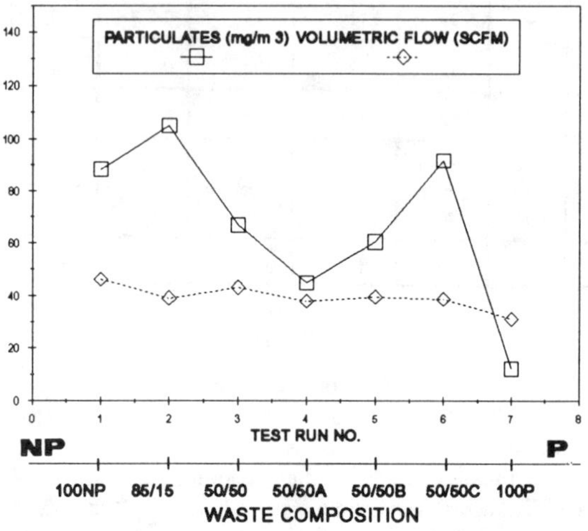


**Fig. 8 Variation of Particulate Matter and Gas Volume with Sample Composition.**

*Emissions of Dioxin and Furan*

The measured concentration levels of polychlorinated dibenzo dioxins (PCDD) measured at ion Nos. 320.2 and 322.4 and polychlorinated dibenzo furans (PCDF) measured at ion Nos. 304.2 and 304.6 are given in table 6 from all the tests.

**Table 6 Concentrations of PCDD/PCDF (ng/m$^3$).**

| RUN No. | 1 | 2 | 3 | 4 | 5 | 6 | 7 |
|---|---|---|---|---|---|---|---|
| TCDF ION No. 304.2 | 90 | 443 | 253 | 255 | 172 | 205 | 174 |
| TCDF ION No. 306.2 | 119 | 619 | 346 | 233 | 124 | 139 | 174 |
| TCDD ION No. 320.2 | 135 | 584 | 317 | 184 | 146 | 119 | 248 |
| TCDD ION No. 322.2 | 91 | 613 | 251 | 144 | 106 | 113 | 230 |

Experimental results show highest concentrations of dioxins and furans from 100% plastic and lowest from 100% non-plastic. This might be due to the presence of precursor compounds such as Polychlorinated biphenyl, formed during the burning of plastic. The concentrations recorded from all the 50/50 mixture samples are between those of pure plastic and non-plastic. Sample#5, which is a combination of 50% plastic B to 50% non-plastic has lower value of PCDD/PCDF than sample#2 (85% non-plastic and 15% plastic). This is mainly due to the absence of polyvinyl chloride in sample#5. This reveals the role of chlorinated hydrocarbons in producing the precursor compounds that are responsible for dioxin and furan formation. The small amount of PCDD/PCDF detected in 100% paper may be due to the trace amount of chlorine in the paper (from the bleaching process) and from the interaction with the refractory material which had trace amounts of chlorine. The residual carbon collected in the region of 300-400 °C are responsible for the formation of PCDD/PCDF by De-nova synthesis[23]. The presence of chlorine and copper in ash, in trace levels, implies the presence of copper chloride which can play a vital role in the formation of PCDD and PCDF through De-nova synthesis[12]. The absolute concentration levels of polychlorinated dibenzo dioxins and furans are plotted against the adiabatic flame temperatures in Figure 9. High temperature has no pronounced effect on the emission of these toxic compounds. The dioxin and furan formation is a weak function of the furnace temperature and strongly depends on waste composition[11]. The concentration of dioxin vary between the various samples tested. Out of the four 50/50 mixture of plastic and non-plastic, PCDD/PCDF emission was found to be lowest from the sample with no polyvinyl chloride (sample# 5).

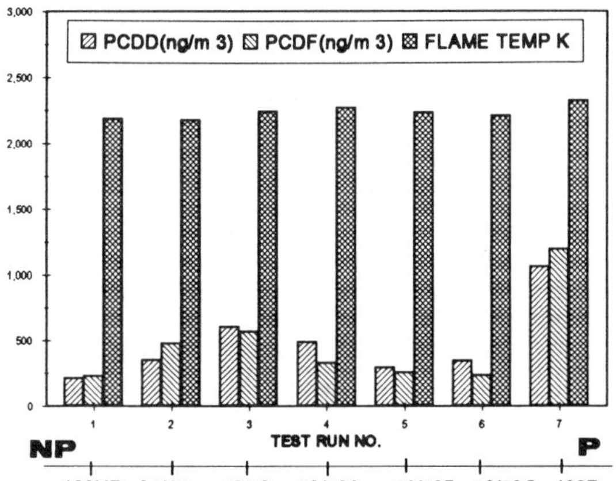


**Fig. 9 Dioxin and Furan Concentration Levels with Adiabatic Flame Temperature.**

The PCDD and PCDF values compared against CO emissions are shown in Figure 10. In the case of 100% non-plastic (sample# 1), 100% plastic (sample# 7) and 85% non-plastic/15% plastic (sample# 2), the dioxin/furan emission is inversely related to CO emission. The relationship between dioxin or furan and CO emission for all of the 50/50 non-plastic and plastic mixtures was found to be poor.

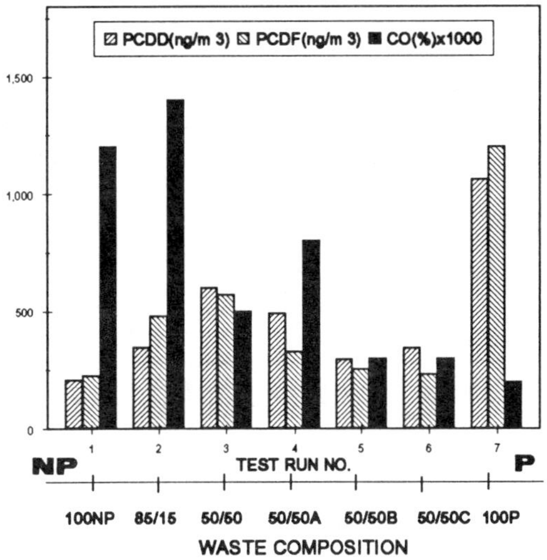


**Fig. 10 Variation of Dioxin and Furan Concentration Levels with CO Emission.**

*Metals in Fly ash*

Particulates from waste incinerators contain oxides of silicon, iron, calcium and aluminum. The other trace metals of importance are antimony, arsenic, barium, beryllium, cadmium, chromium, lead, mercury, silver and thallium. Carcinogenic toxic metals include arsenic, cadmium, chromium and beryllium. Arsenic is only present in sample# 7 (100% plastic) and sample# 6 (50% non-plastic/50% Plastic C) and not present in other samples.

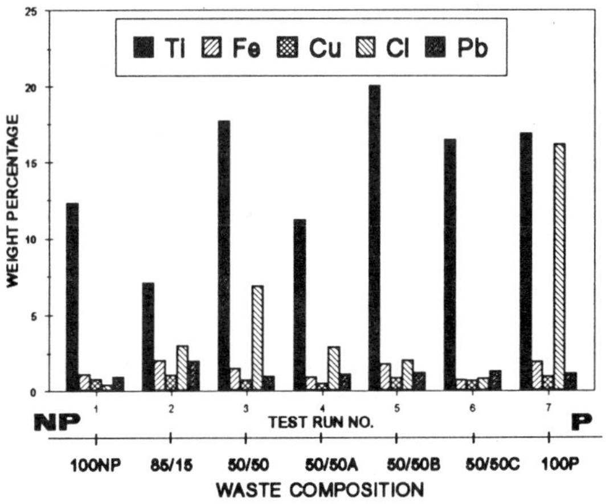


**Fig. 11 Some of the Toxic and Rare Metal Elements Detected by SEM.**

The non-carcinogenic toxic metals are antimony, barium, lead, mercury, silver and thallium. Out of these only lead is invariably present in all the samples. Lead would have come as a contaminant through these samples. However the source of lead contaminants is not clearly understood. The presence of chlorine species can affect metal species and volatility temperature[13]. The chlorine content is highest in 100 % plastic due to the presence of polyvinyl chloride. The 100% non-plastic sample contains trace amounts of chlorine that may be left after the bleaching process in paper making. The other metals found in the fly ash analysis are titanium, iron, and copper. In general metals are present more in plastic than non-plastic. Figure 11 shows the variation of some of the toxic and rare metal elements for all the test runs.

### *Metals in Bottom Ash*

Neutron Activation Method (NAM) is used to analyze metals present in bottom ash. The accuracy is down to one hundredth of a percent. However the limitation is with the number of metals that can be detected. Only short irradiation metal analysis was carried out and metals with longer radioactive decay have to stay longer (months) to complete the test. The presence of rare metals vanadium, barium, indium, magnesium and manganese were detected. The higher vanadium content is found in plastic/non-plastic mixtures than in 100% non-plastic. Barium, which is a non-carcinogenic toxic metal, was found in small percentages in sample# 4 (50% non-plastic/50% plastic A) and # 6 (50% non-plastic/50% plastic C).

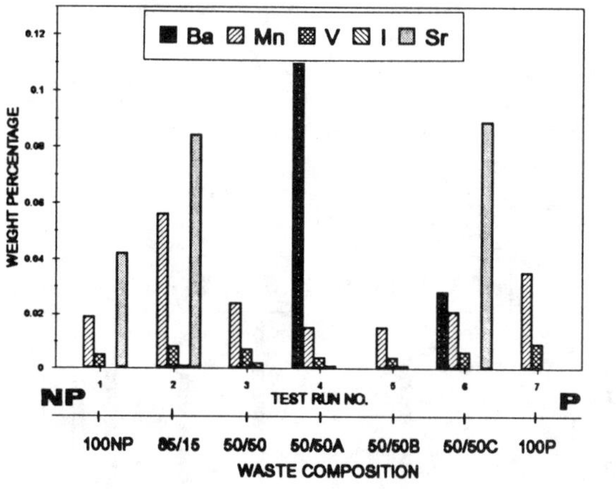


**Fig. 12 Some Toxic and Rare Metal Elements Detected by NAM.**

A small fraction of indium is detected from the non-plastic/plastic sample# 2, 3, 4 and 5. Strontium was detected with sample# 1 (100% non-plastic) and some in sample# 2 (85% non-plastic/15% plastic) and sample# 6 (50% non-plastic/50% plastic C). Figure 12 shows some of the rare and toxic metals presence in bottom ash and its variation with test runs.

## 4.0 CONCLUSIONS

Results of equilibrium thermochemical calculations for the thermal destruction of non-plastic and plastic materials show a significant effect of process configuration on the flame temperature and product gas composition. The effect of waste composition has greater influence on adiabatic flame temperature, combustion air requirement and the evolution of products and intermediate species. The combustion of waste in air produces higher flame temperature for 100% plastic than for non-plastic and mixtures. The 100% plastic requires lower number of moles of oxidant than 100% non-plastic and mixtures. Plastic produces HCl and $H_2S$ with concentration levels ranging from 1000 to 10000 ppm. Emission of NO and $NO_2$ from 100% non-plastic showed an increase with increase in moles of air while that from 100% plastic a slight decrease with increase in moles of air. The higher theoretical flame temperatures predicted with plastic waste corresponds to lower waste feed rate requirement of plastic at constant furnace temperature. This resulted in higher excess air operation with plastic waste and hence lower equivalence ratio. The gas residence time calculated for all the samples was found to be about 1 second. Variation of residence time more or less follows the same trend as excess air for all the samples. The gas volumetric flowrate and residence time stays almost constant for all the samples except for 100% plastic. The experimental data and calculated combustion parameters, show different burning characteristics of plastic and non-plastic. Temperature dependency of CO emission is in good agreement with the theoretical calculations. Higher CO emission was observed for both (100% and 85%) non-plastic than for 100% plastic. Paper burning causes incomplete combustion because of lower heat release than plastic. Concentrations of dioxin vary between the various samples tested. Dioxin and furan levels were found to be high from 100% plastic than 100% non-plastic. Non-plastic waste yields lowest levels of dioxins and furans. Out of four samples of various 50/50 plastic/non-plastic mixtures, PCDD/PCDF emission levels are lowest for the sample without polyvinyl chloride (sample #5). Metals in fly ash and bottom ash did not reveal any specific pattern for their variation. However in general plastic increases metals emission. The presence of lead was found in all samples after combustion although its source is not well understood. The highest amount of chlorine was found in 100% plastic sample. Zinc and arsenic are present only in plastic samples. Sulphur is only detected in trace quantities from 100% non-plastic sample. Bottom ash analysis provided the presence of trace metal elements such as vanadium, barium, indium, magnesium and manganese in ultra trace levels. The results provided here give the insights on the design guidelines for the thermal destruction of plastic containing wastes.

## 5.0 ACKNOWLEDGEMENTS

This research was supported by NSWC, Carderock Division and is gratefully acknowledged. Technical discussions with Messers E. Barry White and Craig Alig of NSWC are greatfully acknowledged.

## 6.0 REFERENCES


1. Pershing, D. W., Lighty, J. S. and Silcox, G. D., "Solid Waste Incineration in Rotary Kilns", Combustion Energy and Science, vol.93, No.1, pp.245-264, 1993.

2. Komatsu, F., Takusagawa, A., Wada, R. and Asahina, K., "Application of Microwave Treatment Technology for Radioactive Wastes", Waste Management, Vol.10, pp.211-215, 1990.

3. Gupta, A. K., Ilanchezhian, E. and Keating, E. L., "Thermal Destruction Behavior of Plastics", Design Technical Conference, International CIE Conference, ASME, Minneapolis, MN, Sep. 11-14, 1994.

4. Oppelt, O. C., "Hazardous Waste Critical Review", Air and Waste Journal, Vol.43, pp.25-73, Jan. 1993.

5. Allen T. D. and Behmanesh, N., "Non-hazardous Waste Generation", Hazardous Waste and Materials, Vol.9, No. 1, 1992.

6. Young P. J. and Parker, A., "The Identification and Possible Environmental Impact of Trace Gases and Vapors in Landfill Gas", Waste Management and Research, No.1, pp .213-226, 1992.

7. Seeker, W. R., "Waste Combustion" Twenty Third Symposium (International) on Combustion, The Combustion Institute, Orleans, France, pp.867-885, 1990.

8. U.S. Environmental Protection Agency in Report., "PCDDs and PCDFs in Stationary Source Emission Air Samples", American Laboratory, pp.33-40, Dec.1991.

9. Reynolds, W. C., "Stanjan Chemical Equilibrium Solver", V.3.89, IBM-PC(c), Stanford University, 1987.

10. Besman, T. M., "SOLGASMIX-PV for the PC", A report published by Oak Ridge National Laboratory, Oct.1989.

11. JANAF Thermochemical Table, 2nd edition, NSRDS-NBS 37, 1971.

12. Williams, P. T., "The Sampling and Analysis of Dioxins and Furans from Combustion Sources", Journal of the Institute of Energy, Vol. 65, pp. 46-54, Mar.1992.

13. Tillman, D. A., "The Combustion of Solid Fuels and Wastes", Academic press, Inc., 1991.

14. Domalski, E. S., Jobe, T. L. and Milne, T. A., "Thermodynamic Data for Biomass Materials and Waste Components", The American Society of Mechanical Engineers, United Engineering Center, NY, 1987.

15. Gupta, A. K., "Critical Analysis on the Thermal Destruction Technologies of Shipboard Surrogate Solid Waste", Annual Research Progress Report submitted to Naval Surface Warfare Center, Carderock Division, Environmental Protection Branch, Dec.1992.

16. Chopra, H., Gupta, A. K. and Keating, E. L. and White, E.B., "Thermal Destruction of Solid Wastes", Proceedings of 27th Intersociety Energy Conversion Engineering Conference, Chapter 1.377, vol.1, 1992.

17. Keating, E. L. and Gupta, A. K., "Pyrolysis and Oxidative Pyrolysis of Cellulose", 16th ASME Annual Energy Sources Technology Conference, Emerging Energy Technology Symposium, Houston, TX, ASME PD Vol.50, pp.109-117, Jan.31-Feb.4, 1993

18. Gupta, A. K.. and Keating, E. L., "Pyrolysis and Oxidative Pyrolysis of Polystyrene", Proceedings of ASME Computers in Engineering Conference, San Diego, CA, Aug.9-12, 1993.

19. Gupta, A. K.. and Keating, E. L., "Pyrolysis and Oxidative pyrolysis of Polyvinyl Chloride", 28th. Proceedings of the IECEC Conference, Atlanta, GA, August 8-13, 1993.

20. Williams, P. T. "A review of Pollution from Waste Incineration", Journal of the Institution of Water and Environmental Management, Vol.4, No.1, Feb.1990.

21. Williams, P. T. and Besler, S., "The Pyrolysis of Municipal Solid Waste", Journal of the Institution of Energy, Vol. No. pp.192-200, 1992.

22. Brunner, C. R., "Handbook of Incineration Systems", McGraw Hill, Inc., 1991.

23. Kun-chieh, L., "Research Areas for Improved Incineration System Performance", JAPCA, Vol.38, No.12, pp.1542-1548, Dec.1988.

24. ASME., " Combustion Fundamentals for Waste Incineration", United Engineering Center, NY, 1974.

25. Seeker, R. W., "Metals Behavior in Waste Combustion Systems", Air Toxic Reduction and Combustion Modeling, ASME , Vol.15, pp.57-62, 1992.

26. Linak, P. W. and Wendt, J. L., "Toxic Emissions from Incineration: Mechanisms and Control", Progress in Energy and Combustion Science, Vol.19, pp.145-185, 1993.

FACT-Vol. 18, Combustion Modeling, Scaling and Air Toxins
ASME 1994

# MATHEMATICAL MODELING OF PULVERIZED COAL COMBUSTION IN AXISYMMETRIC GEOMETRIES

**W. A. Fiveland and J. P. Jessee**
Research and Development Division
Babcock & Wilcox
Alliance, Ohio

## ABSTRACT

Pulverized coal combustion is studied in non-swirling and swirling flames. The various interacting processes which occur during combustion are modeled: turbulent flow, heterogeneous and homogeneous chemical reaction, and heat transfer. Gas phase reactions are based on a two step method with an initial fast step producing a pool of carbon monoxide and products and a final step in which the carbon monoxide is kinetically oxidized, based on the relative turbulence and kinetic time scales. Modeled heterogeneous reactions include evaporation, devolatilization, and char oxidation. Particle dispersion is simulated using a stochastic model. Gas- and particle-phase energy equations are used to determine gas and particle temperatures. Radiation heat transfer in the participating media is modeled by the discrete ordinates method.

Numerical predictions are compared to experimental data for three cases of increasing geometric complexity. Cases include non-swirling and swirling coal combustion in bench scale reactors, and swirling particle-laden flow in a 1/6-scale utility burner. Predictions have been useful in understanding combustion processes and guiding future research to meet long term objectives of using the models as design tools.

## NOMENCLATURE

| | |
|---|---|
| $A$ | area |
| $A_i$ | premultiplication factor |
| $C$ | species concentration |
| $C_\mu$ | constant (= 0.09) |
| $C_D$ | drag coefficient |
| $C_p$ | specific heat |
| $d$ | diameter |
| $E$ | activation energy |
| $g$ | gravity |
| $G$ | incident radiant energy, turbulence production |
| $h$ | enthalpy |
| $h_D$ | mass transfer coefficient |
| $I$ | radiation intensity |
| $k$ | turbulent kinetic energy, conductivity |
| $K$ | chemical kinetic rate |
| $l$ | length scale |
| $m$ | mass |
| $n$ | particle number |
| $Nu$ | Nusselt number |
| $P$ | pressure |
| $Q_a$ | absorption efficiency |
| $r$ | radial position, stoichiometric oxidant to fuel mass ratio |
| $R$ | gas constant, combustion rate |
| $S$ | source |
| $T$ | temperature |
| $t$ | time |
| $u$ | axial component of velocity |
| $v$ | radial component of velocity |
| $V$ | volume |
| $w$ | tangential component of velocity |
| $w$ | discrete ordinate weight |
| $Y$ | mass fraction |
| $z$ | axial position |
| $\alpha$ | volatile yield |
| $\beta$ | extinction coefficient |
| $\phi$ | dependent variable |
| $\Phi$ | scattering phase function |
| $\Gamma$ | effective diffusion coefficient |
| $\mu$ | viscosity |
| $\mu,\xi,\eta$ | direction cosines |
| $\kappa$ | absorption coefficient |
| $\varepsilon$ | emissivity or turbulent dissipation |
| $\rho$ | density |
| $\sigma$ | Prandtl-Schmidt number, or scattering coefficient |
| $\tilde{\sigma}$ | Stefan-Boltzmann constant |
| $\Omega$ | direction of radiation |

## Subscripts

| | |
|---|---|
| b | blackbody |
| c | raw coal |
| ch | char |
| cp | char products |
| CO | carbon monoxide |
| d | drift |
| e | eddy |
| eff | effective |
| f | fuel |
| g | gas |
| h | enthalpy |
| k | turbulent kinetic energy |
| m | mass |
| M | momentum |
| ox | oxidant |
| p | particle |
| r | transit |
| t | turbulent |
| $\phi$ | name of dependent variable |
| l | laminar |
| $\varepsilon$ | turbulent dissipation |
| $\infty$ | bulk fluid |

## Superscripts

| | |
|---|---|
| ' | incoming direction or particle source term |
| d | devolatilization |
| ch | char oxidation |
| CO | carbon monoxide |
| k | kinetic |
| o | initial |
| t | turbulent mixing |
| w | evaporation |

## Miscellaneous

| | |
|---|---|
| $\rightarrow$ | vector |
| - | Reynolds average |
| • | time rate |
| ~ | Favre average |

## INTRODUCTION

There is a need to design more efficient pulverized coal (PC) combustion devices to meet market demands and federal regulations on controlling combustion performance. The existing technology base for designing utility and industrial fossil boilers is primarily empirical. Empirical methods are adequate for the design envelope in which the data were collected, but there is risk in extrapolating the methods outside of the design envelope. Methods are needed to scale designs from the pilot to the full scale and study the various effects of different fuels, burner position and orientation, and geometry on combustion system performance.

Computational methods in heat transfer and fluid flow are being applied to the simulation of complex, yet practical combustion systems. With the development of high speed engineering workstations, numerical simulation now represents a new engineering tool for evaluating designs. These computational tools will eventually be used on a daily basis by the designer to assist in designing combustion systems.

The first comprehensive numerical model for PC combustion was reported by Gibson and Morgan (1970). Since that time, studies by Lockwood, et al., (1980), Fiveland et al., (1984), Boyd and Kent (1986, 1994), Fiveland and Wessel (1988) and Visser (1991) have employed more sophisticated techniques and have demonstrated promise in predicting PC combustor performance. There have also been warnings that computational models have limitations (Weber et al., 1993).

This paper presents predictions for benchscale non-swirling and swirling pulverized coal flames. Model predictions are compared with data to benchmark the model. Although predictions of pulverized coal flames are a topic of current research interest, this work was undertaken to improve several aspects of previous pulverized coal models published in the literature, particularly that of Fiveland, et al., (1984). Models for reaction chemistry of the gas- and solid phases, particle dispersion, radiation heat transfer and numerical solution methods of the transport equations have been improved.

## COMBUSTION MODEL OVERVIEW

Simulation of PC combustion must account for the following physical processes:

- Gas and Solid Phase Flow
- Chemical Reaction of Released Volatiles
- Oxidation of Char
- Heat Transfer

The present combustion model was developed for 2- or 3-dimensional Cartesian or cylindrical geometries. In this paper, the conservation equations of mass, momentum, and energy for the gas and particle phases are solved for axisymmetric geometries. Gas phase transport is modeled using an Eulerian formulation with fixed control volumes in space. Favré-averaged forms of the gas-phase transport equations are used. A dispersed particle phase is modeled by a Lagrangian method using discrete particle streams.

When the coal particle enters an inlet, rapid heat-up causes moisture to evaporate.

$$\textit{Coal} \rightarrow \textit{Dry Coal} + \textit{Water Vapor} \qquad (1)$$

Evaporation is followed by pyrolytic reactions which form volatile fuel and char residue containing inert ash. Coal devolatilizes to a single, gaseous fuel with a fixed composition. The fuel evolution represents the devolatilization and char formation processes.

$$\textit{Dry Coal} \rightarrow \textit{Gaseous Fuel} + \textit{Char} \qquad (2)$$

Combustion of the volatile fuel occurs as a two-step process. The first step proceeds at infinite rate:

$$\textit{Gaseous Fuel} + \textit{Oxidant} \xrightarrow{K = \infty} \textit{CO} + \textit{Products} \qquad (3)$$

The second step involves oxidation of carbon monoxide:

$$\textit{CO} + \textit{Oxidant} \xrightarrow{K^{CO}} \textit{CO}_2 \qquad (4)$$

The rate is limited by either turbulence or kinetic considerations, depending on the respective time scales.

Combustion of the char proceeds at an effective rate determined from the intrinsic rate of the residual char and the physical rate at which oxidant can diffuse to the surface of the char particle.

$$Char + Oxidant \xrightarrow{K^{ch}} CO + Products + Ash \qquad (5)$$

Ash remains after conversion of the char into gaseous products.

An energy equation for the particle models the radiation and convective heat transfer between the particle- and gas-phases. The gas phase energy equation accounts for heat transfer to/from the particle-phase and to domain boundaries.

## GOVERNING EQUATIONS

### Gas Phase Transport Equations

A general steady-state transport equation for an (r-z) geometry can be written for any conserved variable in the form:

$$\frac{\partial}{\partial z}(\bar{\rho}\tilde{u}\phi) + \frac{1}{r}\frac{\partial}{\partial r}(\bar{\rho}r\tilde{v}\phi) = \frac{\partial}{\partial z}\left(\Gamma_\phi \frac{\partial \phi}{\partial z}\right) + \frac{1}{r}\frac{\partial}{\partial r}\left(r\,\Gamma_\phi \frac{\partial \phi}{\partial r}\right) + S_\phi \qquad (6)$$

where $\phi$ is the dependent variable, $\Gamma_\phi$ is the effective diffusion coefficient, and $S_\phi$ is the source of $\phi$ per unit volume. The conservation equations (mass, three components of momentum, and others) are elliptic.

**Gas Phase Fluid Mechanics.** Table 1 lists the values of $\phi$, $\Gamma_\phi$, and $S_\phi$ for the conservation equations of mass and the three components of momentum. Table 2 lists the appropriate modeling constants. The source term, $S'_m$, for the continuity equation represents the mass of gas added due to the mass transfer from the particles. Sources of momentum are also listed in the table and include fluid mechanics terms (e.g., Coriolis forces, pressure terms) and terms due to interaction with the particle phase, $S'_M$.

**Turbulence Model.** In writing Eq. (6) for momentum, we have assumed that the effective diffusion coefficient can be represented by:

$$\mu_{eff} = \mu_l + \mu_t \qquad (7)$$

where $\mu_l$ and $\mu_t$ are the laminar and turbulent viscosities, respectively. Following recommendations by Launder and Spalding (1974), the turbulent viscosity is modeled as:

$$\mu_t = \rho C_\mu \frac{k^2}{\varepsilon} \qquad (8)$$

The standard k-ε model proposed by Launder is used. Values of $\phi$, $\Gamma_\phi$, and $S_\phi$ are listed in Table 1 while the modeling constants are listed in Table 2.

While this model has been widely applied, it suffers from some drawbacks such as under-prediction of the size of wakes and recirculation zones (Jones and Whitelaw, 1982; Sloan et al., 1986). The currently applied model does not account for the effects of particle-phase on turbulence.

**Gas Phase Energy Equation.** A conservation equation for enthalpy (including heats of formation) is solved for the gas phase. Tables 1 and 2 list the appropriate values of $\phi$, $\Gamma_\phi$, and $S_\phi$ and modeling constants, respectively. Sources of enthalpy arise from particle-to-gas and gas-to-gas radiation, and mass transfer from the particles. Upon discretization, the enthalpy sources at boundary cells are further modified to include the effects of heat gains/losses at the walls.

**Gas Phase Chemistry.** The reaction described in equation (3) is modeled by using the eddy dissipation model of Magnussen and Hjertager (1976). This model is formulated by assuming that the rate of combustion is controlled by the rate of turbulent mixing. The effective reaction rate, R, is based on the rate of mixing on a molecular scale of the fuel and oxygen eddies. An effective rate is determined from rates evaluated in fuel-lean, fuel-rich and product eddies. The source of fuel remaining can be written:

$$\bar{S}_f = \bar{S}_f' - \min(R_1, R_2, R_3) \qquad (9)$$

The first term on the right side of the equation is the increase in fuel from the devolatilization, while the second term is the decrease in fuel from combustion. The individual rates are defined by Magnussen and Hjertager (1976). The production source of carbon monoxide, $\bar{S}^o_{CO}$, is related to the effective combustion rate of volatiles and the char oxidation rate of the particles.

The reaction described in Equation (4) is modeled using a hybrid approach. If the kinetic time scale is smaller than the turbulence time scale (i.e. fast chemistry), the reaction is assumed to be limited by turbulent mixing, and the rate is formed using the eddy dissipation model. If the turbulence time scale is smaller than the kinetic time scale, the reaction is assumed to be limited kinetically. The destruction source is thus written as:

$$\bar{S}_{CO} = \min(\bar{S}^t_{CO}, \bar{S}^k_{CO}) \qquad (10)$$

where superscripts t and k denote mixing and kinetic rates, respectively. The mixing rate may be expressed by an equation similar to the destruction portion of Equation (9), whereas the kinetic rate is based on a global reaction proposed by Howard et al. (1972) using average properties and species concentrations:

$$\overline{S^k_{CO}} = -A\, C_{CO}\, (C_{H2O}\, C_{O2})^{1/2} \exp(-E/RT) \qquad (11)$$

where $C_i$ represents the molar concentration. Similar hybrid approaches have been used by Wang et al. (1993) and Azevedo and Carvalho (1993).

### Particle Transport Equations

In this paper, the Lagrangian approach (Medgal and Agosta, 1967; Crowe, 1972) is used to model the particle phase. The flights of the particles are tracked from the entry of the combustion chamber to the exit. The continuous distribution of particle sizes is represented by a number of particle streams that are each of uniform diameter. Each stream is characterized by a

TABLE 1
CONSERVATION EQUATIONS CORRESPONDING TO EQUATION 6.

| $\phi$ | $\Gamma_\phi$ | $S_\phi$ | $S_\phi^P$ |
|---|---|---|---|
| 1 | 0 | 0 | $S_m^{\prime}$ |
| $\tilde{u}$ | $\mu_{eff}$ | $-\frac{\partial P}{\partial z} + \frac{\partial}{\partial z}(\mu_{eff}\frac{\partial u}{\partial z}) + \frac{1}{r}\frac{\partial}{\partial r}(\mu_{eff}\, r\, \frac{\partial v}{\partial z})$ | $S_{M,\tilde{u}}^{\prime}$ |
| $\tilde{v}$ | $\mu_{eff}$ | $\frac{\partial}{\partial z}(\mu_{eff}\frac{\partial u}{\partial r}) + \frac{1}{r}\frac{\partial}{\partial r}(\mu_{eff} r\frac{\partial v}{\partial r}) - 2\mu_{eff}\frac{v}{r^2} + \frac{\rho w^2}{r} - \frac{\partial P}{\partial r}$ | $S_{M,\tilde{v}}^{\prime}$ |
| $\tilde{w}$ | $\mu_{eff}$ | $\mu_{eff}\left[\frac{\partial^2 w}{\partial Z^2} + \frac{1}{r}\frac{\partial}{\partial r}\left(r\frac{\partial w}{\partial r}\right)\right] - w\left[\frac{\mu_{eff}}{r^2} + \frac{\rho v}{r} + \frac{w}{r}\frac{\partial \mu_{eff}}{\partial r}\right]$ | $S_{M,\tilde{w}}^{\prime}$ |
| $k$ | $\frac{\mu_{eff}}{\sigma_k}$ | $G - \bar{\rho}\varepsilon$ | |
| $\epsilon$ | $\frac{\mu_{eff}}{\sigma_\varepsilon}$ | $\frac{\varepsilon}{k}(C_1 G - C_2\bar{\rho}\varepsilon)$ | |
| $\tilde{h}$ | $\frac{\mu_{eff}}{\sigma_h}$ | $-\kappa(G - 4\tilde{\sigma}T^4)$ | $S_h^{\prime}$ |
| $\tilde{Y}_r$ | $\frac{\mu_{eff}}{\sigma_f}$ | $-\min(R_1, R_2, R_3)$ | $S_f^{\prime}$ |
| $\tilde{Y}_{ox}$ | $\frac{\mu_{eff}}{\sigma_{ox}}$ | $-r_f \min(R_1,R_2,R_3) - r_{CO}\bar{S}_{CO}$ | $-\frac{1}{V}\sum_i \dot{n}_i \int_{\Delta t} r_{ch} K^{ch}\, dt$ |
| $\tilde{Y}_{cp}$ | $\frac{\mu_{eff}}{\sigma_{cp}}$ | | $\frac{1}{V}\sum_i \dot{n}_i \int_{\Delta t} (1 + r_{ch}) K^{ch}\, dt$ |
| $\tilde{Y}_{CO}$ | $\frac{\mu_{eff}}{\sigma_{CO}}$ | $\bar{S}_{CO}^{o} - \bar{S}_{CO}$ | |
| $\tilde{Y}_{H_2O}$ | $\frac{\mu_{eff}}{\sigma_{H_2O}}$ | | $\frac{1}{V}\sum_i \dot{n}_i \int_{\Delta t} K^{w}\, dt$ |

$$G = \mu_t\left\{2\left[\left(\frac{\partial u}{\partial z}\right)^2 + \left(\frac{\partial v}{\partial r}\right)^2 + \left(\frac{v}{r}\right)^2\right] + \left(\frac{\partial w}{\partial z}\right)^2 + \left[r\frac{\partial}{\partial r}\left(\frac{w}{r}\right)\right]^2 + \left(\frac{\partial u}{\partial r} + \frac{\partial v}{\partial z}\right)^2\right\}$$

TABLE 2
CONSTANTS FOR THE GAS PHASE EQUATIONS IN TABLE 1

| $C_\mu$ | $C_1$ | $C_2$ | $\sigma_k$ | $\sigma_\varepsilon$ | $\sigma_i^*$ | $\sigma_h$ |
|---|---|---|---|---|---|---|
| 0.09 | 1.44 | 1.92 | 1.0 | 1.21 | 0.9 | 0.9 |

*Value used for all Prandtl-Schmidt numbers not listed in Table 2.

number flow rate of particles and by an initial position, mass, velocity, and temperature. The effect of the particles on the gas phase is included in the gas-phase conservation equations by source terms.

**Particle Momentum Equation.** The equation of motion for a particle may be expressed as:

$$m_p\left(\frac{d\vec{u}_p}{dt}\right) = C_D\,\rho_g\left(\frac{A_p}{2}\right)(\vec{u}_g - \vec{u}_p)\,|\vec{u}_g - \vec{u}_p| + m_p\,\vec{g} \tag{12}$$

where $m_p$ and $A_p$ are the particle mass and cross-sectional area, respectively, $\vec{u}_g$ and $\vec{u}_p$ are the instantaneous gas and particle velocities, $\rho_g$ is the gas density, $\vec{g}$ is the gravitational acceleration, and $C_D$ is the drag coefficient from Crowe (1972). Virtual mass, Basset, Saffman, and Magnus effects have been neglected. The net source of momentum to the gas-phase equations for a single particle stream can be written as follows:

$$\vec{S}_M^{\prime} = \frac{1}{V}\left\{\dot{n}\,\delta(m_p\,\vec{u}_p) - \dot{n}\,\overline{m}_p\,\vec{g}\,\Delta t\right\} \tag{13}$$

where $V$ is the volume of the Eulerian cell, ṅ is the number flow rate of particles in the particular stream and $\delta(m_p\vec{u}_p)$ represents the change of the particle momentum between the inlet and exit of the computational cell. The change in momentum due to gravity ($m_p\,\vec{g}$) does not contribute to the momentum transport of the gas phase.

The Lagrangian stochastic deterministic (LSD) model (Milojevic, 1990) is used to simulate particle dispersion. A Gaussian distribution is assumed for the fluctuating gas velocity components, u', v', w', and the standard deviation is expressed as $\sqrt{2/3k}$. An instantaneous gas velocity, $\tilde{u}_g$, is calculated by adding the fluctuating velocity to the local mean velocity. The instantaneous velocity is used in the particle momentum equation, Eq. (12). This velocity is kept constant during a period of time, $t_i$, which is taken as the minimum of the large eddy lifetime, $t_e$, and the transit time to cross the eddy, $t_r$. The eddy lifetime is calculated from an expression suggested by Hinze (1975):

$$t_e = c_t\,\frac{k}{\varepsilon} \tag{14}$$

where the coefficient, $c_t$, is taken as 0.3 following

recommendations of Milojevic (Milojevic et al., 1987). The transit time is approximated as:

$$t_r = \frac{l_e}{v_d} \quad (15)$$

where $v_d$ is the drift velocity and $l_e$ is the effective eddy size:

$$l_e = t_e \sqrt{\frac{2}{3}k} \quad (16)$$

Because the drift velocity, $v_d$, is not initially known, the relative displacement is monitored, and as soon as this displacement exceeds the eddy length scale, $l_e$, a new eddy is generated. Nonhomogeneous fluid turbulence was considered following the suggestions of Milojevic (1990).

**Particle Energy Equation.** The heat transfer to or from a particle can be represented as:

$$m_p \frac{dh_p}{dt} = \frac{Nu\, k_g}{d_p} A_p(T_g - T_p) + Q_a A_p(G - 4\bar{\sigma} T_p^4) \quad (17)$$

where the terms on the right-hand side represent convective and radiative heat transfer to or from the particle, respectively. The heat of reaction is added entirely to the gas phase. The net source of heat per unit time to the gas phase for a single particle stream can be expressed as:

$$S_h' = \frac{1}{V}\left\{\dot{n}\,\delta(m_p h_p) - \dot{n}\int_{\Delta t} Q_a A_p(G - 4\bar{\sigma} T_p^4)\, dt\right\} \quad (18)$$

**Particle Mass.** Coal particles enter the combustor whereupon the water evaporates, volatiles are released and the remaining char undergoes oxidation. In each case, the mass of the particle changes along its trajectory due to evaporation, devolatilization and char oxidation. The decrease in total mass can be represented, in general, as:

$$\frac{dm_p}{dt} = -K = -(K^w + K^d + K^{ch}) \quad (19)$$

where K is a rate characteristic of evaporation, devolatilization and char oxidation. The source of mass contributed to the gas phase from a single stream of particles becomes an integration over the time step:

$$S_m' = \frac{1}{V}\left\{\dot{n}\int_{\Delta t} (K^w + K^d + K^{ch})\, dt\right\} \quad (20)$$

The total source, $S_m'$, is found by summing over all such particle streams in the cell.

a) Evaporation As the coal heats up, the water in the fuel evaporates. The evaporation rate can be written

$$K^w = -K_D A_p (C_{H2O,p} - C_{H2O,\infty}) \quad (21)$$

The coefficient $K_D$ is a product of a mass transfer coefficient and the molecular weight of water. The parameter, $C_{H2O,p}$, is the concentration of water vapor at the particle surface and $C_{H2O,\infty}$ is the concentration of water vapor in the bulk flow.

b) Devolatilization For the dry coal (see Equation (2)), the use of a single, first-order kinetic expression for pyrolysis is inadequate. Several researchers (Kobayashi, et al., 1976, and Ubhayaker, et al., 1975) have suggested multiple reaction rates. We follow the proposal by Ubhayaker, et al. (1975) in which two first-order reaction rates are expressed as:

$$dry\ coal \xrightarrow{K_i^d} \alpha_i\, gas + (1 - \alpha_i)\, char \qquad for\ i = 1, 2 \quad (22)$$

where $\alpha_i$ is the volatiles' yield, and $K_i^d$ is the kinetic rate of reaction. The kinetic rates are first order in the mass of coal remaining and in an Arrhenius form:

$$K_i^d = m_c A_i e^{-E/RT} \quad (23)$$

The total devolatilization rate becomes:

$$K^d = \sum_i \alpha_i K_i^d \quad (24)$$

The source of volatile fuel is:

$$S_f' = \frac{\dot{n}}{V}\int_{\Delta t} K^d\, dt \quad (25)$$

The total source of volatile fuel is found by summing over all particle streams.

c) Char Oxidation Char oxidation is modeled by Equation (5). The effective char oxidation rate, $K^{ch}$, is a function of the kinetic rate of the chemical reaction and the diffusion rate of the oxidizer to the particle (Field, 1967; Fiveland and Jamaluddin, 1992). The variation of char mass along a trajectory can be written:

$$\frac{dm_{ch}}{dt} = \sum_{i=1}^{2} (1 - \alpha_i) K_i^d - K^{ch} \quad (26)$$

where the first term is the gain of mass due to devolatilization while the second term is the loss of mass by oxidation.

## Radiative Transport Equations

In furnaces and combustion chambers, radiative heat transfer is an important mode of heat transfer. We apply the discrete-ordinates method (Fiveland, 1982, and Fiveland and Jamaluddin,

1991) to the radiative transport equation (RTE) for an emitting-absorbing and scattering gray medium. The discrete ordinates method approximates the RTE in a finite number of ordinate directions and replaces integrals by quadratures summed over the ordinate directions:

$$\frac{\mu_m}{r}\frac{\partial(rI_m)}{\partial r} + \xi_m \frac{\partial I_m}{\partial z} - \frac{1}{r}\frac{\partial \eta_m I_m}{\partial \varphi} = - \beta I_m + \kappa I_b + \frac{\sigma}{4\pi}\sum_{m'} w_{m'} \Phi_{m',m} I_{m'} \quad (27)$$

The resultant equations are integrated over each control volume. The source term for the energy equation can be found from summing Eq. (27) over all directions:

$$S_h = \kappa \sum_{m'} w_{m'} I_{m'} - 4\, \kappa \bar{\sigma} T^4 \quad (28)$$

The radiative transport equation requires absorption and scattering cross sections for particles and an absorption cross section for the gas mixture. In each computational cell the amounts of coal, char, and ash are determined from particle calculations, and the particle absorption and scattering coefficients are computed using Mie theory (Van De Hulst, 1957; Dave, 1969). The absorption coefficients of the gas mixture are found using Edwards' wide band models (Edwards, 1976).

## SOLUTION METHODS

Gas- and solid-phase equations are solved iteratively. The gas phase equations are discretized using the standard control volume method (Patankar, 1980). The mass and momentum equations are solved using the SIMPLE algorithm (Caretto, et al., 1972) on a nonstaggered grid. The coupling of the pressure and velocity fields is achieved using the pressure weighted interpolation method of Rhie and Chow (1983) with an extension by Majumdar (1988) to provide solutions independent of relaxation factors. The discrete, gas-phase conservation equations are solved using an algebraic multigrid solver which is based on the work of Ruge and Stüben (1987) and Hutchinson and Raithby (1986).

Particle phase equations are solved by Lagrangian methods in which the motion of individual particle streams is tracked through the combustor. A predictor-corrector algorithm with an adaptive time-step (Radakrishnan and Pratt, 1988) is used to integrate the stiff ordinary differential equations which govern the particle mass, momentum and energy transport. The sources of mass, momentum and energy for the gas phase are determined during particle tracking using a method based on that of Crowe (1979). A radiation solution is found using the standard streaming technique of the discrete ordinates method (Fiveland, 1982, 1991). The $S_4$ approximation was used for all cases presented in this paper.

A segregated solution procedure using Picard iteration is used to converge the coupled, nonlinear equations. Various linearization and relaxation techniques are used to avoid divergence. In particular, the source terms for the gas phase equations due to the particle phase are relaxed. To make the most efficient use of computer resources, particle and radiation calculations are performed less frequently than the gas phase calculations (i.e. a set of particle calculations are performed every 5 global gas phase iterations).

## RESULTS AND DISCUSSION

Three test cases - each of increasing geometric complexity - were considered. These cases include:

1) Non-swirling pulverized coal combustion in a bench scale research reactor.

2) Swirling pulverized coal combustion in the CANMET tunnel furnace.

3) Particle-laden jet flow in a 1/6-scale utility burner.

The following sections detail test conditions and present computational results.

### Non-swirling Pulverized Coal Combustion

Non-swirling pulverized coal combustion was modeled and compared with the experimental data collected by Thurgood (1979). The configuration of the combustor is shown in Figure 1. Coal particles and air are injected through a central primary nozzle at a rate of 0.0038 kg/s and 0.0056 kg/s, respectively. The entrance length of the primary nozzle is long enough to insure no slip between the particles and air which both enter at a velocity of 33.51 m/s in the axial direction. The oxidant streams through an annulus at a mass rate of 0.0361 kg/s corresponding to a velocity of 34.1 m/s.

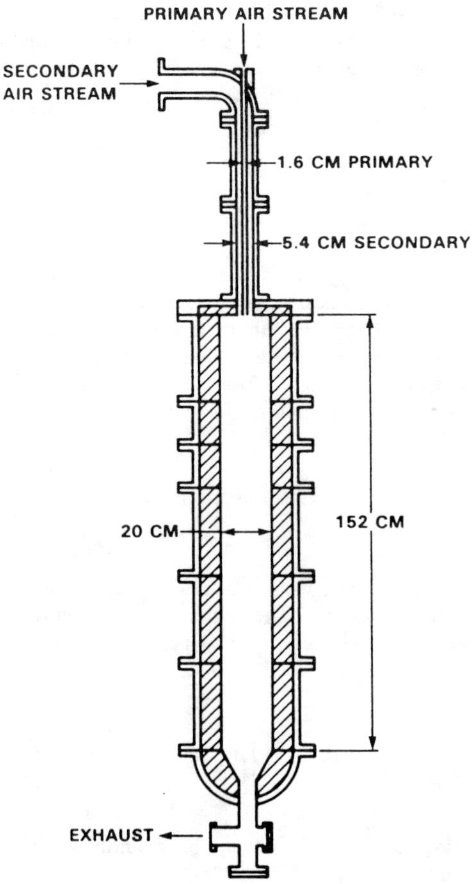


FIGURE 1. COMBUSTOR GEOMETRY FOR NON-SWIRLING PC COMBUSTION CASE.

The domain was modeled using a nonuniform 30x31 grid as shown in Figure 2. Inlet velocity, turbulence energy, and dissipation profiles (see Table 3) were those recommended by Thurgood (1979). Walls were assumed to be adiabatic with an emissivity of 0.7. The inlet particle conditions were obtained from test conditions and assumptions of no slip and uniform spatial distribution. The particle size distribution was based on that reported by Thurgood. Particle size and spatial distributions at the inlet are listed in Table 4.

TABLE 3
PRIMARY AND SECONDARY INLET CONDITIONS FOR NON-SWIRLING PC COMBUSTION CASE

| | Primary Air | Secondary Air |
|---|---|---|
| Axial velocity (m/s) | 33.6 | 34.1 |
| Radial velocity (m/s) | 0.0 | 0.0 |
| Tangential vel. (m/s) | 0.0 | 0.0 |
| Turbulent energy ($m^2/s^2$) | 16.8 | 17.4 |
| Dissipation rate ($m^2/s^3$) | 268 | 926 |
| Temperature (K) | 356 | 589 |
| $O_2$ (kg/kg air) | 23.1% | 23.1% |
| Coal flow rate (kg/s) | 0.00121 | - |

TABLE 4
COAL SIZE AND SPATIAL DISTRIBUTIONS FOR NON-SWIRLING PC COMBUSTION CASE

| Particle Radial Position (m) | Fraction Mass Flow at Each Radial Position | Particle Size (μm) | Fraction Mass Flow for Each Particle Size |
|---|---|---|---|
| $3.988 \times 10^{-4}$ | $8.333 \times 10^{-3}$ | 15 | 0.00615 |
| $1.1906 \times 10^{-3}$ | $3.333 \times 10^{-2}$ | 25 | 0.2 |
| $1.9843 \times 10^{-3}$ | $5.555 \times 10^{-2}$ | 40 | 0.3 |
| $2.7781 \times 10^{-3}$ | $7.777 \times 10^{-2}$ | 55 | 0.3 |
| $3.5718 \times 10^{-3}$ | 0.1 | 80 | 0.18375 |
| $4.3656 \times 10^{-3}$ | 0.1222 | | |
| $5.1594 \times 10^{-3}$ | 0.1444 | | |
| $5.9312 \times 10^{-3}$ | 0.16667 | | |
| $6.7469 \times 10^{-3}$ | 0.1889 | | |
| $7.5406 \times 10^{-3}$ | 0.1 | | |

Particle Temperature - 356 K

A vector plot of the flow field is shown in Figure 3. Arrows on the figure are proportional to the flow velocity. The strong inlet jets of fuel and oxidant dissipate quickly, and a one-

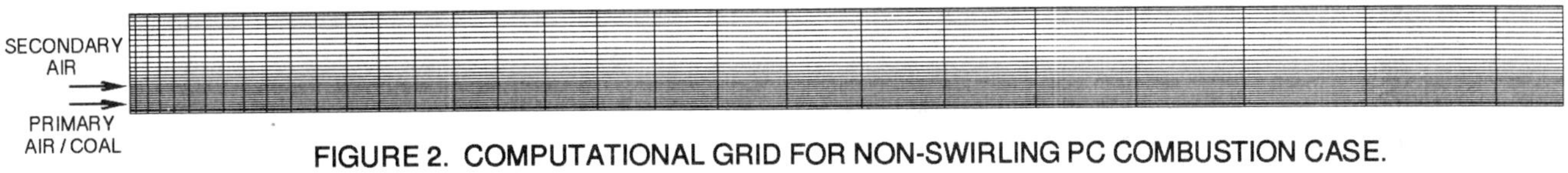


FIGURE 2. COMPUTATIONAL GRID FOR NON-SWIRLING PC COMBUSTION CASE.

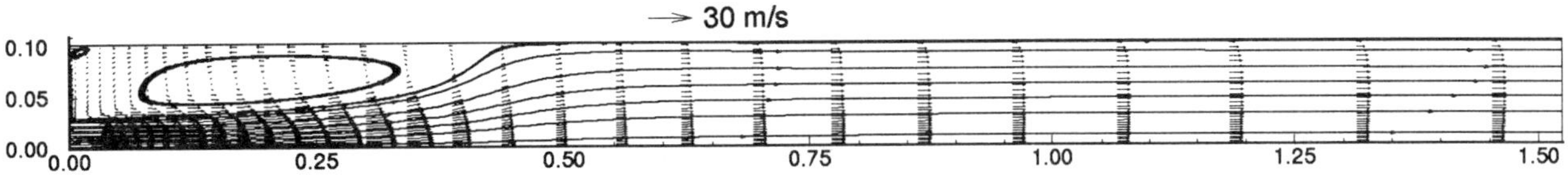


FIGURE 3. VELOCITY VECTORS FOR NON-SWIRLING PC COMBUSTION CASE.

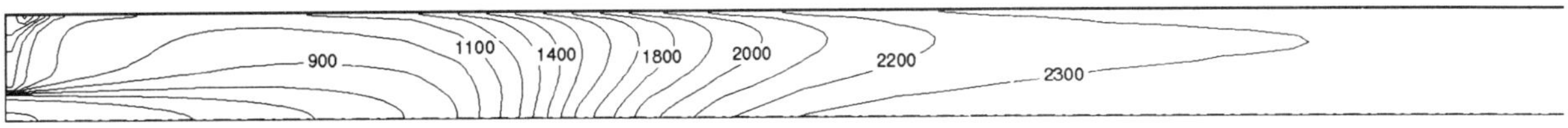


FIGURE 4. TEMPERATURE (K) FOR NON-SWIRLING PC COMBUSTION CASE.

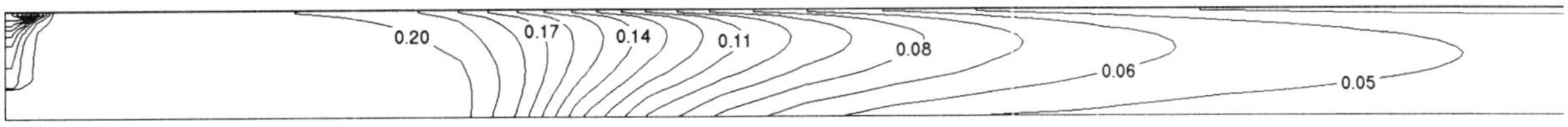


FIGURE 5. $O_2$ CONTOURS (MOLE FRACTION) FOR NON-SWIRLING PC COMBUSTION CASE.

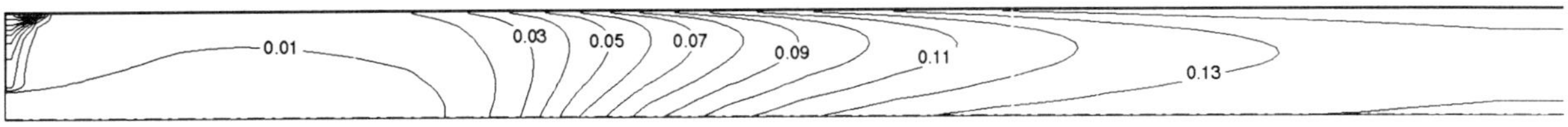


FIGURE 6. $CO_2$ CONTOURS (MOLE FRACTION) FOR NON-SWIRLING PC COMBUSTION CASE.

dimensional flow is established about half way down the combustor. Combustion occurs in this one-dimensional region; thus, the weak recirculation near the inlet does not significantly affect the combustion results.

Figure 4 shows a contour map for isotherms. The temperature is low at the inlet, but rises rapidly as the particles devolatilize and the volatiles burn. In Figures 5 and 6, contour maps for $O_2$ and $CO_2$ concentration are shown. As volatiles and released and burned, $O_2$ concentration decreases. $CO_2$ concentration increases as the volatiles are burned. Predicted and measured lift-off distances agree within 30%while other centerline profiles agree qualitatively. Sensitivity to the grid was investigated by doubling the number of grid lines in each direction. Although the original grid was relatively coarse, the grid refinement shifted the lift-off distance by less than 1%. In Figure 7, radial molar concentrations are presented. At the exit, concentration profiles agree fairly well with data collected by Thurgood (1979).

## Swirling Pulverized Coal Combustion

Swirling pulverized coal combustion was modeled and compared with the experimental data collected by Hughes (1987). The tests were conducted in the cylindrical tunnel furnace of the Canada Centre for Mineral and Energy Technology (CANMET). The case corresponds to Trial D reported by Hughes (1987). The tunnel configuration is shown in Figure 8. Air and fuel enter the furnace via coaxial streams. Coal and primary air enter with zero tangential velocity through a center zone while swirling secondary air enters through a annular zone. The burner has an approximate swirl number of 0.1. The test conditions are documented by Hughes (1987).

The furnace was modeled using a 64x40 nonuniform axisymmetric grid. The inlet velocity and turbulence profiles were obtained using inlet cold-flow from measurements and

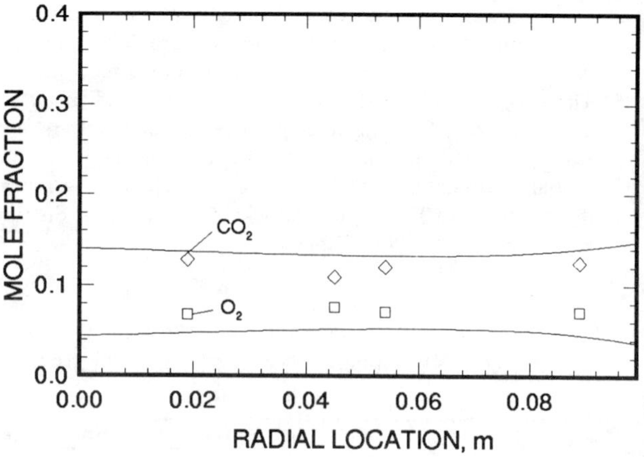


FIGURE 7. EXIT RADIAL PROFILES OF SPECIES CONCENTRATIONS FOR NON-SWIRLING PC COMBUSTION CASE ( x=1.39m ).

scaling recommendations of Holloway and Tavoularis (1989). The inlet particle conditions were obtained from test conditions and assumptions of no slip and uniform spatial distribution. The size distribution and starting locations are displayed in Table 5.

Predictions of gas temperature are compared to measurements in Figure 9. Radial profiles are shown at six axial locations. At each location, two data sets are shown - one for the near-side of the furnace and one for the far-side. At the first and second axial locations, the temperature is significantly underpredicted near the centerline. Better agreement is obtained at the remaining axial locations. Likely sources for the discrepancies between the

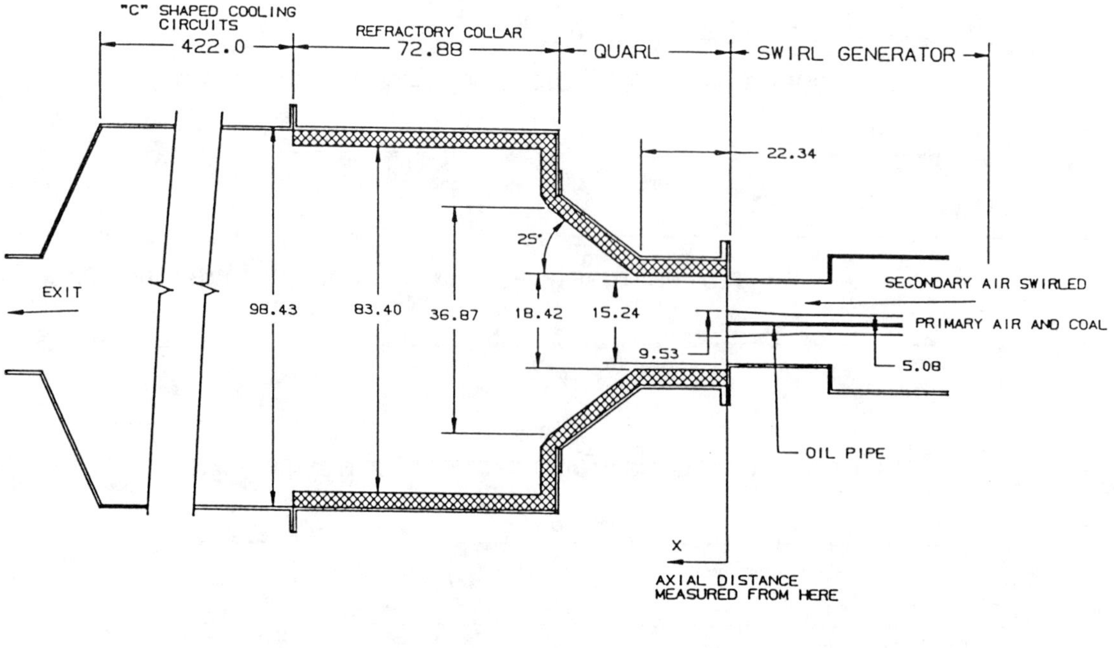


FIGURE 8. SCHEMATIC OF THE INTERIOR OF THE CANMET TUNNEL FURNACE (HUGHES, 1987).

TABLE 5
COAL SIZE AND SPATIAL DISTRIBUTIONS
FOR CANMET CASE

| Particle Radial Position (mm) | Fraction Mass Flow at Each Radial Position | Particle Size (μm) | Fraction Mass Flow for Each Particle Size |
|---|---|---|---|
| 2.987 | 0.0156 | 13.3 | 0.6635 |
| 8.934 | 0.0468 | 47.9 | 0.2272 |
| 14.89 | 0.0781 | 82.8 | 0.0743 |
| 20.84 | 0.1093 | 109.7 | 0.0301 |
| 26.80 | 0.1406 | 123.5 | 0.0046 |
| 32.75 | 0.1719 | 137.4 | 0.0003 |
| 38.71 | 0.2032 | | |
| 44.47 | 0.2345 | | |

Particle Temperature - 363 K

predictions and data at the first two locations include uncertainties in radiation properties (soot), devolatilization rates, and/or the combustion model. The comparison of species profiles in Figure 10 indicates inaccuracies in the homogeneous and heterogeneous chemistry models. The underprediction in centerline gas temperature at x=0.91m is mirrored by an overprediction of $O_2$ and an underprediction of $CO_2$. Although model changes could be made to produce a better fit of predictions to data, none were made because more fundamental experiments and sensitivity studies are need to justify such changes. Predictions for $O_2$ and $CO_2$ compare better to data at locations further downstream. Similar results were obtained during an AEA round-robin benchmark.

## Particle-laden Jet Flow

Numerical predictions of a isothermal, particle-laden flow were compared to experimental measurements taken from one/sixth-scale model of the utility burner as shown in Figure 11. Although this case is non-reacting, it tests the flow, turbulence, and particle flow modules for a practical burner geometry.

A particle-laden jet issues from a central tube, while secondary air exits the burner through two annular registers. Two tests were conducted: one with small particles (geon) and one with large particles (glass). Particle measurements were taken with a LDA at three axial locations - 1.28, 22.9, and 45.7 cm downstream from the burner exit. The latter two locations correspond to 1 and 2 burner diameters from the exit, respectively.

Two axisymmetric analyses - one for each particle class - were performed to model the dispersed two-phase flow downstream of

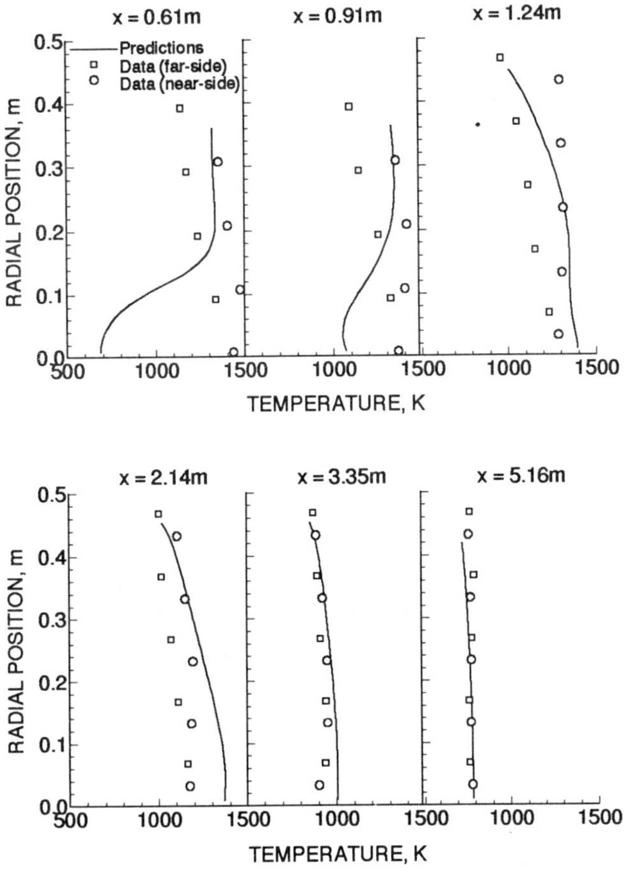


FIGURE 9. RADIAL GAS TEMPERATURE PROFILES FOR CANMET CASE.

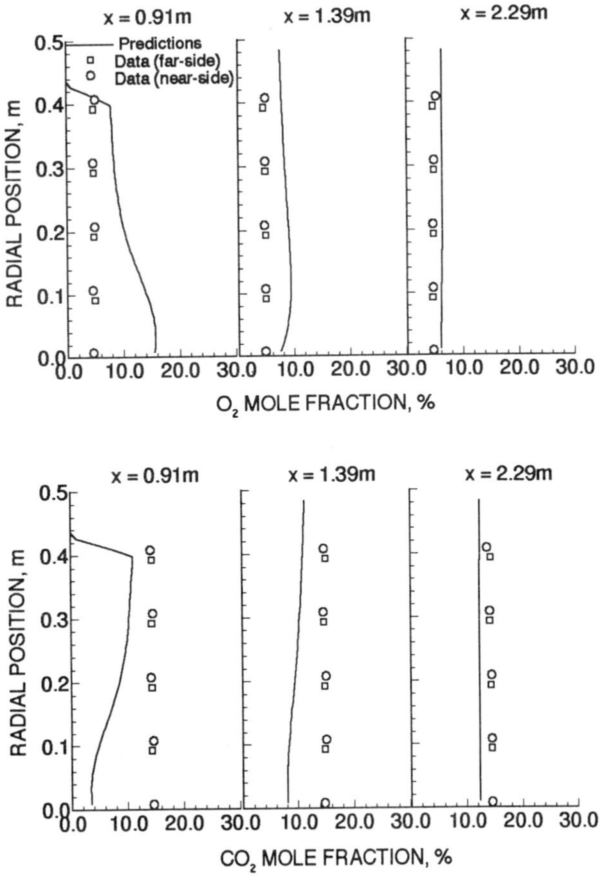


FIGURE 10. RADIAL GAS SPECIES PROFILES FOR CANMET CASE.

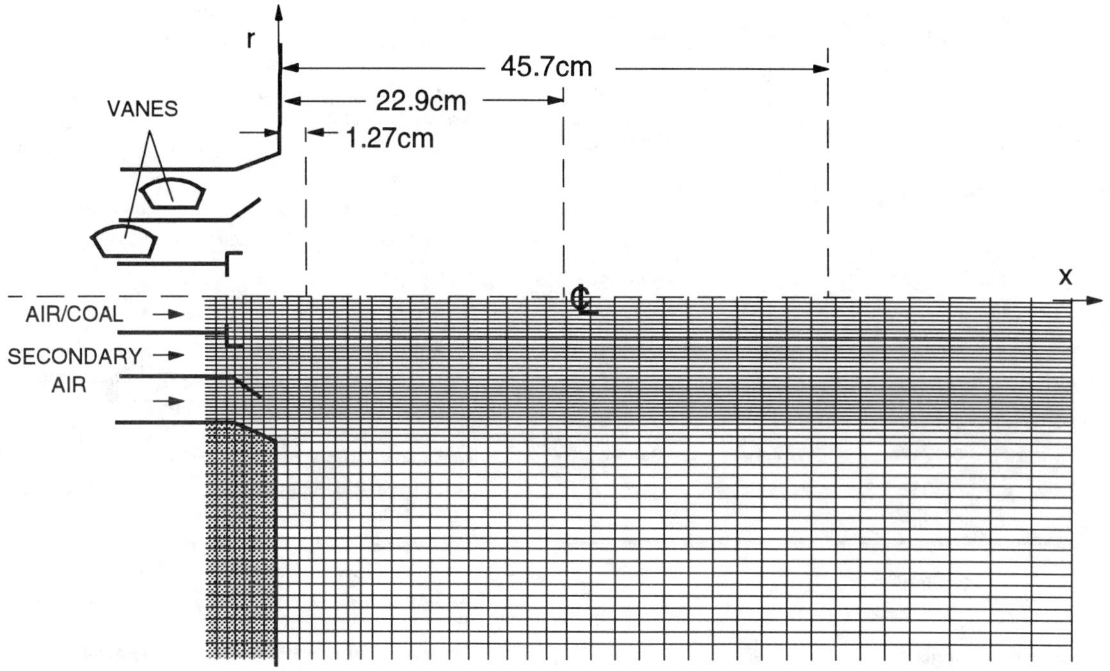


FIGURE 11. SCHEMATIC OF 1/6-SCALE BURNER AND IDENTIFICATION OF MEASUREMENT PLANES.

the throat. The analyses began several centimeters upstream of the primary air exit. A 53x57 nonuniform grid was used to represent the domain. A portion of the grid is shown in Figure 11. Inlet conditions were obtained using a 3-dimensional cylindrical model of the burner internals. The primary and secondary airflow rates, and the particle flow rates were based on measured values and are included in Table 6. The entrained flow entering the model domain at large radius was modeled using a correlation from Beér and Chigier (1972). For each case, a mono-dispersed size distribution was assumed corresponding to the number mean diameter. The spatial particle distributions were assumed uniform within the primary tube, and a no-slip condition was used to provide the inlet particle velocities.

The predictions were compared to measurements taken at the three axial planes described above. Quantities compared included mean axial and tangential particle velocities. A sample number of predicted trajectories is shown in Figure 12. The small geon particles are dispersed much more than the large glass particles.

The predicted and measured geon particle velocity profiles at the three planes are shown in Figures 13 and 14. As seen in Figure 13, the geon particles exit the burner with a high axial velocity, but decelerate rapidly due to drag forces from the gas phase. For all planes, the locations and magnitudes of peaks and valleys are well predicted. As seen at plane 1, the recirculation of particles behind the flame stabilizing ring appears to be underpredicted. The discrepancy may be attributed to the inability

TABLE 6
DESCRIPTION OF PARTICLE-LADEN JET CASE

| | | |
|---|---|---|
| Properties (air): | | |
| Density | 1.12 kg/m$^3$ | |
| Temperature | 297.9 K | |
| Reference Pressure | 98.7 kPa | |
| Inlet conditions: | | |
| Primary mass flow rate | 0.0495 kg/s | |
| Secondary mass flow rate | 0.4121 kg/s | |
| Swirl number | ~0.45 | |
| | Geon | Glass |
| Particle flow rate (kg/s) | 3.64x10$^{-3}$ | 8.71x10$^{-3}$ |
| Particle properties: | | |
| | Geon | Glass |
| Number mean diameter (μm) | 72 | 128 |
| Density (kg/m$^3$) | 1400 | 2500 |

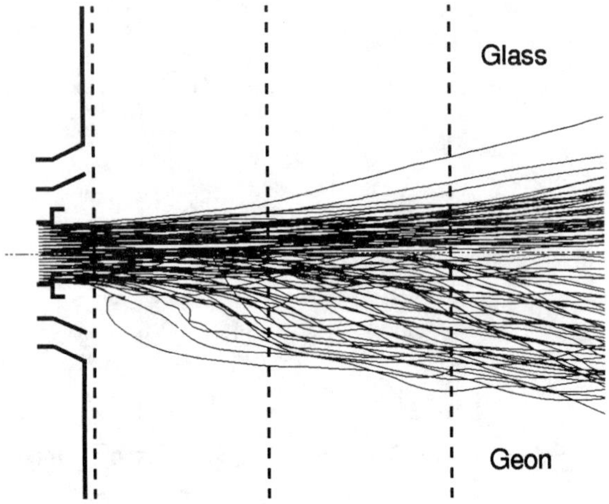


FIGURE 12. SAMPLE GEON AND GLASS PARTICLE TRAJECTORIES.

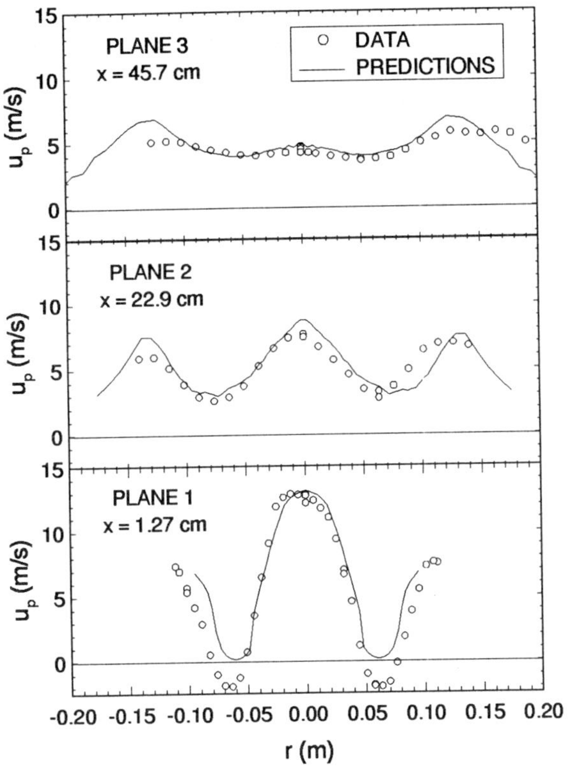


FIGURE 13. MEAN AXIAL VELOCITY PROFILES FOR GEON PARTICLES.

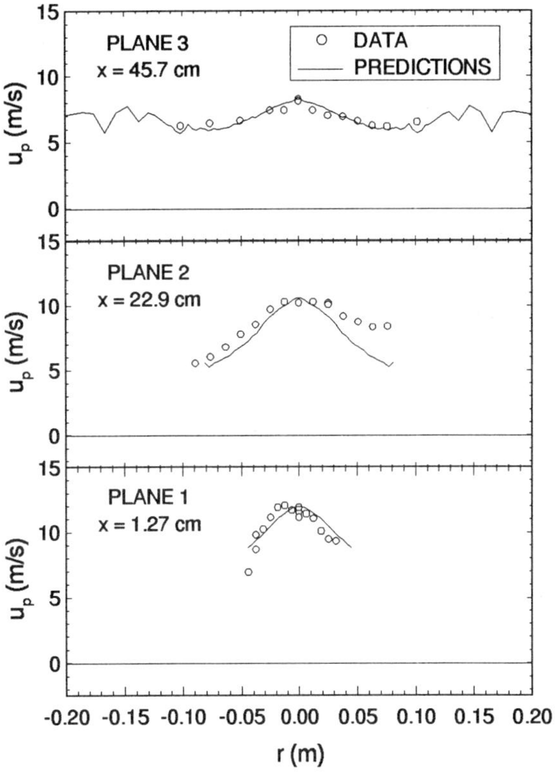


FIGURE 15. MEAN AXIAL VELOCITY PROFILES FOR GLASS PARTICLES.

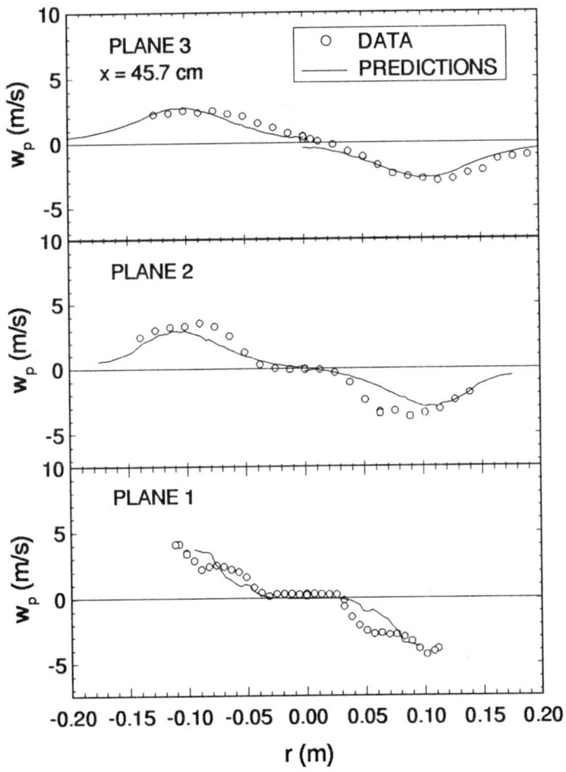


FIGURE 14. MEAN TANGENTIAL VELOCITY PROFILES FOR GEON PARTICLES.

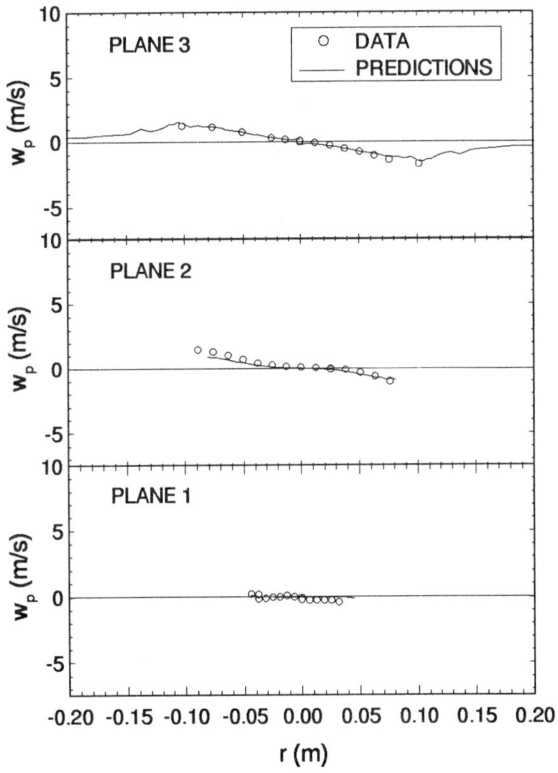


FIGURE 16. MEAN TANGENTIAL VELOCITY PROFILES FOR GLASS PARTICLES.

of the k-ε turbulence model to predict the size and shape of the recirculating zone.

The tangential velocities of the geon particles are shown in Figure 14. At the burner exit (plane 1), the particles in the center of the coal pipe region have nearly zero tangential momentum, but the particles on the edge of the region have been accelerated tangentially by the swirling secondary flow. At a burner diameter downstream (plane 2), the particles have been noticeably captured by the secondary stream. The spreading continues as the particles move to plane 3.

The predicted and measured glass particle velocity profiles at the three planes are shown in Figures 15 and 16. All predictions compare very well to the data. Plane 1 in Figure 15 shows the particles exiting the burner with a high axial velocity. The particles decelerate as they flow downstream; however, the glass particles maintain higher axial velocities than the geon particles -- a trend that may be expected since the glass particles are both larger and more dense than the geon particles. Also, the glass particles are not as strongly entrained by the secondary flow. Figure 16 displays the corresponding tangential velocity profiles for the glass particles. As the particles exit, they have negligible tangential velocity. The particles are slowly accelerated tangentially by the swirling secondary flow as they are carried downstream.

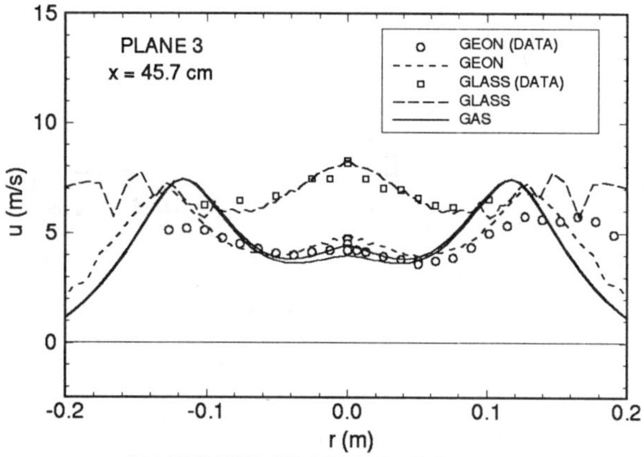


(a) MEAN AXIAL VELOCITY PROFILES AT PLANE 3.

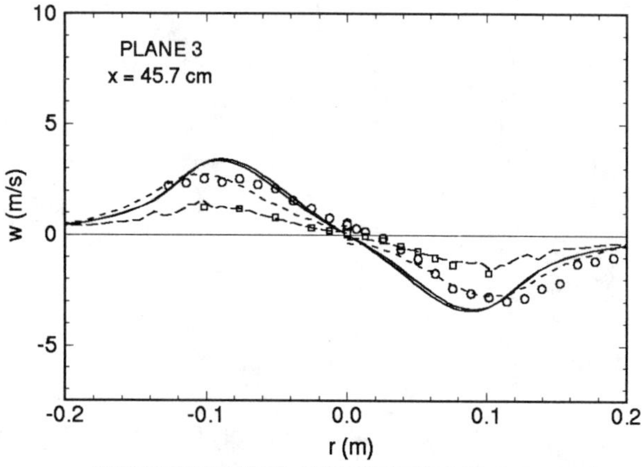


(b) MEAN TANGENTIAL VELOCITY PROFILES AT PLANE 3.

FIGURE 17. COMPARISON OF MEAN GAS AND PARTICLE VELOCITIES AT PLANE 3 .

In Figure 17, the mean velocity profiles for the gas and geon and glass particles are compared at plane 3. Figure 17a shows the mean axial velocities. As expected, the smaller and lighter geon particles follow the gas flow whereas the larger and heavier glass particles tend to maintain their exit velocities. The geon particles are accelerated faster in the tangential direction by the swirling secondary flow than are the glass particles, as seen in Figure 17b. In both Figure 17a and 17b, two predicted gas velocity profiles are shown. The two curves correspond to the gas velocities from the geon and glass particle analyses, and indicate the influence of the particle phase on the gas phase. The curve with higher axial velocity in Figure 17a and the curve with lower tangential velocity in Figure 17b correspond to the glass particle case.

## CONCLUSIONS

An Axisymmetric model has been presented for pulverized coal applications. The methodology presented is easily extended for three-dimensional applications.

Three cases of increasing complexity show the models capability to replicate some experimental trends, albeit developmental work is far from complete. Isothermal gas phase flow with a disperse particle phase compares excellently with measured axial and tangential velocities, and the models show promise for use as scaling tools for such applications.

However, even in simple geometries, predictions for combustion remain a significant challenge. The models for turbulence and chemistry which are used in today's practical codes are not adequate to achieve quantitative agreement with data. Studies, using simplified kinetics, are now underway. In addition, validation data is needed to benchmark the models. The models have limitations, but with care they can be used to qualitatively assess new burner and furnace designs.

Future work should focus on improving the chemistry and turbulence models. Stochastic PDF methods hold promise for improving the chemistry predictions. However, much development work is still needed in this area, and parallel computation must become more prevalent before such models may be applied on a daily basis to practical combustion systems.

## ACKNOWLEDGEMENTS

The authors would like to thank Pat Hughes of CANMET and Phil Stopford of AEA for providing the experimental data and modeling conditions, respectively, for the CANMET case. The authors also acknowledge Mort Licht and Ruth Ilan of B&W/R&DD for making the measurements on the 1/6-scale burner.

## REFERENCES

Azevedo, J.L.T and Carvalho, M.G., "Study of the Influence of Flue Gas Recirculation in Modeling Pulverized Coal Combustion," Second International Conference on Combustion Technologies for a Clean Environment, Lisbon, Portugal, July 19-22, 1993.

Beér, J.M. and Chigier, N.A., Combustion Aerodynamics, Wiley, 1972.

Boyd, R.K. and Kent, J.H., "Three-Dimensional Furnace Computer Modeling," *21st Symposium (International) on Combustion*, The Combustion Institute, p. 1245, 1986.

Boyd, R.K. and Kent, J.H., "Comparison of Large Scale Boiler Data with Combustion Model Predictions," *Energy and Fuels*, Vol. 8, pp. 124-130, 1994.

Caretto, L.S., Gosman, A.D., Patankar, S.V., and Spalding, D.B., "Two Calculation Procedures for Steady, Three-Dimensional

Flows with Recirculation," *Proc. 3d Int. Conf. Num. Methods Fluid Dyn.*, Paris, Vol. II, p. 60, 1972.

Crowe, C.T., "A Computational Model for the Gas Droplet Flow Field in Vicinity of an Atomizer," Models of Turbulence, Academic Press, New York, NY; 1972.

Crowe, C.T., "Gas-Particle Flow" - Pulverized Coal Combustion and Gasification, Edited by L. D. Smoot and D. T. Pratt, Plenum Press, New York, NY; 1979.

Dave, F.J., "Subroutine for Computing the Parameters of Electromagnetic Radiation Scattered by Spheres," 360 D-17, 4.007, IBM Corporation, 1969.

Edwards, D.K., "Molecular Gas Band Radiation," *Advances in Heat Transfer*, Vol. 12, pp. 115-193, 1976.

Field, M.A., Gill, D.W., Morgan, B.B., and Hawksley, P.G.W., Combustion of Pulverized Coal, British Coal Utilization Research Association, Leatherhead, 1967.

Fiveland, W.A., "A Discrete Ordinates Method for Predicting Radiative Heat Transfer in Axisymmetric Enclosures," ASME Paper No. 82-HT-20, 1982.

Fiveland, W.A., Cornelius, D.K., and Oberjohn, W.J., "COMO: A Numerical Model for Predicting Furnace Performance in Axisymmetric Geometries," ASME Paper No. 84-HT-103, 1984.

Fiveland, W.A., and Jamaluddin, A.S., "Three-Dimensional Spectral Radiative Heat Transfer Solutions by the Discrete Ordinates Method," *Journal of Thermophysics and Heat Transfer*, Vol. 5, No. 3, pp. 335-339, 1991.

Fiveland, W.A., and Jamaluddin, A.S., "An Efficient Method for Predicting Unburned Carbon in Boilers," *Combustion Science and Technology*, Vol. 81, pp. 147-167, 1992.

Fiveland, W.A., and Wessel, R.A., (1988). Numerical Model for Predicting Performance of Three-Dimensional Pulverized-Fuel Fired Furnaces, *ASME J. of Engng. for Gas Turbines and Power*, 110, 117-126.

Gibson, M.M. and Morgan, B.B., "Mathematical Models of Combustion of Solid Particles in a Turbulent Stream With Recirculation," *J. Inst. Fuel*, 43, pp. 517-523, 1970.

Hinze, J.O., Turbulence, 2nd Edition, McGraw-Hill Book Co., 1975.

Holloway, A.G.L. and Tavoularis, S., "Turbulence Measurements in the Swirling Flow at the Entrance to the CCRL Tunnel Furnace Facility," University of Ottawa Report UOME-FM-89-06, 1989.

Howard, J.B., Williams, G.C., and Fine, D.H., "Kinetics of Carbon Monoxide in Postflame Gases," Proceedings of the 14th International Symposium on Combustion, 1972.

Hughes, P.M.J., "A Data Package for the Validation of a Computer Model of the CCRL Tunnel Furnace Facility," CANMET Report 87-4E, 1987.

Hutchinson, B.R. and Raithby, G.D., "A Multigrid Method Based on the Additive Correction Strategy," *Numerical Heat Transfer*, vol. 9, pp. 511-537, 1986.

Jones, W.P., and Whitelaw, J., "Calculation Methods for Reacting Turbulent Flows: A Review," *Combustion and Flame*, Vol. 48, pp. 1-26, 1982.

Kobayashi, H.J., Howard, J.B., and Sarofim, A.F., "Coal Devolatilization at High Temperatures," *Sixteenth Symposium International) on Combustion*, The Combustion Institute, pp. 411-425, 1976.

Launder, B.E. and Spalding, D.B., "The Numerical Computation of Turbulent Flows," *Computer Methods in Applied Mechanics and Engineering*, Vol. 3, 1974.

Lockwood, F.C., Salooja, A.P., and Syed, S.A., "A Prediction Method for Coal-Fired Furnaces," *Comb. and Flame*, 38, pp. 1-15, 1980.

Magnussen, B.F., and Hjertager, B.H., "On Mathematical Modeling of Turbulent Combustion with Emphasis on Soot Formation and Combustion," *16th Symposium (International) on Combustion, The Combustion Institute*, 719-729, 1976.

Majumdar, M., "Role of Underrelaxation in Momentum Interpolation for Calculation of Flow with Nonstaggered Grids," *Numer. Heat Transfer*, vol. 13, pp. 125-132, 1988.

Medgal, D., and Agosta, V.D., *Journal of Applied Mechanics*, Vol. 35, pp. 860-865, 1967.

Milojevic, D., Borner, T, Wennerberg, D., Berlemont, A., Desjonqueres, P., and Avila, R., "A Comparision of Six Two-Phase Flow Models for the Prediction of Turbulent Gas Particle Flows in Jets and Channels," Int. Sem. Transient Phenomena in Multiphase, Int. Center for Heat and Mass Transfer, Dubrovnik, May 1987.

Milojevic, D., "Lagrangian Stochastic-Deterministic (LSD) Predictions of Particle Dispersion in Turbulence," *Journal of Particles and Particle Systems Characterization*, Vol. 7, pp. 181-190, 1990.

Patankar, S.V., Numerical Heat Transfer and Fluid Flow, Hemisphere Publishing Corporation, New York, 1980.

Radakrishnan, K., and Pratt, D.T., "Fast Algorithm for Calculating Chemical Kinetics in Turbulent Reacting Flow," *Combustion Science and Technology*, Vol. 58, pp. 155-176, 1988.

Rhie, C.M., and Chow, W.L., "Numerical Study of the Turbulent Flow Past an Airfoil with Trailing Edge Separation, *AIAA J.*, vol. 21, pp. 1525-1532, 1983.

Ruge, J.W. and Stüben, K., "Algebraic Multigrid", in Multigrid Methods, ed. by S. McCormick, Volume 3 of the SIAM Frontiers Series, SIAM, Philadelphia, 1987.

Sloan, D.G., Smith P.J., and Smoot, L.D., "Modeling of Swirl in Turbulent Flow Systems," Progress in Energy and Combustion Science, Vol. 12, pp. 163-250, 1986.

Thurgood, J.R., "Mixing and Combustion of Pulverized Coal," Ph.D. Thesis, Brigham Young University, Provo, UT, 1979.

Ubhayaker, S.K., Stickler, D.B., Van Rosenburg, C.W., and Gannon, R.E., "Rapid Devolatilization of Pulverized Coal in Hot Combustion Gases," *International Symposium on Combustion*, The Combustion Institute, 1975.

Van De Hulst, H.C., Light Scattering by Small Particles, Wiley, New York, NY, 1957.

Visser, B.M., "Mathematical Modeling of Swirling Pulverized Coal Flames," Delft Technical University, Ph.D. Thesis, January 1991.

Wang, D.M., Watkins, A.P., and Cant, R.S., "Three-Dimensional Diesel Engine Simulation with a Modified EPISO Procedure," *Numerical Heat Transfer*, vol. 24, pp. 249-272, 1983.

Weber, R., Peters, A.A.F., Breithaupt, P.P., "Mathematical Modeling of Swirling Pulverized Coal Flames: What can Combustion Engineers Expect from Modeling?," ASME Paper Fact-Vol. 17, pp.71-86, 1993.

FACT-Vol. 18, Combustion Modeling, Scaling and Air Toxins
ASME 1994

# NUMERICAL SIMULATION OF COAL GASIFICATION REACTORS

**G. J. Kovacik**
Alberta Research Council
Devon, Alberta, Canada

**K. J. Knill**
Advanced Scientific Computing Limited
Waterloo, Ontario, Canada

## ABSTRACT

A computational fluids code has been developed for simulating coal gasifiers. The code includes physical models for simultaneous coal devolatilization, char oxidation and gasification and gas phase product redistribution. The gasification of a subbituminous coal in a 10 kg/h oxygen blown laboratory coal gasifier, operated under various conditions was then simulated.

Predictions of major gas species compared quite well with the experimental data, and provided much more insight regarding temperature, flow and concentration distributions within the gasifier than was possible experimentally. In experimental tests with high coal burnout, the gas species approached chemical equilibrium at the exit of the gasifier and were well predicted by the model. The numerical model was then used to simulate the flow and reaction behaviour in a generic, full scale (800 tonne/day) entrained flow gasifier. These results demonstrate that the processes occurring at full scale are similar to those observed in the laboratory gasifier.

## INTRODUCTION

Coal gasification reactor modeling is considered a priority by Canadian utility companies because Integrated Gasification Combined Cycle (IGCC) technology is regarded as the most likely process for future coal based electricity generating plants Macdonald et. al. (1992). As IGCC technology nears commercialization, utilities will need to understand the processes involved in a gasifier in order that they can procure, build and operate new power plants. However, there is very little information available regarding the processes occurring in a gasifier. Developers still consider the process technology and gasifier design to be proprietary. This makes it difficult for utilities to consider and evaluate process options and design details.

The potential for using computational techniques to aid in the understanding, selection, and scale up of coal gasification technologies to specific situations has been recognized. A consortium project involving seven utility companies and two research organizations was established to develop computational fluid models for coal gasification. In this paper, the coal gasification model is presented and simulations conducted on a laboratory scale gasifier are presented. The model is then applied to a generic full scale gasifier to determine the influence of scale on the processes occurring.

## MODEL DETAILS

A commercial computational fluids package, TASCflow3D, was used to model the gasifier. TASCflow3D contains several novel methods to solve a variety of fluid flows in a robust, efficient and accurate manner and has been described by Knill *el al.* (1993). Specific elements of the model related to coal gasification are described here.

### Particle Tracking

Particle tracking is calculated using a Lagrangian stochastic model (Faeth, 1987) of a particle equation of motion:

$$m_p \frac{dv_p}{dt} = 3\pi\mu d_p C_{cor} (v_f - v_p). \quad (1)$$

The change in particle velocity, $v_p$, is a function of particle mass, $m_p$, particle diameter, $d_p$, fluid velocity, $v_f$, and viscosity, $\mu$, as well as a drag correction term, $C_{cor}$. The instantaneous fluid velocity, $v_f$, is divided into a mean, $\overline{v}_f$, and a fluctuating component, $v'_f$. The fluctuating velocity is re-calculated each time the particle travels through an eddy based on a normally distributed random number and the turbulent kinetic energy.

Individual particles are tracked through the domain until they burn out or escape through the outlet. Several particles (typically 250) are injected each iteration to represent the tracks of all particles. The sources generated during the particle flight are added to the appropriate source terms of the mass, momentum, energy and species equations.

## Coal Gasification

Coal gasification is modeled as simultaneous devolatilization and char gasification processes. The coal volatiles are assumed to be a hydrocarbon mixture containing all of the coal hydrogen with the remaining mass being carbon. The volatiles evolve at a rate expressed in Arrhenius form:

$$\frac{dV}{dt} = A_v \exp(-E_v/T_p)(V - V_{\max}) \quad (2)$$

where $A_v$ and $E_v$ are kinetic rate constants, $V$ is the fraction of coal evolved as volatiles, and $V_{max}$ is the maximum volatile yield. Values of these parameters are shown in Table 1. Knill et al. (1989) showed that devolatilization occurs during particle heating and is nearly instantaneous for particle temperature greater than 1300 K. Thus, the kinetic constants are chosen to ensure that devolatilization is complete in 1 ms at 1300 K.

Table 1: Model Parameters

| Parameter | Units | Value |
|---|---|---|
| $A_{c,c}$ | [$s^{-1}$] | 32.7 |
| $A_{c,h}$ | [$s^{-1}$] | 0.5 |
| $A_{\mathrm{edm}}$ | | 1.3 |
| $A_{\mathrm{eq}}$ | | 40.7 |
| $A_{\mathrm{shift}}$ | [$s^{-1}$] | 8.1e-9 |
| $A_v$ | [$s^{-1}$] | 8.4e5 |
| $E_{c,c}$ | [$K$] | 28867.0 |
| $E_{c,h}$ | [$K$] | 28867.0 |
| $E_{\mathrm{eq}}$ | [$K$] | 4217.5 |
| $E_{\mathrm{shift}}$ | [$K$] | 10926.5 |
| $E_v$ | [$K$] | 8900.0 |
| $V_{\max}$ | | 42.6 |

The remaining char is gasified with $CO_2$ and $H_2O$:

$$\begin{array}{lllll} C & + & CO_2 & \rightarrow & 2CO \\ C & + & H_2O & \rightarrow & CO + H_2 \end{array} \quad (3)$$

Both reactions are first order in the $CO_2$ and $H_2O$ partial pressures, $P_{CO_2}$ and $P_{H_2O}$, and they proceed in parallel. The total char gasification rate, $dm_c/dt$, is calculated from:

$$\frac{dm_c}{dt} = \frac{k_{c,c}\, k_{d,c}}{k_{c,c} + k_{d,c}} P_{CO_2} + \frac{k_{c,h}\, k_{d,h}}{k_{c,h} + k_{d,h}} P_{H_2O} \quad (4)$$

where $k_{c,c}$ and $k_{c,h}$ are the chemical reaction rates for $CO_2$ and $H_2O$ gasification, respectively, and $k_{d,c}$ and $k_{d,h}$ are the corresponding reactant diffusion fluxes to the particle. The chemical reaction rates are expressed in Arrhenius form:

$$k_{c,i} = A_{c,i} \exp\left(\frac{-E_{c,i}}{T}\right) \quad (5)$$

where the subscript $i$ represents either $c$ or $h$. The gasification reaction parameters were obtained from Baxter (1987). The diffusion rates are expressed as:

$$k_{d,i} = \frac{2\Gamma_i}{d_p\, P_f} \quad (6)$$

where $\Gamma_i$ is the diffusion coefficient of $CO_2$ and $H_2O$ and $P_f$ is the total pressure of the fluid.

## Gas Phase Reactions

As the volatiles are released from the coal, they react with oxygen to form complete combustion products.

$$C_xH_y + \left(x + \frac{y}{4}\right) O_2 \rightarrow x\, CO_2 + \frac{y}{2} H_2O \quad (7)$$

The volatile reaction rate, $R_v$, is controlled by the mixing of fuel and oxidant according to the Eddy Dissipation Model (EDM) of Magnussen and Hjertager (1976). In the EDM, reaction rate is defined as the product of a characteristic eddy lifetime, $k/\epsilon$ and the minimum of the volatile and oxygen mass fractions, $Y_v$ and $Y_o$, respectively:

$$R_v = A_{\mathrm{edm}}\, \rho \frac{\epsilon}{k} \min\left(Y_v,\ \frac{Y_o}{r_o}\right) \quad (8)$$

where $r_o$ is the stoichiometry (mass $O_2$/mass volatile) in Reaction (7).

The volatile combustion and char gasification products are redistributed in the gas phase according to the water gas shift reaction:

$$CO_2 + H_2 \rightleftharpoons CO + H_2O \quad (9)$$

The water/gas shift reaction may progress in either direction depending on gas composition and temperature as determined by equilibrium, $K_{\mathrm{eq}}$:

$$K_{\mathrm{eq}} = \frac{[P_{CO}][P_{H_2O}]}{[P_{CO_2}][P_{H_2}]} = A_{eq} \exp\left(\frac{-E_{eq}}{T}\right) \quad (10)$$

The rate of the shift reaction, $R_{\mathrm{shift}}$, is expressed as the minimum of the mixing rate, given by the Eddy Dissipation Model and the kinetic rate of the reaction, $R_{\mathrm{eq}}$. If the forward reaction predominates, then the shift rate is calculated from:

$$R_{\mathrm{shift}} = \min\left[A_{\mathrm{edm}} \rho \frac{\epsilon}{k} \min\left(Y_{CO_2},\ \frac{Y_{H_2}}{r_f}\right),\ R_{\mathrm{eq}}\right] \quad (11)$$

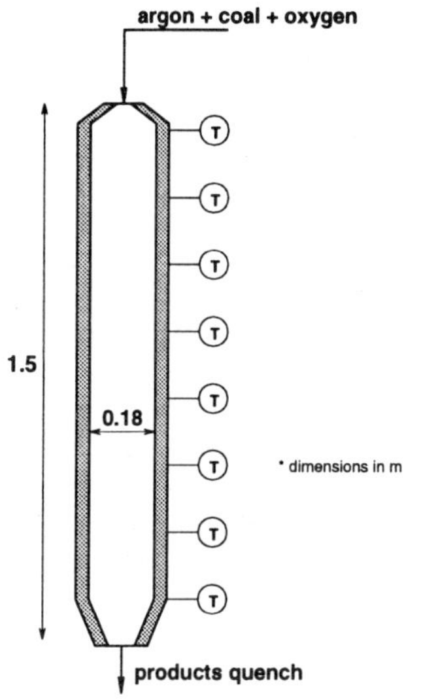


Figure 1: Schematic of the gasifier

where $r_f$ is equal to 22 and the shift kinetic rate is calculated from Chen *et al.* (1987):

$$R_{eq} = k_{shift} \left[ P_{CO} P_{H_2O} - K_{eq} P_{CO_2} P_{H_2} \right] \quad (12)$$

$$k_{shift} = A_{shift} \exp\left( \frac{-E_{shift}}{T} \right) \quad (13)$$

## TEST PROBLEM

### Laboratory Gasifier

The laboratory coal gasifier simulated in this work is the experimental unit at CANMET's Bell's Corner Research Facility and is shown schematically in Fig. 1. The gasifier is a down-fired, dry feed, cylindrical reactor with an inside diameter of 0.18 m and an internal length of 1.51 m. The gasifier is lined with a castable ceramic refractory for insulation. Coal is transported in argon through a central 4 mm pipe. Oxygen is either injected in a 10 mm annulus around the coal pipe or is premixed with the coal and fired through the coal pipe. In the test reported here, the oxygen, coal and argon are all premixed. The reactor has a converging outlet section which dumps the gasification products into a quench pot. A water spray quenches the gasification products as they enter the quench pot.

The reactor is instrumented with thermocouples along the inner face of the insulation lining. A gas sampling line is connected to the quench pot to analyze the exit dry gas composition. There is no instrumentation inside the reactor to sample intermediate gas species. The only method of identifying reaction behaviour is through visual observation ports along the length of the reactor.

The mass inputs to the laboratory gasifier are shown in Table 2. A Canadian low sulphur subbituminous coal is fired at a rate of 9.9 kg/h. with oxygen fired at a rate of 9.5 kg/h. This results in a C/O ratio (carbon/oxygen in coal and $O_2$) of 0.56.

### Generic Full Scale Gasifier

The generic full scale coal gasifier simulated here is a scaled-up version of the laboratory gasifier based on information regarding a 800 tonne/day coal feed, oxygen blown, entrained flow unit obtained from Blamire, 1993. The process and design information generally available on any commercial scale coal gasification reactor is extremely vague so that the values assumed here are only best estimates.

This gasifier is a down-fired, coal/water slurry feed, cylindrical reactor with an inside diameter of 1.5 m and an internal length of 10.0 m. This reactor is refractory lined and the internal wall temperature is maintained at 1500 K to allow ash melting.

The coal/water slurry is preheated to evaporate the water and mixed with oxygen to form a single jet. The mass inputs to the reactor were estimated from Kovacik, et al (1988) and are shown in Table 2. The feed rates result in a C/O ratio of 0.61.

Table 2: Reactor Inputs

| Mass flows | Laboratory (kg/h) | Full scale (kg/h) |
|---|---|---|
| $O_2$ | 9.5 | 28,608 |
| Ar | 0.5 | nil |
| Coal | 9.9 | 33,333 |
| $H_2O$ | | 33,516 |
| Coal analysis | (wt% as fed) | |
| C | 62.8 | |
| H | 4.3 | |
| O | 17.2 | |
| N, S * | 1.8 | |
| Moisture | 3.5 | |
| Ash | 10.4 | |

* - not included in simulation

### Numerical Modeling

In both gasifiers, the simulation geometry consists of a 10° section and is modeled using a 150 × 30 × 3 grid. The three velocity components and pressure are solved using an implicit coupled multigrid solver. Turbulence is treated using the $k/\epsilon$ turbulence model and radiation is handled using a Gibb's diffuse radiation model. Separate scalar equations are formulated for each of the gas phase fluid components: volatile, $O_2$, $CO_2$, CO, $H_2$, $H_2O$. The argon is calculated as the difference of the other components.

The coal, argon and oxygen enter the reactor at 300 K. In the laboratory reactor, the inner lining is approximated as a constant temperature surface of 1200 K (based on measurements) with an emissivity of 0.8. In the full scale gasifier, the wall temperature is assumed to be 1500 K to allow ash melting.

Fifty iterations were completed without the gasification model activated to develop a reasonable cold flow. Then, the flow was seeded with a starting temperature of 1200 K to initiate the reactions. The energy, scalars, particle tracking and gasification model were all activated at once. Particles were injected before the first hydrodynamic iteration and every iteration thereafter for the first 100 iterations. The high frequency of particle injection initially helps to obtain a stable gasification. The diffusion radiation model was activated after 50 iterations. The delay was introduced to help let the coal react in the early reaction process. If the radiation model was active from the first iteration, the coal was difficult to ignite in some cases.

The time step was maintained at 0.05 s to help come to a steady state solution. After the above 100 iterations the injection frequency was reduced to once every 10 iterations and a further 50 iterations were required to converge the solution. The total time to reach the final converged solution (from initial cold flow to final hot convergence) on the SGI personal IRIS work station was approximately 20 CPU hours.

## RESULTS AND DISCUSSION

### Laboratory Gasifier

A streakline plot of the flow patterns in the laboratory gasifier is shown in Fig. 2 and the predicted temperature profile is shown in Fig. 3. The coal/$O_2$ jet expanded to fill the reactor 0.75 m downstream from the inlet, generating a large external recirculation zone. This zone was observed through the viewports of the experimental gasifier.

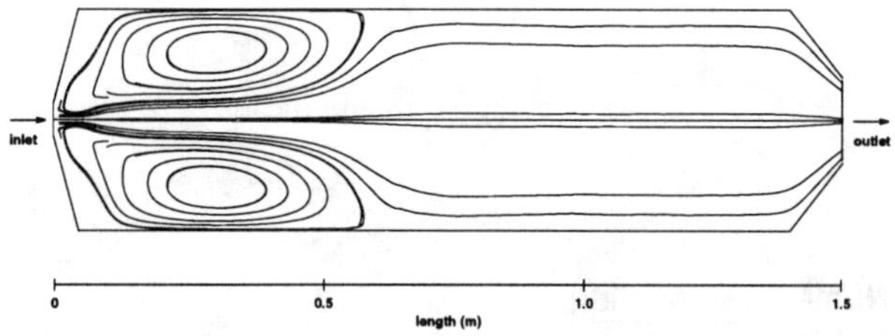


Figure 2: Flow streaklines in the laboratory reactor

Two fairly distinct reaction zones were prevalent in the reactor. In the opinion of the authors, the coal jet rapidly entrained hot reaction products from this external recirculation zone which helped to stabilize the devolatilization zone 0.3 m downstream of the inlet. The volatiles reacted with the $O_2$ to form $CO_2$ and $H_2O$ accounting for the high temperatures observed in the upstream part of the reactor.

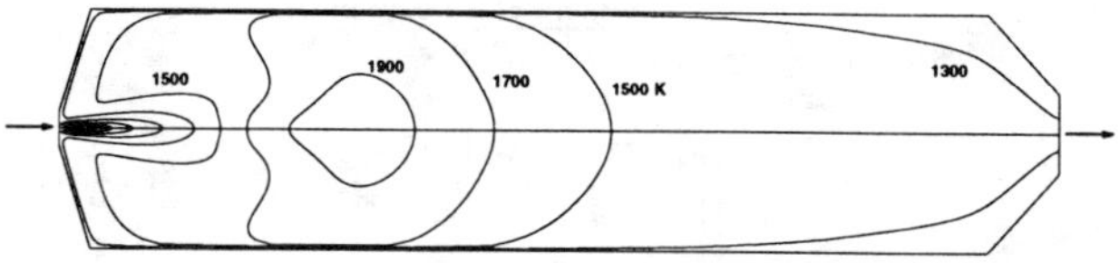


Figure 3: Temperature in the laboratory gasifier

Downstream of the devolatilization zone, the char reacted with $CO_2$ and $H_2O$. These reactions helped to drop the temperature from 1970 K to 1300 K at the outlet. The maximum fluid temperature of 1970 K was predicted one-third of the distance down the reactor. This was where the volatiles combustion was most intense and compared qualitatively with visual observations (radiative intensity) of the gasifier.

In the char reaction region, the gas temperature was too cold to ensure complete conversion of the char. The predicted carbon conversion (including carbon in the volatile and char) was only 82 wt%. The measured carbon conversion for this test was 96 wt%. This difference may be attributable to both model assumptions and experimental measurement error. The model neglects an important reaction of char with $O_2$ which may occur in the upstream part of the jet as the coal is devolatilizing. Increasing the char conversion in the early part of the reaction zone will tend to increase the overall carbon conversion. Also, typical experimental carbon mass balance closures ranged from 90 to 105% indicating that an error on the order of 10% was not uncommon.

Predicted CO concentration distributions within the gasifier are shown in Fig. 4. The CO concentration is increasing along the length of the reactor up to 1.0 m and then decreases toward the outlet. The CO concentration increases due to the char gasification reactions. The water gas shift reaction reduces CO (reforming $CO_2$) at the low gas temperatures around the outlet. Once the volatile reactions are complete, the water gas shift redistributes the products as they are formed in char gasification reactions. As shown in Fig. 5, the gas composition approaches equilibrium after 0.75 m although the shift is too slow to reach equilibrium in the reactor at temperature less than 1500 K.The CO concentration increases due to the char gasification reactions. The water gas shift reaction reduces CO, reforming more $CO_2$, at the low gas temperatures around the outlet. Once the volatile reactions are complete, the water gas shift redistributes the products as they are formed in char gasification reactions. As shown in Fig. 5, the gas composition approaches equilibrium after 0.75 m although the shift is too slow to reach equilibrium in the reactor at temperature less than 1500 K.

The experimentally measured and predicted, dry basis,

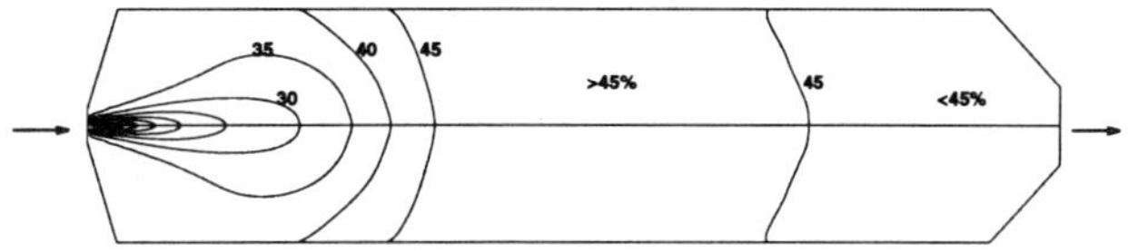


Figure 4: CO (vol% dry) in the laboratory gasifier

gasifier outlet gas concentrations are given in Table 3. The good agreement between predictions and experiment indicate that the carbon conversion is reasonable and that the predicted exit temperature is probably close to the outlet temperature. However, without direct measurement, it is difficult to conclude on the validity of the gasification model.

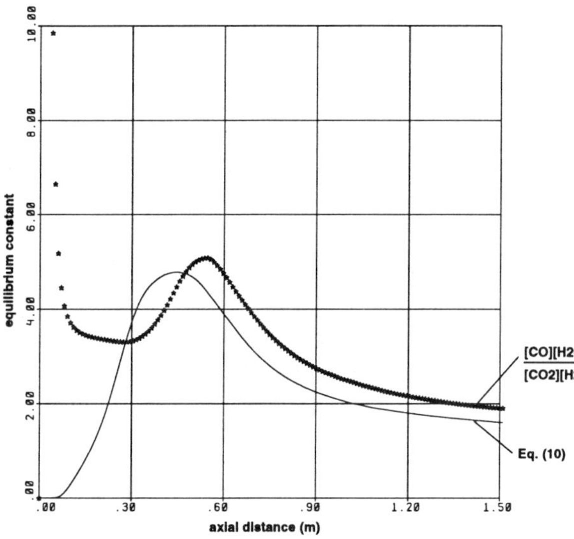


Figure 5: Gas equilibrium in the laboratory reactor

## Full Scale Gasifier

The coal/steam/$O_2$ jet expanded to fill the reactor 5 m downstream from the inlet, generating a large external recirculation zone similar to that observed in the experimental gasifier.

Temperature contours in the full scale gasifier are shown in Fig. 6. A cold central jet persists for almost half the distance of the reactor. This was due to the large quantity of water in the feed that must be evaporated before the coal can react. There is a rapid temperature rise after 4 m where devolatilization and char gasification begin almost simultaneously. The maximum temperature of 2000 K is reached after 7 m where volatile combustion is complete. Unlike the laboratory unit the temperature only decreases slightly, to 1970 K at the outlet due to the much higher mass through put and the lower wall heat loss.

Table 3: Outlet Composition of Gaseous Products

| Gas (mol% dry) | Laboratory Exp. | Laboratory Predicted | Full scale Reported | Full scale Predicted |
|---|---|---|---|---|
| $CH_4$ | 0.9 | 0.0 | 0.6 | 0.0 |
| CO | 43.1 | 43.5 | 38.8 | 44.9 |
| $CO_2$ | 26.7 | 25.3 | 26.0 | 29.0 |
| $H_2$ | 15.2 | 17.4 | 34.6 | 26.1 |
| Inert | 13.1 | 13.7 | – | – |
| Other | 1.0 | – | – | – |

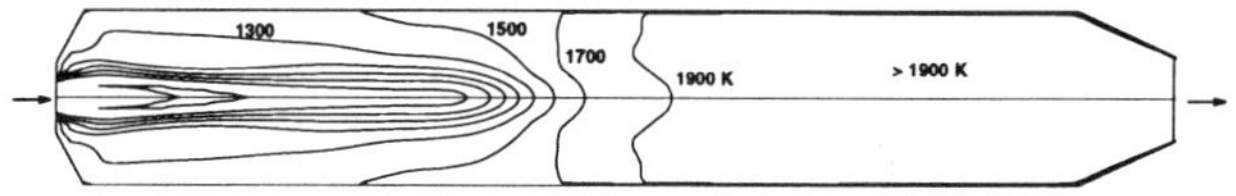


Figure 6: Temperature in the full scale gasifier

The char reaction proceeds to completion in the large scale simulation due to the high gas temperature. Although the C/O feed ratio was higher in the full scale, the higher temperatures, lower heat loss and much higher water vapour concentrations compensate to increase the carbon conversion. The CO concentration increases up to the outlet as shown in Fig. 7. The low concentration of CO in the upstream half of the reactor demonstrates that the preheat and char reaction zones are completely separated in the reactor. Compared to the laboratory gasifier, a relatively small fraction of CO and high fraction of $H_2$ is produced in the final gas due to the high concentration of water. The reported (Kovacik, 1988) and predicted, dry basis, gasifier outlet gas concentrations are given in Table 3 and the calculated and theoretical equilibrium gas compositions are compared in Fig. 8. The gas composition started far from equilibrium due to the high concentration of injected water. However, the gas composition rapidly approached equilibrium toward the outlet at the high gas temperatures ($> 1900$ K).

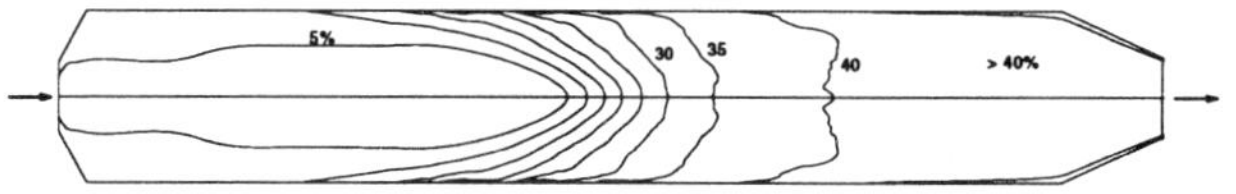


Figure 7: CO in the full scale gasifier

The predicted gas composition is in reasonable agreement with reported values for commercial gasifiers firing subbituminous coal. However, the reported values are those expected after gas cooling and are not actual gasifier outlet concentrations. Commercially, this gas cooling is achieved through a combination of radiative and convective heat exchange and finally a rapid quench with water. The water gas shift reaction would still be significant in the relatively long residence time radiation and convective coolers. A cooler exit temperature in the simulation would bring the predicted values in line with the reported results. Also, it is known that commercial coal gasifiers operate at pressures from 15 to 35 atmospheres. This simulation was performed at atmospheric pressure as the actual reactor pressure was not known. Considering the limited information available for commercial scale coal gasification in terms of reactor geometry and boundary conditions the model seems to provide a reasonable insight as to the conditions that could be expected inside a large scale gasification reactor. To improve the relevance of the modeling to full scale gasifiers, inputs and internal temperature and gas composition data are required.

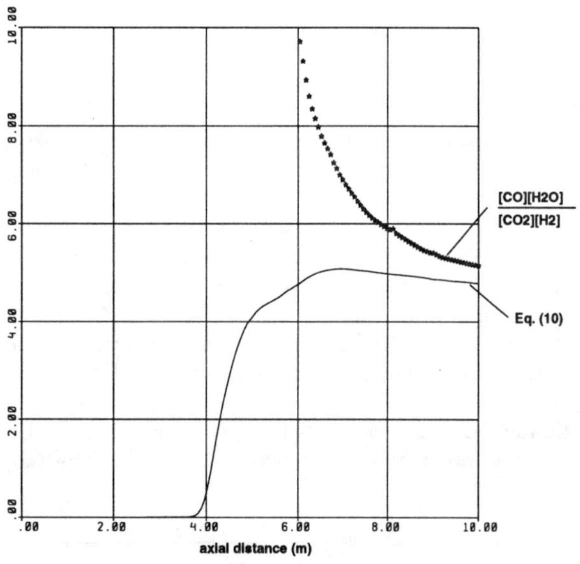


Figure 8: Gas equilibrium in the full scale reactor

## SUMMARY AND CONCLUSIONS

A coal gasification prototype model was developed and implemented in a commercial software package and used to predict a laboratory scale gasifier. The model was able to predict the observed heating, devolatilization and outlet gas composition in the reactor. The model has provided an insight into the processes occurring during gasification which can be used to independently evaluate new gasification technologies.

Future modifications to the model are planned to model char oxidation and volatile gasification. Both processes are considered necessary to model gasification processes occurring over a wider range of C/O ratios. The chemistry will also be extended to include higher reactor pressures and N and S species to complete the prediction of the gas composition.

## ACKNOWLEDGEMENTS

The authors acknowledge the support of Alberta Power, Edmonton Power, Nova Scotia Power, Ontario Hydro, Sask Power, TransAlta Utilities, the governments of Alberta Nova Scotia and Canada and to E. Furimsky of CANMET for providing experimental data.

## REFERENCES

Macdonald, D. E., Kleta, J. K. and MacKenzie R. C., 1992, "Clean-coal Technology Developments in Alberta, Canada" *International Journal of Environmental Issues in Minerals and Energy Industry*, pp. 69-67.

Blamire, K., Nova Scotia Power, Personal communication, 1993.

Kovacik, G. J., Chambers, A. K., 1988, "Evaluation of Canadian Coals for Gasification with Texaco, British Gas, and Shell Technologies", contract report to the Canadian Coal Gasification Technical Committee.

Knill, K. J., Chui, E. and Hughes, P. M., 1993, "Application of Grid Refinement in Modeling a Tangentially Fired Coal Boiler", *1993 International Joint Power Generation Conference*, Kansas City, October 17-20, 1993.

Faeth, G. M., (1987), "Mixing, Transport and Combustion in Sprays", *Prog. Energy Combust. Sci*, Vol 13, pp. 293-345.

Knill, K. J., Maalman, T. F. J. and Morgan, M. E., 1989, "Development of a Characterization Technique for High Volatile Bituminous Coals", Doc. No. F 88/a/10, International Flame Research Foundation, IJmuiden, the Netherlands.

Magnussen, B. F. and Hjertager, B. H., 1976, "On Mathematical Modeling of Turbulent Combustion with Emphasis on Soot Formation and Combustion", *Sixteenth Symposium (International) on Combustion*, The Combustion Institute, Pittsburgh, pp. 719-729.

Baxter, L. L., (1987) "Condensed Phase Behaviour in Combustion Environments", Ph.D. Dissertation, Brigham Young University, Provo, Utah.

Chen, W. J., Sheu, F. R. and Savage, R. L. (1987) "Catalytic Activity of Coal Ash on Steam Methane Reforming and Water-Gas Shift Reactions", *Fuel Processing Technology*, Vol 16, pp. 279-288.

FACT-Vol. 18, Combustion Modeling, Scaling and Air Toxins
ASME 1994

# MODELLING OF A 275 kW GENERIC COAL FIRED BURNER: TWO AND THREE DIMENSIONAL PREDICTIONS COMPARED TO EXPERIMENTAL DATA

**C. N. Eastwick, A. P. Manning, S. J. Pickering, and A. Aroussi**
Department of Mechanical Engineering
University of Nottingham
Nottingham, United Kingdom

## ABSTRACT

The design of pulverised coal Low NOx burners increasingly incorporates some mathematical modelling, although 'rule-of-thumb' design and experimental work exceeds this. Due to time and computing constraints industry models these burners by two dimensional predictions, however the burners often have three dimensional geometry. These two dimensional predictions are often used to compute NOx emissions and are then denigrated because they do not match experimental data.

This paper describes the comparison of two and three dimensional computational fluid dynamics predictions with experimental data for a generic three dimensional laboratory scale burner.

The numerical simulation was performed using a commercial CFD code. The modelled burner has coal segregators in the coal laden primary air inlet and this causes angular concentrations of both coal particles and air velocities.

Comparisons of axial velocity, temperature and oxygen concentrations were made at several ports for two and three dimensional predictions and experimental data.

The conclusion of this paper is that the inclusion of three dimensional inlet conditions causes significant differences in the temperature and oxygen concentrations between different azimuthal planes in the near burner region. The axial velocity displays significant axial variance in the first 0.35 burner diameters downstream of the quarl exit.

As the oxygen concentration and temperature fields are used to calculate NOx emissions any inaccuracy in their prediction is important. Therefore the use of two dimensional calculations to predict NOx for three dimensional burners is suspect.

## INTRODUCTION

The design of pulverised coal Low NOx burners increasingly incorporates some mathematical modelling, although 'rule-of-thumb' design and experimental work exceeds this. Due to time and computing constraints industry models these burners by two dimensional predictions, however the burners often have three dimensional geometry. These two dimensional predictions are often used to compute NOx emissions and are then denigrated because they do not match experimental data.

This paper describes work that was carried out to validate two and three dimensional models of a single swirl stabilised burner firing into a cylindrical furnace. This work was part of a collaborative project on mathematical modelling of power station boilers. The project was funded by the Department of Trade and Industry and involved the UK power generation companies PowerGen and National Power, with the Harwell Combustion Centre and the Universities of Nottingham and Wales, College Cardiff.

Previous work pertinent to this study was recently reviewed by Brewster et al (1993). Notable contributors to this field include Lockwood and Mahmud (1988), Visser and Weber, (1990) Smoot et al (1984), Truelove and Williams (1988) and the combustion centre at AEA Harwell (Stopford and Marriott, 1990).

The commercial computational fluid dynamics code FLOW3D, (CFDS, 1992) and its radiation counterpart RAD3D, (Guilbert, 1989), were used for the mathematical modelling, while the experimental data was supplied by PowerGen, (Cunningham et al, 1991). The experimental test rig considered was the Combustion Test Facility, a cylindrical furnace at a now closed PowerGen laboratory.

This paper describes the computational and experimental method and compares experimental and predicted results.

The burner investigated is a generic swirled wall fired burner (Fig 1). The burner has three dimensional geometric features in the primary air stream, which carries the coal, and this leads to a three dimensional distribution of coal particles and primary air velocity.

## EXPERIMENTAL DETAILS

The burner modelled is a swirling pulverised coal burner (Fig 1). The burner has three air inlet streams, primary, secondary and tertiary. The primary air stream carries the pulverised coal. In the primary pipe of the burner axial swirl vanes are used to ensure that the coal particles are thoroughly mixed with the primary air. The swirl is then removed by coal segregators close to the burner exit. These coal segregators take the form of 'cups' every 90° as shown in Fig 2. The coal segregators cause the coal to be concentrated in four streams, this creates fuel rich areas and hence reduces NOx emissions. The secondary and tertiary flows are swirled by axial vanes set at 68° and 41° (swirl nos 2.18 and 0.78) from equation of swirl from a nozzle (Gupta et al, 1984). A central blockage, which includes an oil injector for ignition, is present. The quarl is conical with an opening angle of 25°.

This burner was fired singly in the CTF with a thermal load of 275 kW. The CTF furnace was a roughly cylindrical refractory lined furnace (Fig 3). Measurements of the inlet velocities and temperatures were taken together with the coal's proximate and ultimate analysis. These are listed in Tables 1 and 2. The particle size distribution used for this experiment is typical of pulverised coal fired in large UK utility boilers, ie 70% passing through a 76 micron mesh. Data within the furnace was obtained through flame sampling access ports. Axial velocity and temperature were recorded for five ports and gas composition for six. Data was recorded along a two dimensional traverse from the furnace wall to the flame centreline. The angular position of the burner with respect to this traverse was not recorded. Therefore, the relation of the coal segregators and particle concentration to the experimental results is unknown. The axial velocity measurements were made using a water cooled, double-headed pitot probe, whilst the flame temperatures were recorded by a water-cooled suction pyrometer probe. The gas analysis was performed by the use of a special spray-cooled probe which quenched reactions at the entry to the probe.

Unfortunately no data for the flue gas or wall temperatures were recorded for this experiment.

The axial position of the ports were as follows: 0.35, 1.3, 2.26, 3.21, 4.16 and 6.82 burner diameters downstream of the quarl exit, where the burner diameter was 179 mm.

## COMPUTATIONAL METHOD

FLOW3D (CFDS, 1992) is a fully viscous Navier-Stokes solver which uses a non-staggered grid to solve Eulerian flow equations. The turbulence closure model used for this work was the k-$\epsilon$ two equation model. The coal particle tracks were computed using Lagrangian co-ordinates. The grids used were body-fitted.

Coal devolitilisation was modelled using a single reaction model, (Badzioch and Hawskley, 1970) and char oxidation by a simple surface reaction model (Field et al, 1967). The parameters used for these models are shown in Table 3. Values for devolatilisation constants were taken from Truelove and Jamaluddin (1986) and the char constants from Wall et al (1986). These were values recommended by Harwell. As coal specific data for these constants were unavailable they were treated as computing parameters. The gaseous combustion was modelled by a conserved scalar approach (Bilger, 1980) with the use of a Beta probability density function.

Radiation was computed by RAD3D (Guilbert, 1989), a separate code that interfaces to FLOW3D during calculations. RAD3D allows the user to calculate radiation by either the Monte Carlo, (Howell, 1968), or discrete transfer method, (Lockwood and Shah, 1981). For this work the Monte Carlo method was employed due to code restrictions at the time of calculation. The code problems have now been solved and present calculations use the discrete transfer method in preference to the Monte Carlo method because of time savings.

Both two and three dimensional computations were undertaken for this work. The inlet conditions used for the models are listed in Table 1, whilst Table 2 gives details of the coal fired. The profiles for primary, secondary and tertiary axial velocities were obtained from previous validated work, (Eastwick, 1992) and are shown in Fig 4. The swirl velocity profiles for the secondary and tertiary air streams were assumed to be plug flow because axial swirl vanes had been used (Gupta et al, 1984). The three dimensional model used primary inlet conditions computed by PowerGen (Read, 1992), for the primary axial velocity, as well as particle distribution and velocities.

The axis of the burner was modelled as a symmetry plane and periodic planes used at the azimuthal boundaries to account for swirl. The exit of the furnace (Fig 3) was assumed to be a symmetrical radial exit. This exit was described by a mass flow (Neumann) boundary. The flue gas and wall temperatures were not recorded during the experimental work, therefore the flue gas temperature was taken from a similar experiment in the same furnace. The wall temperatures were then calculated from the known thermal rating of the burner and regarded as a computing parameter.

The gradient and absolute value of the heat flux along the furnace wall is vital in establishing the flame shape. Too much heat removed close to the burner can lead to the delay of ignition and alter the flow field. Although the total flux was known, the gradient of the heat flux had to be assumed, and this may have caused inaccuracies in the prediction of the temperature field.

FLOW3D uses a stochastic particle turbulence model which requires the use of a very large number of particles. Due to time and computing restraints it was not possible to use particle turbulence for these calculations. For the same reason a small representative number of particles were used for the calculations. The grids used in this work were investigated for grid independence during previous validation work (Eastwick, 1992) and can therefore be assumed to be grid independent. The grids used were; for the two dimensional case 3,173 cells and for the three dimensional case 25,384 cells. As the grids were body fitted the division of cells between burner, quarl and furnace are described in greater detail in Table 4.

## THREE DIMENSIONAL EFFECTS

The three dimensionality present in the burner is due to the coal segregators in the primary air inlet alone. This causes azimuthal distributions of coal particles and velocities. The distributions of particles and velocities were not measured but were computed by PowerGen (Read, 1992). The distributions were concentrated approximately in 30° sectors every 90°. Three dimensional computations using the inlet values obtained by PowerGen were calculated. The distributions of velocities and particles used are shown in Figs 5 and 6.

For the three dimensional model a 90° sector was modelled with nine cells for the three dimensional predictions. Due to time restrictions only one three dimensional calculation was possible, so eliminating any possibility for grid independency tests or investigation of three dimensional inlet profiles. This is a recognised limitation in the work published here.

## RESULTS AND DISCUSSIONS

Comparisons between two and three dimensional predictions and experimental data were made. For clarity only two planes of the three dimensional predictions have been presented here. These planes are at 45° and 80° and represent the extremes of predictions.

The position of the planes in relation to the coal segregators are shown in Fig 2.

### Comparison of Two and Three Dimensional Predictions

The axial velocity predictions for the two and three dimensional results are very similar for all ports, with the only difference being seen in the first port. The three dimensional axial velocity contours for the first port are shown in Fig 7. This figure represents a 90° sector looking on to the end of the burner. Small differences in magnitude are discernable through the field, however the same plot for port two (Fig 8) exhibits very little difference in magnitude through the 90° sector.

Figs 9 and 10 display contours of temperature and oxygen concentration looking onto a 90° sector of port one. Radial profiles of temperature and oxygen concentrations for port two are shown graphically in Figs 11 and 12, together with two dimensional and experimental results. From these figures the differences between two and three dimensional results are apparent. The temperature contours at ports three and five (Figs 13 and 14) show the gradual smoothing of the large azimuthal differences seen in port one, with the temperature contours at port five showing no significant three dimensional effects. However, the oxygen concentration contours at port five still show some three dimensionality (Fig 15).

The graphical comparisons of two dimensional and three dimensional at port three for temperature and port four for oxygen concentrations (Figs 16 and 17) show the differences between two and three dimensional predictions in greater detail. The greatest differences are seen between the two dimensional and three dimensional (at 45°) predictions with differences in the IRZ for temperature being of the order 40-50 K.

From these figures it can be concluded that the inclusion of three dimensional inlet conditions has a significant effect on the temperature and oxygen concentration field in the near burner region. This presence of three dimensionality for temperature and gas composition in the near field has an important impact on NOx predictions. Both the temperature and oxygen concentration fields are used in the computation of NOx levels, therefore their accurate calculation is necessary for NOx predictions. Unfortunately NOx calculations could not be made at the same time as these computations due to code restrictions. However, it is hoped that these will be published at a later date.

### Comparison of Predictions with Experimental Results

The experimental axial velocity measurements were recorded using a pitot probe and because of problems encountered during the experiment were known to be unreliable (Cunningham et al, 1991). For this reason no comparison between predictions and experimental results for axial velocity are presented here.

Fig 11 shows the temperature predictions and experimental data. From this figure it can be concluded that the two dimensional results are not in good agreement with experimental data. The best agreement is achieved by the three dimensional result at 80°. This is also true for the oxygen concentrations (Fig 12). The temperature plot at port three, Fig 16, shows a deterioration of both two and three dimensional predictions with experimental data. This could be for a number of reasons, not least the assumed wall temperature profile. There could also be experimental error as acknowledged in the experimental report (Cunningham et al, 1991). The oxygen concentration plot in Fig 17 emphasises the experimental uncertainty.

From these results it can tentatively be concluded that the two dimensional traverse was closer to the 80° plane than 45°. Fig 17 does not agree with this conclusion as the three dimensional result at plane 45° shows better agreement with the experimental data. However the experimental scatter is significant. It can also be concluded that the two dimensional modelling of a burner that has three dimensional inlet conditions is inadequate.

## CONCLUSIONS

From this work it can be concluded that three dimensional effects in coal and velocity inlet distributions can lead to three dimensionality in the near burner region. For the burner investigated here the angular differences extended to approximately five burner diameters downstream of the quarl exit and affected both the temperature and oxygen concentration field. The presence of three dimensional effects in the temperature and gas composition in the near burner region would have a significant impact on the prediction of NOx. Unfortunately NOx predictions were unavailable due to code restrictions but may be published at a later date.

Experimental data was available at one plane whose angular position with respect to the coal segregators was unknown. It is therefore difficult to validate the three dimensional predictions and draw a firm conclusion as to the angular position of the experimental plane. However, it can be noted that the closest agreement to experimental data was achieved by the three dimensional plane at 80°.

Possible reasons for the discrepancies between experimental and predicted results are the assumed heat flux boundary conditions, the assumed flue gas temperatures, the constants used in the devolatilisation model and experimental error. With greater experimental detail it would be worthwhile investigating prediction discrepancies, however with the limited experimental information this work is restricted to an indication of trends.

The most important conclusion of this work is that the two dimensional modelling of a burner with three dimensional inlet conditions is insufficient. This is of particular importance as industrial burner manufacturers often perform two dimensional calculations to compute NOx emissions. A two dimensional model can lead to inaccuracies in predicted temperature and gas composition field and therefore in predicted NOx.

## REFERENCES


Badzioch S and Hawskley P G W, 1970, 'Kinetics of thermal decomposition of pulverised coal particles', *Industrial Engineering Chemical Process Design and Development*, Vol 9, No 4, pp 521-530.

Bilger R W, 1980, 'Turbulent flows with non-premixed reactants', *Turbulent Reacting Flows*, Eds R A Libby and F A Williams, Spring-Verlag, 1980.

Brewster B S, Hill S C, Radulovic P T and Smoot, L D, 1993, 'Comprehensive modelling', *Chapter 8 of Fundamentals of Coal Combustion*, ed L D Smoot, Elsevier.

CFDS, 1992, FLOW3D Release 3.2 User Manual.

Cunningham A T S, Hoadley D, Matthews K J, 1991, 'FMF Low NOx Burner Tests in The Combustion Test Facility', *TR/91/22001/M*.

Eastwick C N, 1992, 'Mathematical modelling of power station boilers', *Technical Report*, University of Nottingham.

Field M A, Gill D W, B Morgan, Hawksley P G W, 1967, 'Combustion of Pulverised Coal', *BCURA*.

Guilbert P, 1989, 'Computer program RAD3D for modelling thermal radiation', *Technical Report AERE-R13534, HCCP/R31/1989*.

Gupta A K, Lilley D G and Syred N, 1984, 'Swirl Flows', Abacus Press.

Howell J R, 1968, 'Monte Carlo application to heat transfer', *Advances in Heat Transfer*, Vol 5, pp 2-54.

Lockwood F C and Mahmud T, 1988, 'The prediction of swirl burner pulverised coal flames', *22nd Symp (Int) on Combustion,* The Combustion Institute, pp 165-173.

Lockwood G C and Shah N G, 1981 'A new radiative solution method for incorporation in general combustion prediction procedures', *18th Symp (Int) on Combustion*, The Combustion Institute, Pittsburgh, pp 1405-1414.

Read A J, 1992, 'Calculation of the pf distribution at the exit of an FMF low NOx burner', *PT/92/220059/M*.

Stopford P J and Marriott N, 1990, 'Modelling the CANMET and Marchwood Furnaces using the PCCC code', *Proc of the Seminar on Harwell Coal Combustion Program*, p 59.

Truelove J S and Jamaluddin A S, 1986, 'Models for rapid devolatilisation of pulverised coal', *Combustion and Flame*, Vol 64, pp 369-372.

Truelove J S and Williams R G, 1988, 'Coal combustion models for flame scaling', *22nd Symp (Int) on Combustion*, The Combustion Institute, pp 155-161.

Smoot L D, Hedman P O and Smith P J, 1984, 'Pulverised coal combustion research at Brigham Young University', *Prog Energy Comb Sci*, Vol 10, pp 359-441.

Visser B M and Weber R, 1990, 'Predictions of near burner zone properties of six swirling pulverised coal flames', *IFRF Rep No F36/y/14.*

Wall T F, Phelan W J and Bortz S, 1986, 'Coal burnout in the IFRF No 1 furnace', Combustion and Flame, Vol 66, No 2, pp 137-156.

| PARAMETER | VALUE |
|---|---|
| COAL FLOW (kg/h) | 35.7 |
| PRIMARY AIR TEMPERATURE (C°) | 100 |
| SECONDARY AIR TEMPERATURE (C°) | 300 |
| EXCESS OXYGEN IN FLUE (%) | 3 |
| AIR FUEL RATIO OVERALL | 10.02 |
| TOTAL AIR FLOW (kg/h) | 357.7 |
| PRIMARY AIR/FUEL RATIO | 1.76 |
| PRIMARY AIR FLOW (kg/h) | 62.8 |
| TOTAL SECONDARY AIR FLOW (kg/h) | 294.4 |
| TERTIARY AIR FLOW (kg/h) | 224.8 |
| SECONDARY AIR FLOW (kg/h) | 70.1 |
| GROSS CALORIFIC VALUE OF COAL (MJ/kg) | 27.73 |
| OVERALL THERMAL INPUT (kW) | 275 |

Table 1 Inlet Parameter

| | |
|---|---|
| ASH DRY % | 18.9 |
| VOLATILES DRY % | 29.6 |
| VOLATILES DAF % | 36.5 |
| SULPHUR DRY % | 1.11 |
| CARBON DRY % | 67.64 |
| HYDROGEN DRY % | 4.32 |
| OXYGEN DRY % | 6.25 |
| CHLORINE DRY % | 0.33 |
| NITROGEN DRY % | 1.45 |
| NITROGEN DAF % | 1.78 |
| GROSS CV MJ/kg (dry basis) | 27.73 |

Table 2 Coal Analysis

| | | |
|---|---|---|
| Devolatilisation | Pre-exponential Constant | 20,000.0 $s^{-1}$ |
| Devolatilisation | Activation Temperature | 5941.0 K |
| Char oxidation | Rate Constant | 497.0 $kg/m^2s$ atmos |
| Char oxidation | Activation Temperature | 8540.0 K |

Table 3 Model Constants

| | Grid | | | | | |
|---|---|---|---|---|---|---|
| | 2D | | | 3D | | |
| | NI | NJ | NK | NI | NJ | NK |
| Burner | 27 | 28 | 1 | 27 | 28 | 8 |
| Quarl | 8 | 28 | 1 | 8 | 28 | 8 |
| Furnace | 51 | 43 | 1 | 51 | 43 | 8 |
| Total | 4073 | | | 25384 | | |

Table 4 Grid Parameters

* Where NI, NJ and NK are the number of cells in the axial, radial and azimuthal direction respectively

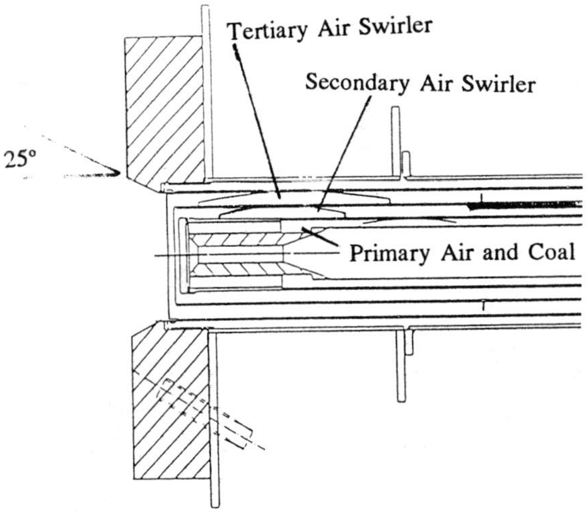


Figure 1 Burner Cross-Section

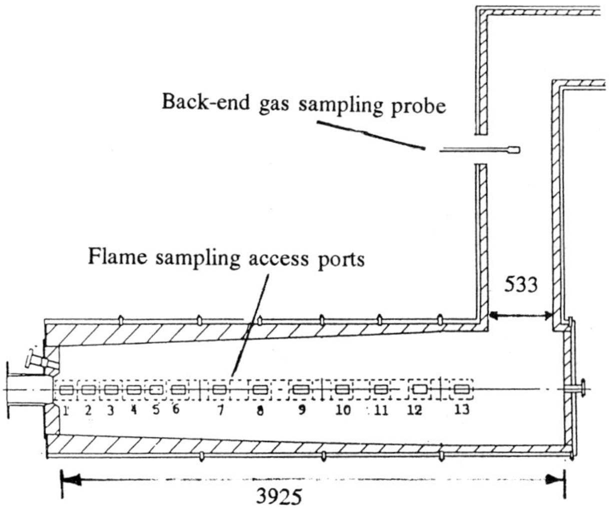


Figure 3 The Combustion Test Facility

All dimensions in mm

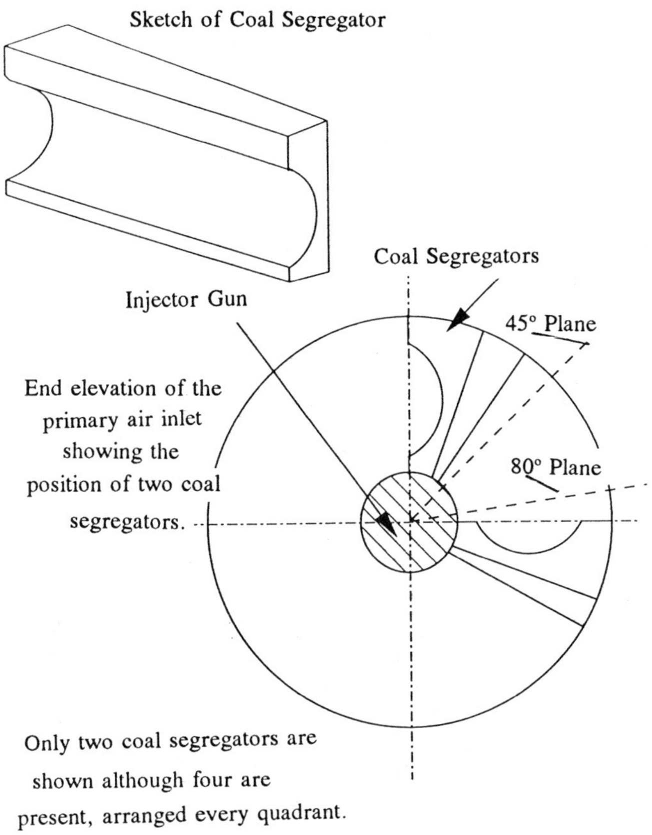


Figure 2 Coal Segregators

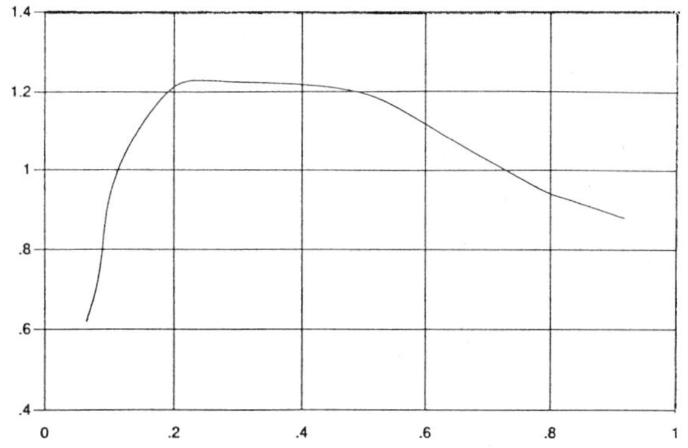


Figure 4 Axial Velocity Inlet Profiles for Primary, Secondary and Tertiary Air Streams

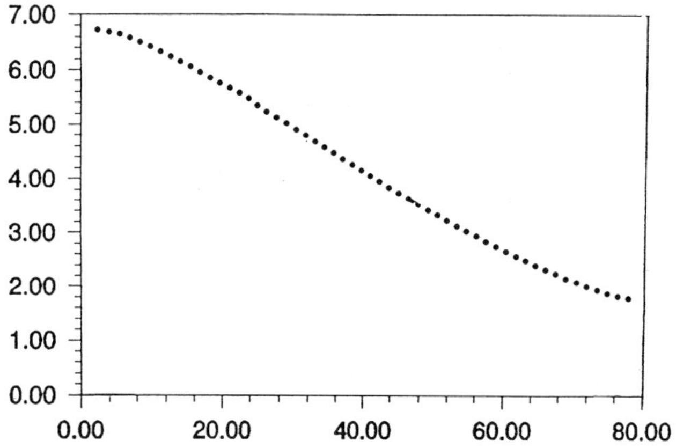


Fig 5 Primary Axial Velocity at Burner Exit

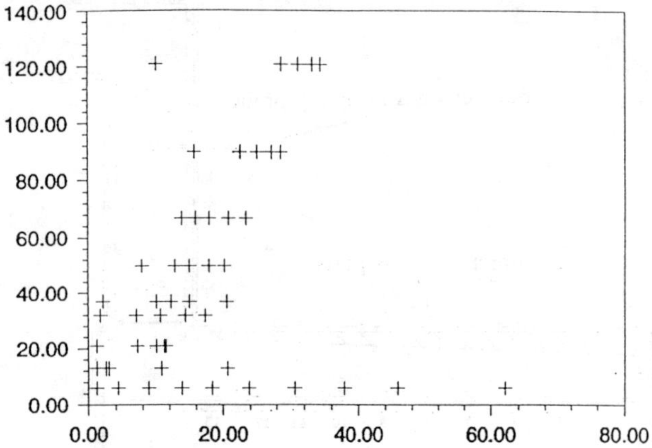


Figure 6 pf Size Distributions at Burner Exit

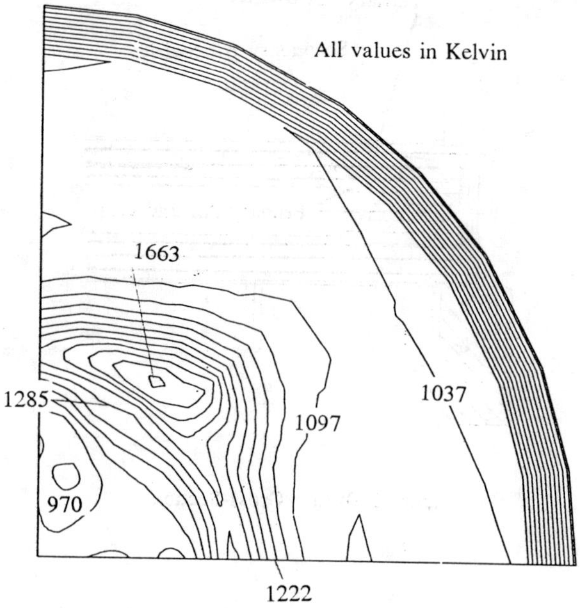


Figure 9 Temperature Contours Port 1

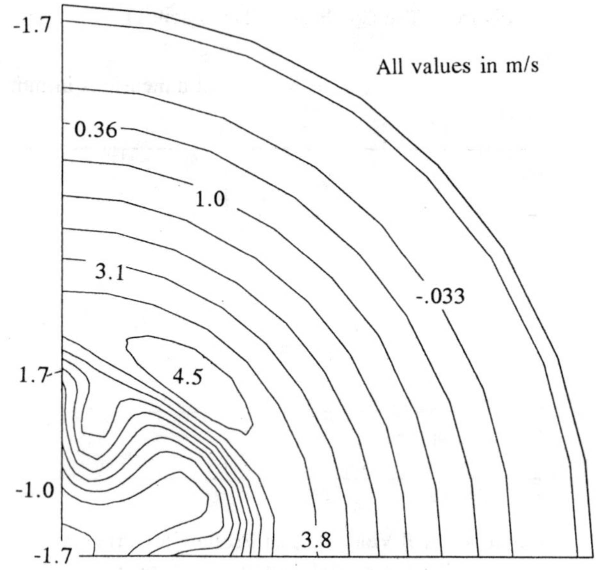


Figure 7 Axial Velocity Contours Port 1

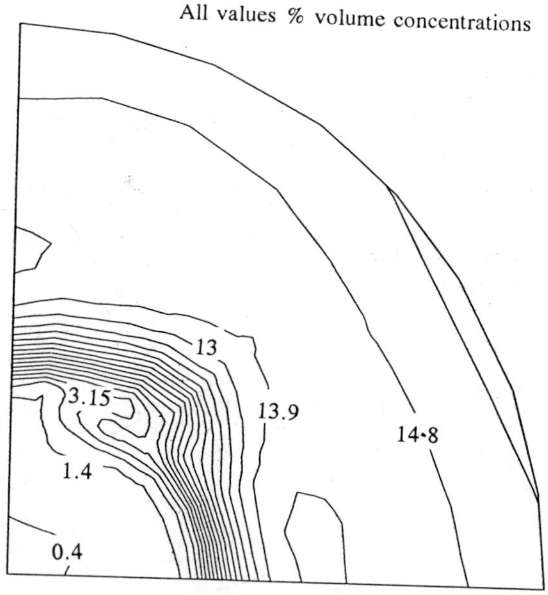


Figure 10 Oxygen Concentration Contours Port 1

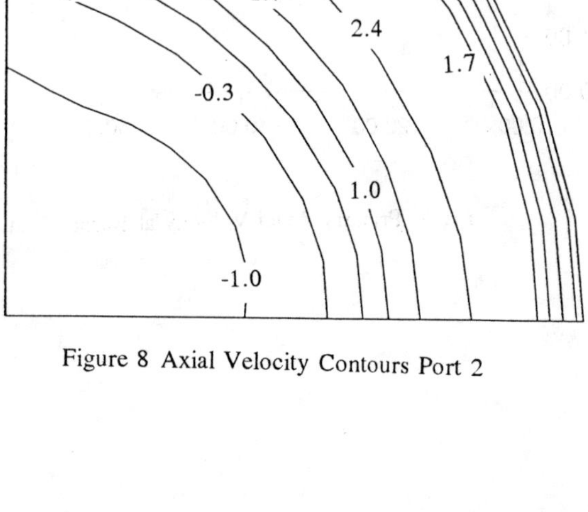

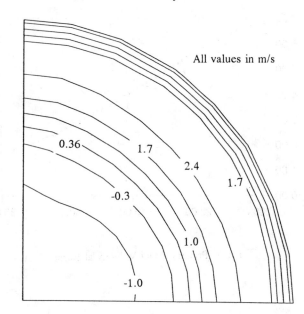


Figure 8 Axial Velocity Contours Port 2

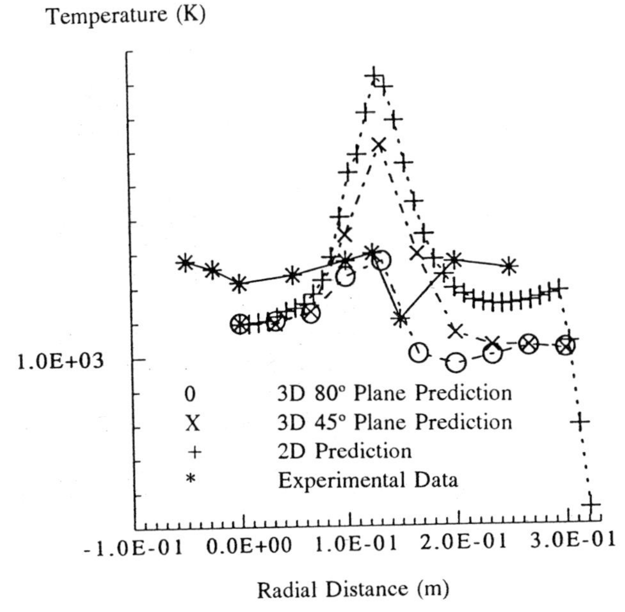


Figure 11 Radial Profiles of Temperature at Port 1

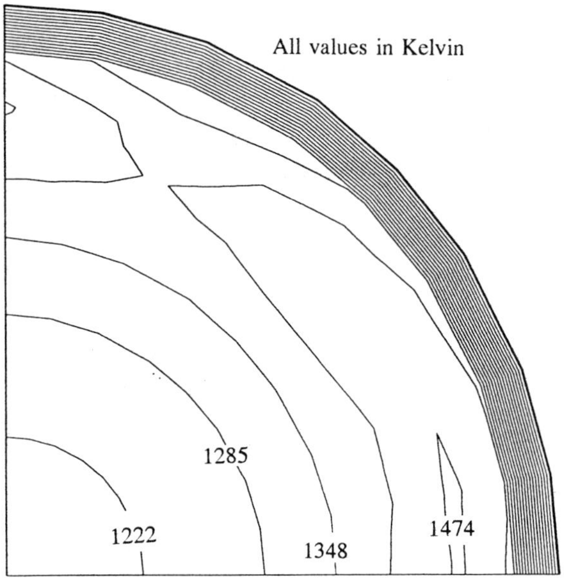


Figure 13 Temperature Contours Port 3

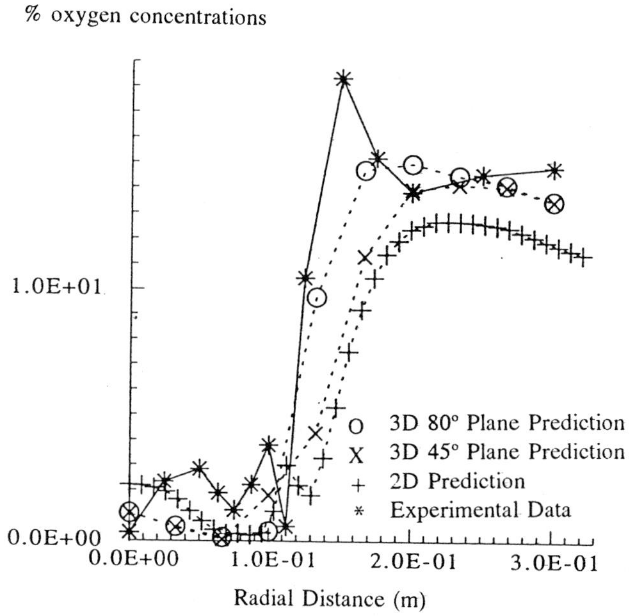


Figure 12 Radial Profiles of Oxygen Concentrations at Port 1

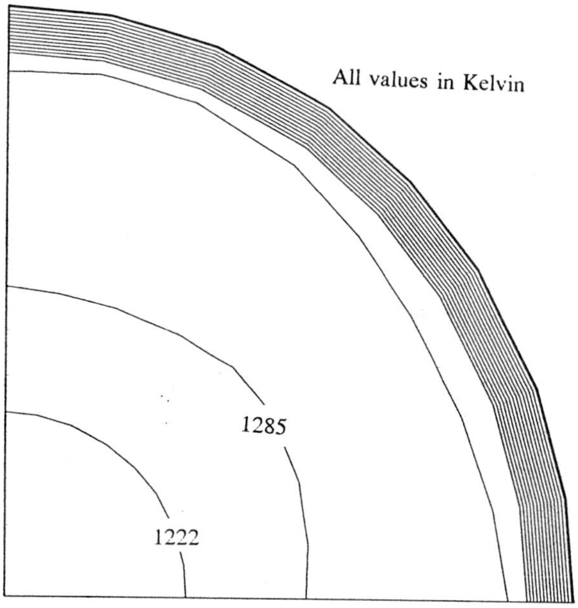


Figure 14 Temperature Contours Port 3

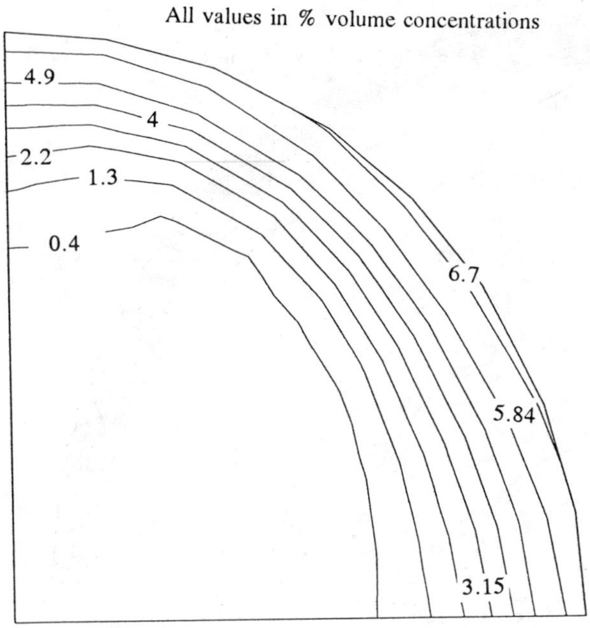


Figure 15 Oxygen Concentration Port 5

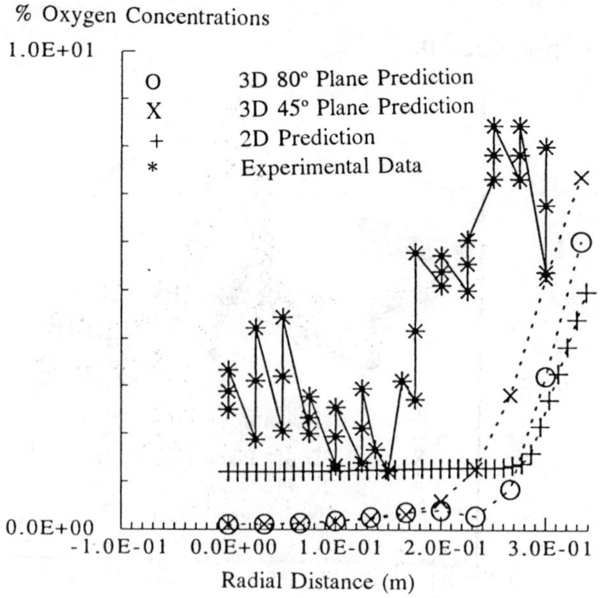


Figure 17 Radial Profiles of Oxygen Concentrations at Port 4

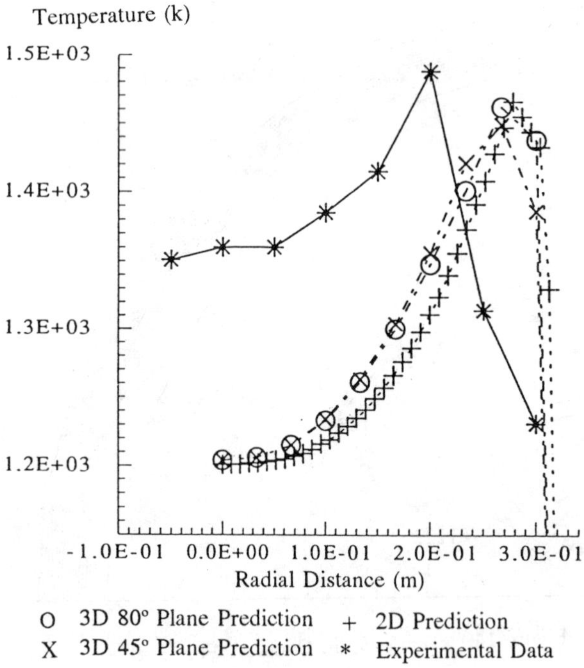


Figure 16 Radial Profiles of Temperature at Port 3

FACT-Vol. 18, Combustion Modeling, Scaling and Air Toxins
ASME 1994

# SCALE UP AND MODELLING OF GAS REBURNING

Roy Payne
David K. Moyeda
Energy and Environmental Research Corporation
Irvine, California

## ABSTRACT

The supplemental use of natural gas in a reburning mode has been developed as a retrofit technology for the control of $NO_x$ emissions from coal fired utility boilers. The technique involves combustion staging, achieved through the injection of natural gas at an elevation above the main burner zone, followed by the subsequent use of overfire air to complete burn out. In recent demonstrations of the technology on full–scale utility boilers, $NO_x$ reduction levels greater than 70% have been reported with the use of up to 20% natural gas. Designs for the full–scale application of gas reburning have been based largely on an extensive experimental data base, coupled with the application of a variety of process and performance modelling techniques. This paper summarizes the data obtained in recent demonstrations of gas reburning on utility boilers, and compares results with expectations derived from experimental data. In general, when site specific constraints and boundary conditions are taken into account, a good correspondence between experimental and boiler data has been found. In addition, various modelling techniques have shown some success in predicting full–scale performance; although the development of improved models for coupling turbulent mixing and reaction would lead to a better understanding of some results observed in the full–scale trials.

## INTRODUCTION

As a result of the 1990 Amendments to the Clean Air Act, operators of utility boilers in the United States are implementing various strategies to control $NO_x$ emissions. Part of this strategy can involve the selective cofiring of natural gas. When the natural gas is injected in a "reburning" mode, $NO_x$ emissions generated in the main firing zone can be significantly reduced. Full–scale demonstrations of this technology have shown that the emissions reductions achievable are higher than those typically attainable with other technologies, such as low $NO_x$ burners and selective non–catalytic reduction.

Designs for the full–scale application of gas reburning have been based largely on an extensive small–scale experimental data base, coupled with the application of a variety of process and performance modelling techniques. Experimental studies at bench and pilot scales have served to elucidate the important process controlling parameters, whist various modelling techniques have allowed extrapolation of the data base to full–scale boiler performance.

This paper reviews the reburning process, the results of three demonstrations of the process at full scale, and compares the full–scale results with expectations derived from experimental data. The role of computational modelling in the scale up and design of full–scale boiler systems is also presented and discussed, with reference to techniques for the modelling of process chemistry, mixing, and boiler thermal performance.

## REBURNING PROCESS

Reburning is a combustion modification technology which removes $NO_x$ from combustion products by using fuel as a reducing agent. This technology, based on the principle of Myerson (1957) that CH fragments can react with NO, is alternatively referred to as "in-furnace $NO_x$ reduction" or "staged fuel injection" and has been appreciated for over two decades. The term "reburning" was coined by Wendt, Sternling, and Matovich (1973) who reduced NO produced in a laboratory scale flat flame by adding methane or ammonia to the combustion products. Japanese industry has demonstrated $NO_x$ reduction at full scale with reburning (Takahashi et al., 1983), and several studies in the United States have recently been directed towards its application to coal–fired utility boilers (Folsom, Sommer, and Payne, 1991; Borio et al., 1991).

An overview of the reburning process as applied to a utility boiler is shown in Figure 1. In this process, a portion of the fuel is diverted from the main firing zone and is injected at an elevation above this zone. The injected fuel forms a slightly fuel rich "reburn" zone. In the reburn zone, hydrocarbon radicals, primarily CH, generated during breakdown of the reburn fuel react with NO molecules from the primary zone to form other nitrogenous species such as HCN. The HCN then decays through several reaction intermediates, NCO

$\rightarrow$ NH $\rightarrow$ N, and ultimately reaches $N_2$ via the reverse Zeldovich reaction:

$$N + NO \rightarrow N_2 + O$$

Following the reburn zone, overfire air is added to complete combustion of the reburn fuel and to bring the boiler to its normal operating stoichiometry.

The parameters which control the performance of the reburn process have been defined though an extensive series of experimental studies which are summarized by Chen et al. (1986), Greene et al. (1986), and Overmoe et al. (1986). These studies were performed on both a bench and a pilot scale reactor set up to simulate the typical thermal history in a coal fired utility boiler. The geometry of the reactors is illustrated in Figures 2 and 3. The bench scale reactor is a refractory lined combustion tunnel. The reactor is downfired with a total capacity of 25 kW. Natural gas can be fired in refractory channels surrounding the main chamber to control the temperature profile of the furnace. The pilot scale reactor was specifically designed for scale up and evaluation of the reburning process. The walls on this reactor are refractory lined and jacketed for water cooling. This reactor is downfired with a total capacity of 3.0 MW. Special cooling panels located on the walls of the reactor provide the ability to manipulate the temperature profile. Details of the experimental set up for the bench and pilot scale reactors and fuels tested can be found in the above references.

The studies of Greene et al. (1986), and Overmoe et al. (1986) have shown that the most critical parameters controlling the reburn process are the reburn zone stoichiometric ratio and the initial $NO_x$ level entering the reburn zone. The impacts of these variables on the $NO_x$ reduction achievable with reburning in the bench and pilot scale facilities are shown in Figures 4 and 5. As shown in Figure 4, most fuels exhibit a maximum $NO_x$ reduction at a reburn zone stoichiometry of about 0.9. Here, reburn zone stoichiometry is defined as the ratio of the total air supplied to the primary and reburn zones to the total stoichiometric air requirements of the primary and reburn fuels. Studies of the chemistry involved in the process suggest that the optimum occurs at this stoichiometry for gaseous fuels since this is the point at which the generation of CH radicals is highest. Driving the reburn zone more fuel rich does not generate additional radicals since insufficient oxygen is available to begin breakdown of the reburn fuel. For fuels containing fuel bound nitrogen, the optimum occurs at a stoichiometry of 0.9 due to tradeoffs between primary NO destruction and formation of HCN from the fuel–bound nitrogen. Figure 5 shows that higher initial $NO_x$ emissions lead to higher $NO_x$ reductions with reburning, and that final $NO_x$ emissions are a function of the initial $NO_x$ independent of the fuel type.

The reburn zone residence time and temperature are also important parameters. Figure 6 shows the influence that the injection temperature of the reburning fuel has on the process effectiveness (Chen et al., 1988). In general, higher temperatures and longer residence times are preferred. The practical implications of these requirements are that the reburning fuel must be placed as close as practical to the main flame, allowing for some separation between the main flame and point of reburning fuel injection, to ensure that the reburn fuel is injected at the highest possible temperature but not injected into the primary flame, and that the burnout air must be placed high enough in the furnace to provide for adequate residence time in the reburn zone, but low enough to permit complete combustion of the unburnt materials exiting the reburn zone. Generally, reburn zone residence times between 250 to 500 milliseconds at temperatures above 1371°C (2500°F) are preferred for most boiler applications.

Small scale studies have also shown that rapid and complete mixing of the reburning fuel and overfire air is also important to optimizing the process. Figure 7 presents the results of pilot–scale experiments where the reburning fuel injector type was varied from an in–furnace distributed injector, which provided rapid mixing and excellent distribution of the reburning fuel, to a single axial jet, which provided both slow mixing and poor distribution of the reburning fuel (Chen et al., 1988). As shown in Figure 7, significant variations in performance were observed for these two injection systems. A system simulating a typical full–scale installation, four cross flow jets, was also tested. Its performance was better than the single jet, but not as good as that observed with the in–furnace distributed injector, which illustrates the importance of good reburning fuel mixing and dispersion.

In summary, the fundamentals of the reburn process are well understood, and a large body of data exists concerning controlling parameters and their relative impacts on performance. In addition, to the data cited above, the existing database includes the effects of parameters such as: fuel type and composition; the distribution of residence times and temperatures, which represent the boundary conditions for process application; and the importance of mixing and mechanical features, which are key to scale up. As a result of recent full–scale boiler demonstrations, much of the understanding gained at laboratory and pilot experimental scales has been verified and refined, and reliable methodologies now exist for the design of retrofit applications and for performance prediction. Such methodologies relate not only to $NO_x$ reduction, but also to reciprocal impacts on boiler performance and overall operability. In this context, it is noteworthy that the nature of the reburning process, with its requirement for strict zone separation, coupled with definable zonal residence times and temperature distributions, lends itself more readily to computational modeling than do other combustion modification techniques for $NO_x$ control. Computational models therefore form a key and integral part of current process design methodologies.

## PROCESS SCALE UP

To apply the process to full–scale boilers, a scale up methodology involving the use of computational and physical models was developed (Payne et al., 1988). The first step of the process is the construction of a 1/10 to 1/15 scale physical flow model of the full–scale boiler. The physical model is used to study the aerodynamics of the boiler flow field and to determine mixing characteristics which are key to process performance. Preliminary designs for the injection systems are developed using empirical correlations for the jet trajectory and entrainment. These designs are studied in the flow model and optimized to provide effective dispersion and rapid mixing of the reburn fuel and overfire air with the main flow exiting the primary zone. This modelling component is particularly important in the design of full–scale retrofit application. Although computational fluid dynamics modelling can be applied to model the overall boiler flow field, physical flow modelling is the only technique sufficiently reliable for the determination of injection requirements (injector number, location and velocities) for complex three-dimensional flow situations such as those encountered in practice.

The next step in the design methodology is to evaluate the potential impacts of the reburn system on the boiler thermal performance. This is accomplished with the use of furnace heat transfer models which are used to predict boiler specific heat absorption and gas temperature profiles. Details of the thermal performance models are provided by Li (1993). The furnace heat transfer model is linked to a boiler cycle model to simulate the steam side performance. The models allow an assessment of the impacts of reburning application on boiler performance parameters such as steam generation, efficiency and gas and surface temperatures, and may be used to establish optimum locations for injection of the various streams. The furnace model also defines the temperature-time distributions which are key elements in the prediction of reburning performance. It also incorporates combustion sub-models and provides for the prediction of carbon burn-out in the case of coal or fuel oil, whether this is introduced as the main burner fuel or as the reburning fuel.

The final step in the design process consists of estimating the $NO_x$ reduction effectiveness of the reburn process. Several methods for estimating process performance have been developed. First, an empirical approach is to use the small–scale data accounting for the temperature and residence time in the reburn zone and the full–scale system baseline $NO_x$ data. The $NO_x$ reduction estimate obtained in this manner represents the theoretical reduction which is possible in the absence of mixing effects. A second approach utilizes a computational model of the reburn process described by Wu, Payne and Nguyen (1991). The computational model is a two-phase modular model which incorporates detailed gas-phase and coal devolatilization chemistry packages, and is centered around a kinetic mechanism which simulates the complex chemistry which occurs in the reburn zone. The kinetic mechanism consists of 201 reactions in 43 species. Plug flow and well stirred reactors are set up to simulate the various steps in the reburn process. The heat transfer model results are used to specify the residence time and thermal profile for each reactor. By varying the reactor set up and the contacting rate of reactants in a plug flow reactor, an ad hoc simulation of reburning fuel mixing can be accomplished. Computational results obtained with the model described by Wu, Payne and Nguyen (1991) are presented in this paper.

## FULL SCALE RESULTS

Gas reburning has been applied to four utility boilers in the United States as summarized in Table 1. The Lakeside and Niles boilers are both cyclone boilers manufactured by Babcock and Wilcox. The Hennepin boiler is a tangentially fired boiler manufactured by ABB Combustion Engineering. Cherokee is a wall fired boiler manufactured by Babcock and Wilcox. The Cherokee boiler was retrofit with low $NO_x$ burners prior to installation of the gas reburning system. The emissions reductions observed at each site are summarized in Table 1. Overall, emissions reductions between 50 and 60 percent were obtained for natural gas flows between 15 to 25 percent of the total boiler heat input.

Shown in Figures 8 and 9 are data from the Hennepin, Lakeside, Cherokee, and Niles utility boiler demonstration projects, where final $NO_x$ emission levels are plotted as a function of the natural gas addition rate and reburn zone stoichiometry, respectively. These plots show the general variability in the recorded data, and indicate the different levels of overall performance achieved on the different units. Of particular interest in Figure 8 is the observation that small percentages of gas addition can lead to significant reductions in $NO_x$ emissions. Such results cannot readily be predicted based on small–scale data or modelling alone, and, through analysis, it is clear that other parameters are important in establishing overall performance.

In Figure 9 averaged data are plotted against reburn zone stoichiometry, rather than gas addition rate. For the most part these data appear to be in general agreement with small scale test results, in that decreasing stoichiometry reduces $NO_x$ emissions at each of the sites, and that $NO_x$ levels reach a minimum at a stoichiometry close to 0.9. Due to lack of a wide range of full–scale data, it is difficult to determine whether stoichiometries below 0.9 result in higher reductions; although the general trend in the data appears to suggest that reburn zone stoichiometries below this value do not result in incrementally lower $NO_x$ emissions.

Close examination of the data suggests that a number of effects combine to influence the $NO_x$ emission level achievable in a specific application. Mixing and the distribution of reburning fuel and overfire air are important variables; whilst reducing the primary zone heat input and stoichiometry can lower the $NO_x$ starting point for the reburning system. Such effects tend to be site specific, but can be very important to the designer of the reburn system.

Analyses of the full–scale results have been conducted in order to assess the importance of the various parameters observed to be significant in small scale tests. One of the more significant results is shown in Figure 10 where full–scale data from all of the various boiler projects are plotted against experimental data generated during various process development studies. Lines of constant reburn zone stoichiometry are also indicated for the full–scale data. In this plot, the independent variable is the initial $NO_x$ actually entering the reburning zone. This parameter is estimated from the full–scale data, using empirical and computational models of the impacts of load reduction and stoichiometry variations on the primary $NO_x$ emissions. As shown in Figure 10, the full–scale results are generally in good agreement with experimental results. This is particularly encouraging given the large range of scale involved (from 20 kW thermal to 150 MW electric), and suggests that the underlying mechanisms controlling the process also govern the performance at full scale. However, the fact that the full–scale data are consistently above the experimental line may be due to the effects of scale or mixing.

## MIXING IMPACTS

Although some of the full–scale results suggest that mixing is important, the precise impact of mixing on full–scale $NO_x$ reduction has been difficult to quantify. For example, all of the full–scale demonstrations have utilized the recirculation of flue gas to the reburn nozzles to promote mixing of the natural gas with the gases exiting the primary zone. However, different results have been achieved by varying the quantity of flue gas supplied to the reburn nozzles. More recent full–scale results (unpublished) have shown that it may be possible to eliminate or reduce flue gas recirculation to the reburn nozzles by using advanced reburn nozzle designs.

To assess the impacts of flue gas recirculation on the 34 $MW_e$ cyclone fired boiler, the process modelling approach described in the preceding section has been applied to the results of the field demonstration. This is a particularly challenging situation because of the complex flow field in this unit, which is characterized by strongly three dimensional effects and reverse flow zones in and around the locations of both reburn fuel and overfire air injection. Furnace heat transfer models were used to define the overall thermal characteristics of the boiler under reburning conditions, and an isothermal flow model was used to predict the mixing patterns. A kinetic model of the reburn zone in the cyclone boiler was set up to simulate the reburn process following the methodology described by Wu, Payne and Nguyen (1991). Performance predictions were developed assuming that the reburn process followed a range of time temperature histories with flow paths representing: perfect mixing of the reburn fuel and flue gas, mixing of the reburn fuel through the center of the boiler, and mixing of the reburn fuel along the cooler water–wall surfaces. Each of these paths represented jet mixing patterns which were observed in the isothermal flow model. Time–temperature histories for each path were determined based upon the results of the furnace heat transfer model.

Figure 11 compares the predicted and observed impact of flue gas recirculation on reburn performance. The prediction band represents the variation in performance observed for the various mixing profiles. Comparison of the field and predicted results indicates that the model was able to simulate the general trend observed in the field data, but not the exact results. Considering the assumptions made in the model and the scatter in the field data, the comparison shown in Figure 11 is thought to be quite encouraging, though it is still somewhat surprising that the steepness of the full–scale data trend was not more closely simulated.

To further assess the ability of the model to predict full–scale performance, additional calculations were performed for the cyclone reburning system by varying the mean reburn stoichiometry in the model, in addition to the local stoichiometries represented by the various mixing paths. The predicted and observed results are summarized in Figure 12. Two sets of model predictions were performed: one set in which it was assumed that some of the overfire air was entrained into the reburning zone as observed during the isothermal flow model studies, and one set where the entrained overfire air was neglected. Once again, the trend predicted by the model is encouraging, but the model tends to over predict the field results.

The above results demonstrate that the coupling of a detailed chemistry model with a simplified mixing model based upon the results of thermal and physical models of the full–scale system can predict the general performance trends observed in full–scale systems. Therefore, models such as these can be used to design and evaluate full–scale reburn systems. However, simplified mixing models cannot predict some details of full–scale reburn performance. To achieve better agreement between full–scale results and model predictions, and to explain some trends which have been observed at full scale, it is believed that more accurate models that couple the complex interactions between turbulence and chemistry which occur in the turbulent mixing process are needed. One such model is currently under investigation, and is an extension of the concept of turbulent mixing described by Broadwell et al. (1983).

## CONCLUSIONS

Reburning is a combustion modification technology which has been well–characterized at small scales and has recently been demonstrated at full scale. Scale up and successful application of reburn technology to utility boilers involves the use of a variety of physical and computational modelling techniques. For the most part, the results of these modelling techniques has shown good agreement with full–scale results. However, the current suite of modelling techniques are not able to explain all of the trends observed at full scale. To more accurately predict full–scale performance, computational models which couple chemistry with turbulent mixing are needed. The use of these models can help with improved prediction of full–scale performance and the impacts of various potential injection configurations, as well as the optimization of installed systems and the development of improvements in reburn technology.


## ACKNOWLEDGMENTS

The authors recognize the assistance of Quang Nguyen, Bruce Li, and Tony Marquez in collecting and analyzing the full–scale data, and in performing the various modelling studies.


## REFERENCES


Borio, R. W., et al., 1991, "Reburn Technology for $NO_x$ Control on a Cyclone–Fired Boiler; An Update," Presented at the ASME 1991 International Joint Power Generation Conference & Exposition, October 6–10, San Diego, CA.

Broadwell, J. E., et al., 1983, "The Structure of Turbulent Diffusion Flames and Nitric Oxide Formation," *Proceedings of the 1982 Joint Symposium on Stationary Combustion $NO_x$ Control, Volume 1: Utility Boiler Applications*, Electric Power Research Institute, Palo Alto, CA.

Chen, S. L., et al., 1986, "Bench and Pilot Scale Process Evaluation of Reburning for In–Furnace $NO_x$ Reduction," *Twenty–First Symposium (International) on Combustion*, The Combustion Institute, pp. 1159–1169.

Chen, S. L. et al., 1988, "Studies on Enhancement of Reburning for Advanced $NO_x$ Control in Coal–Fired Boilers," Presented at the 1988 AFRC Fall International Symposium, October 4–6, 1988, Pittsburgh, PA.

Folsom, B. A., Sommer, T. M., and Payne, R., 1991, "Demonstration of Combined $NO_x$ and $SO_2$ Emission Control Technologies Involving Gas Reburning," Presented at the AFRC/JFRC International Conference on Environmental Control of Combustion Processes, October 7–10, Honolulu, Hawaii.

Greene, S. B., et al., 1986, "Bench Scale Process Evaluation of Reburning for In–Furnace $NO_x$ Reduction," *ASME Journal of Engineering for Gas Turbines and Power*, Volume 108, pp. 450–454.

Li, B. W., et al., "Use of Computer Models for Reburning/Cofiring Boiler Performance Evaluations," *ASME FACT Volume 17, Cofiring and $NO_x$ Control*, A. K. Gupta et al., editors, 1993.

Myerson, A. L., 1957, "Ignition Limits and Products of the Multistage Flames of Propane–Nitrogen Dioxide Mixtures," *Sixth Symposium (International) on Combustion*, The Combustion Institute, pp. 154–163.

Overmoe, B. J., et al., 1986, "Pilot Scale Evaluation of $NO_x$ Control from Pulverized Coal Combustion by Reburning, " Proceedings of the 1985 Joint Symposium on Stationary Combustion $NO_x$ Control, Volume 1: Utility Boilers Applications, Electric Power Research Institute, Palo Alto, CA.

Payne, R., et al., 1988, "Demonstration of Gas Reburning/Sorbent Injection $NO_x/SO_2$ Control Technology on Three Utility Boilers," Presented at the AIChE 1988 Summer National Meeting, August 21–24, Denver, Colorado.

Takahashi, Y., et al., 1983, "Development of "MACT" In-Furnace $NO_x$ Removal Process for Steam Generators," *Proceedings of the 1982 Joint Symposium on Stationary Combustion $NO_x$ Control, Volume 1: Utility Boiler Applications*, Electric Power Research Institute, Palo Alto, CA.

Wendt, J. O. L., Sternling, C. V., and Matovich, M. A., 1973, "Reduction of Sulfur Trioxide and Nitrogen Oxides by Secondary Fuel Injection," *Fourteenth Symposium (International) on Combustion*, The Combustion Institute, pp. 897–904.

Wu, K. T., Payne, R., and Nguyen Q. H., 1991, "Development and Application of a Gas Reburning Process Model for the Design of Boiler $NO_x$ Reduction," ASME Paper No. 91-JPGC-FACT-24.

TABLE 1. RESULTS OF FULL SCALE REBURN APPLICATION.

| Unit | Size (MW) | Initial $NO_x$ (lb/MMBtu) | Controlled $NO_x$ (lb/MMBtu) | $NO_x$ Reduction (%) |
|---|---|---|---|---|
| Illinois Power Hennepin | 71 | 0.75 | 0.243 | 68 |
| P.S. Colorado Cherokee (post LNB retrofit) | 158 | 0.50 | 0.20 | 60 |
| CWLP Lakeside | 33 | 0.95 | 0.36 | 62 |
| Ohio Edison Niles | 108 | 0.90 | 0.45 | 50 |

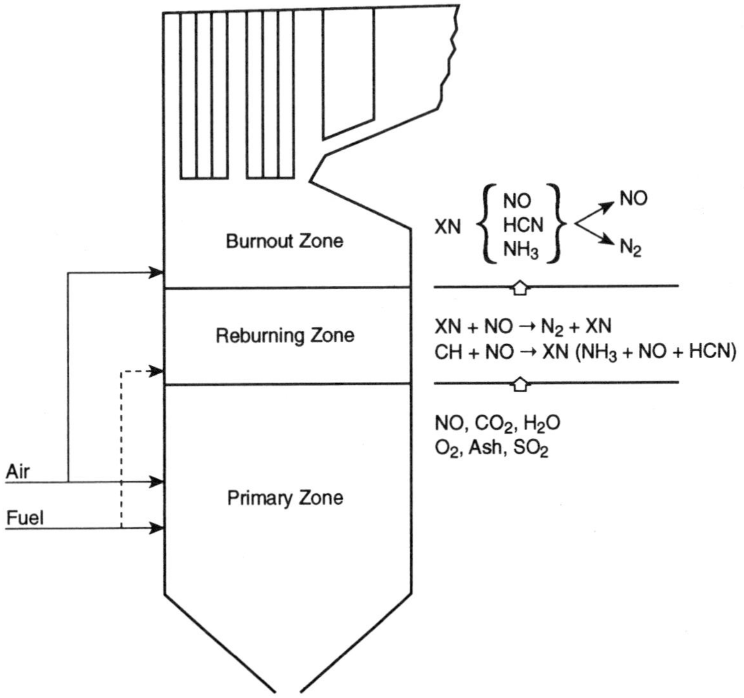


FIGURE 1. SCHEMATIC OF THE REBURNING PROCESS.

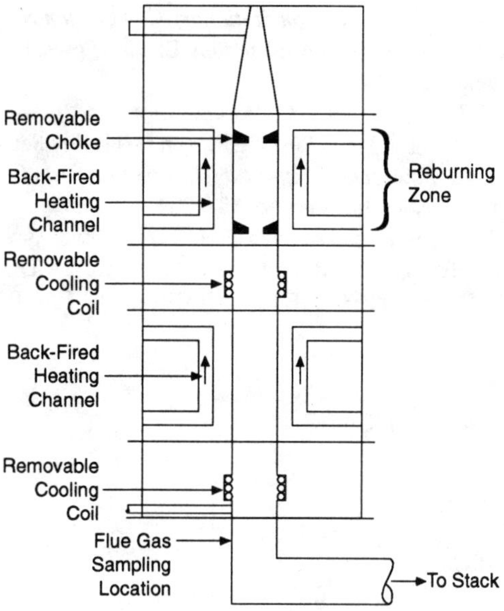


FIGURE 2. BENCH SCALE REACTOR.

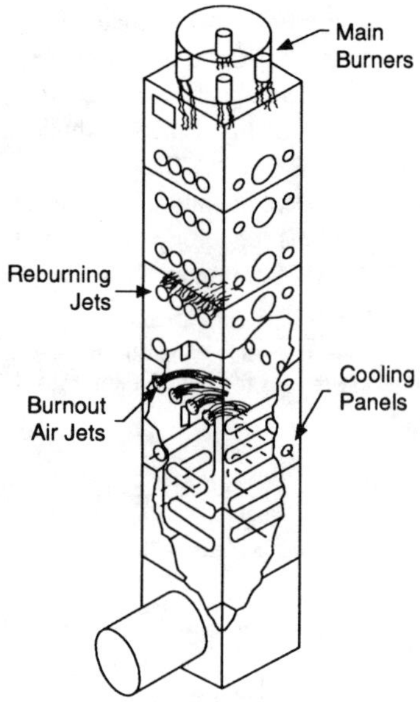


FIGURE 3. PILOT SCALE REACTOR.

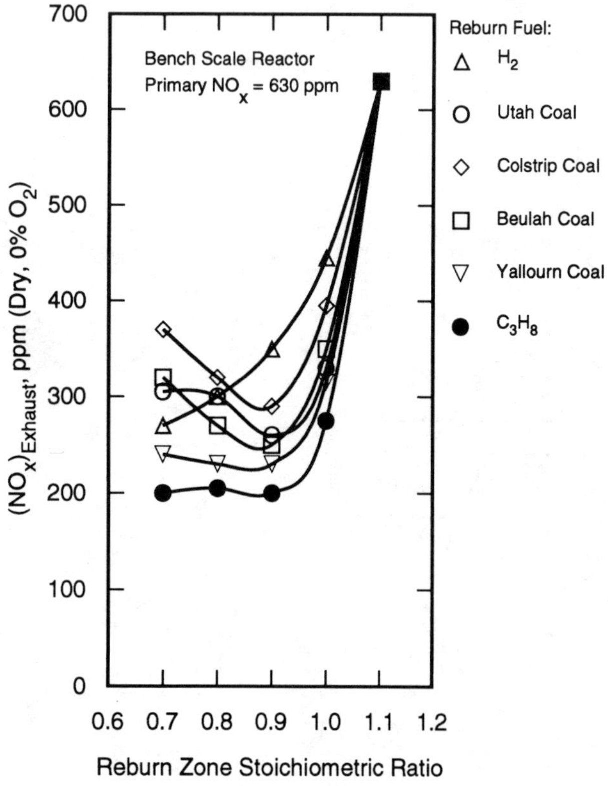


FIGURE 4. IMPACT OF REBURN ZONE STOICHIOMETRY.

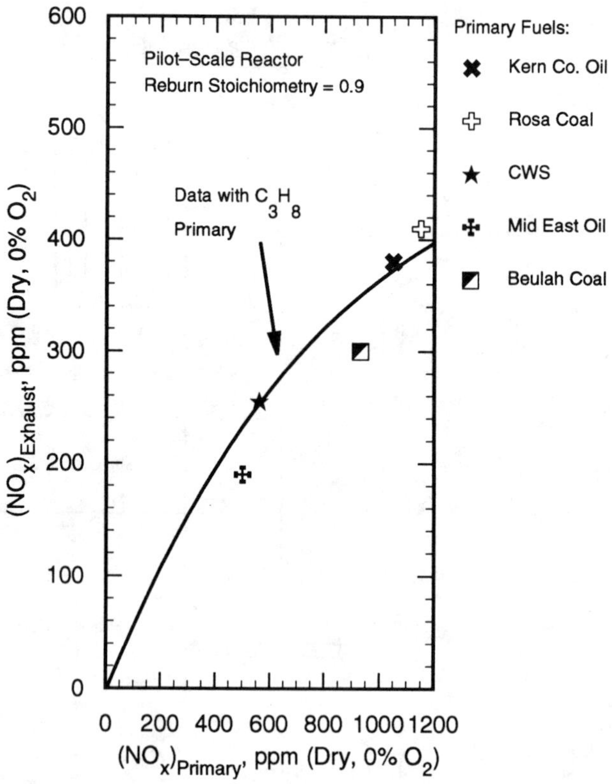


FIGURE 5. IMPACT OF INITIAL $NO_x$.

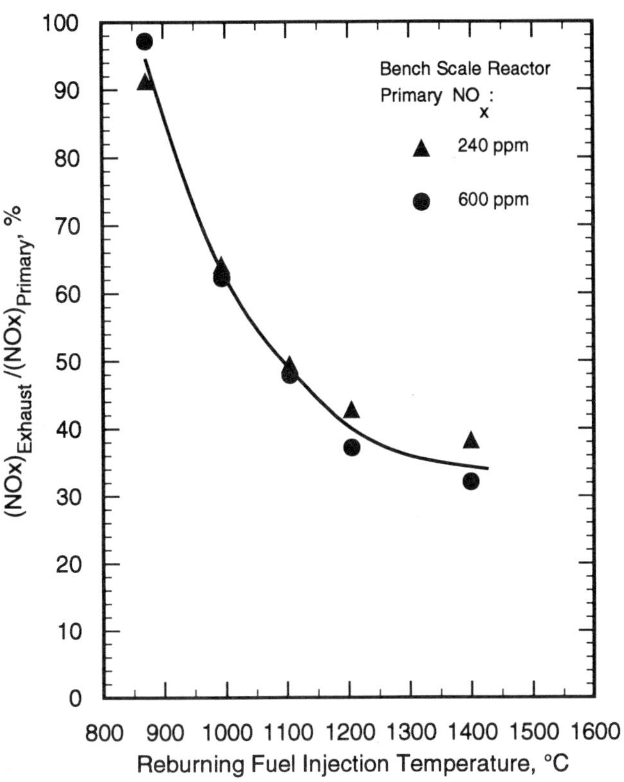


FIGURE 6. IMPACT OF TEMPERATURE.

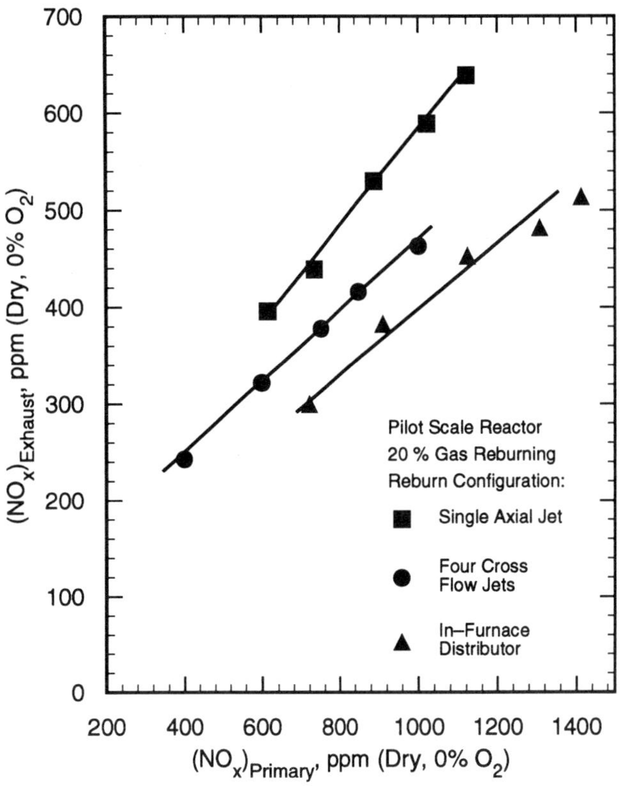


FIGURE 7. IMPACT OF REBURNING FUEL MIXING.

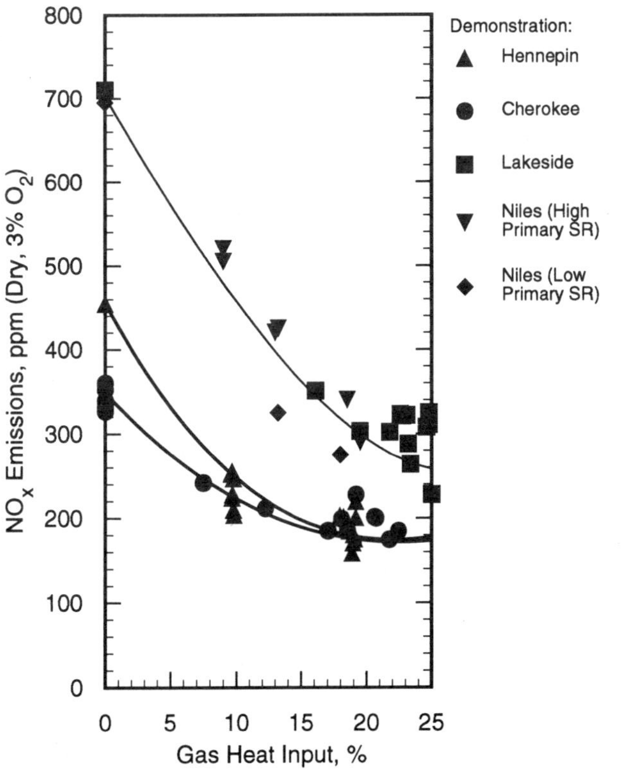


FIGURE 8. FIELD RESULTS VERSES % GAS INPUT.

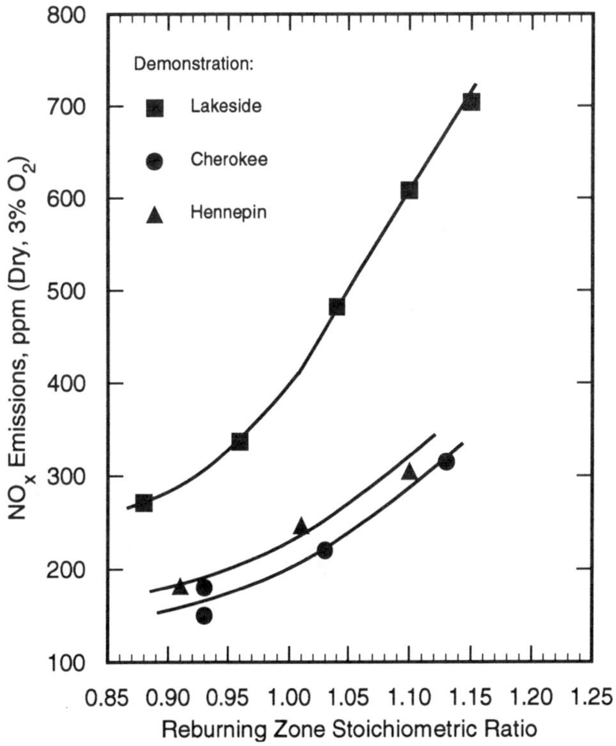


FIGURE 9. FIELD RESULTS VERSES REBURN SR.

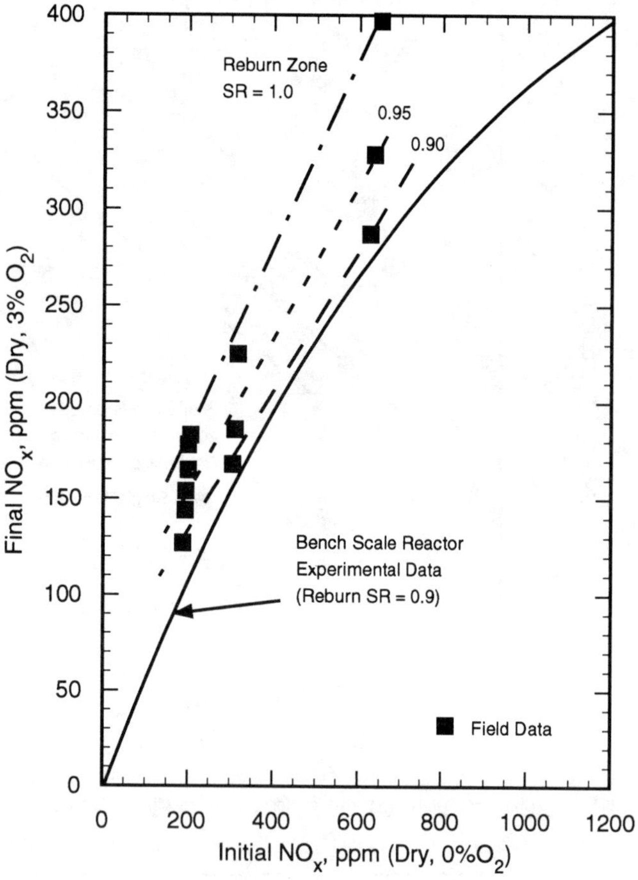


FIGURE 10. COMPARISON OF REBURNING DATA.

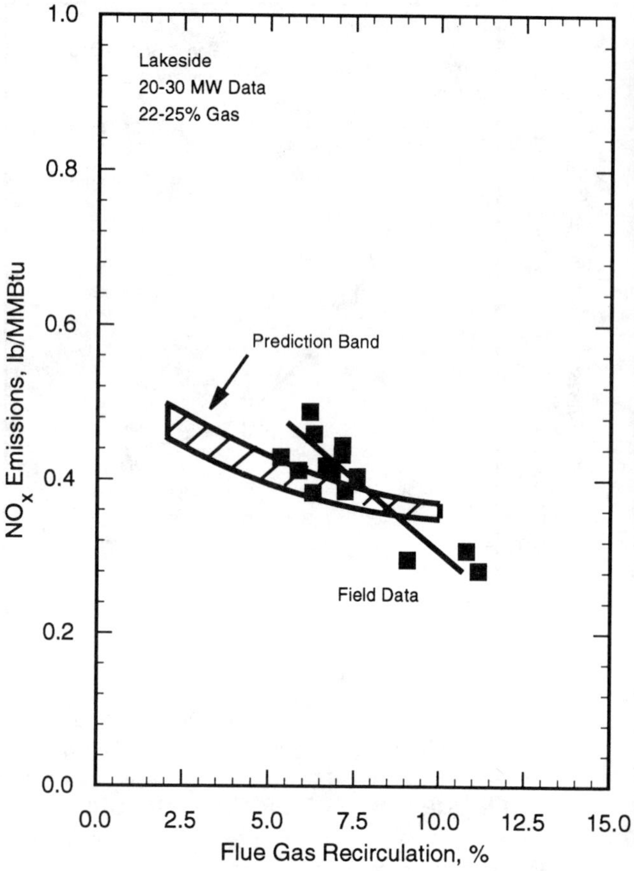


FIGURE 11. PREDICTION OF FGR IMPACT.

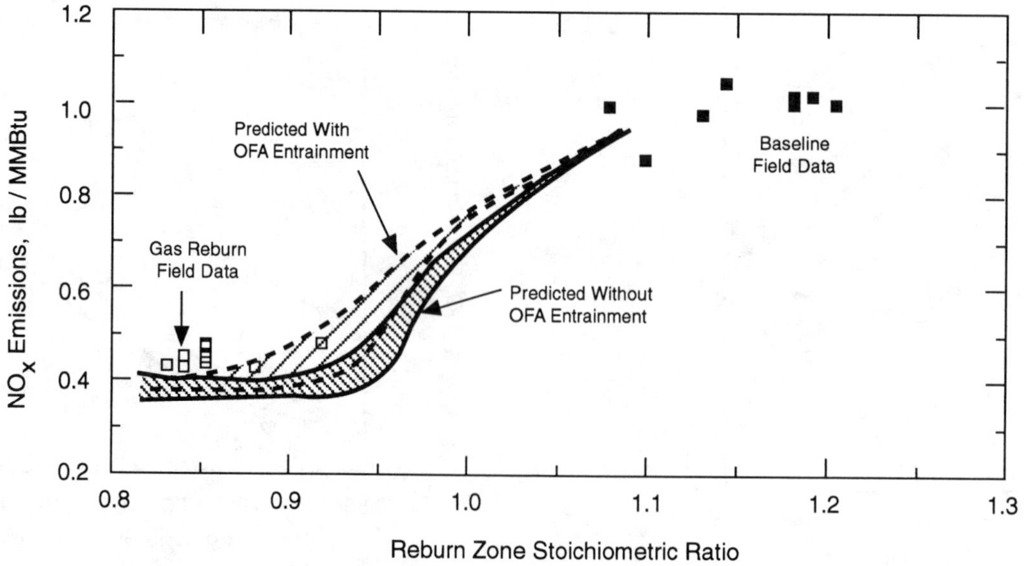


FIGURE 12. PREDICTION OF CYLONE GAS REBURN PERFORMANCE .

FACT-Vol. 18, Combustion Modeling, Scaling and Air Toxins
ASME 1994

# MATHEMATICAL MODELING AND SCALING OF FLUID DYNAMICS AND $NO_x$ CHARACTERISTICS OF NATURAL GAS BURNERS

André A.F. Peters and Roman Weber
International Flame Research Foundation
IJmuiden
The Netherlands

## ABSTRACT

Predictions of two unstaged (high-$NO_x$) swirling natural gas flames of 4 MW and 12 MW thermal input have been performed and compared with the measurements. In scaling the burners, the constant-velocity-scaling criterion has been applied together with geometrical similarity. The applied mathematical models for main flame properties and nitric oxide predictions have been discussed. The two-step-reaction turbulent combustion model is based on the eddy-break-up concept, i.e. mixing. In the $NO_x$-postprocessor, thermal-NO and prompt-NO formation rates are obtained by means of statistically averaging of the corresponding chemical reaction rates over the fluctuating temperature. In the predictions the two flames scale perfectly whereas the measurements indicate some departures from thermal similarity in particular. It was observed that the degree of combustion for both flames was dependent on the residence time of fluid pockets in the reacting shear layer. As the turbulent combustion model has been based solely on mixing, i.e. the turbulent flow field, the non-similarity of the flames could not be predicted. Possible future improvements to the turbulent combustion model are addressed. 3D-modeling, i.e. modeling of individual natural gas jets, shows a significant improvement and is therefore to be preferred over 2D-modeling.

## INTRODUCTION

Designing and developing combustion equipment requires careful consideration on how the burner, developed either at laboratory scale or at semi-industrial scale, would perform when scaled to full-industrial size. The scaling problem has received relatively little attention, particularly with respect to scaling low-$NO_x$ burner designs. This paper is part of a larger project on scaling of natural gas burners.

Within the SCALING 400 study (Weber et al., 1993a), the objective is to provide scaling flame data over the widest possible range of thermal inputs and to extract from these data a set of reliable scaling laws relating small scale burner performance to the performance at full industrial scales. A series of five tests have been performed on five versions of a generic natural gas burner, designed using the constant-velocity-scaling criterion together with geometric similarity. These tests have been carried out in a set of five thermally (reasonably) similar furnaces over a range of thermal inputs spanning from the smallest laboratory scale (30 kW) to full industrial scale (12 MW). These limits span a factor of 400 in thermal input and the aim is to determine scaling laws that relate burner performance over this entire range. Flame data obtained at a selected set of crucial intermediate scales (300 kW, 1.3 MW and 4 MW) will facilitate accurate identification of scaling trends over this wide range of thermal inputs.

As part of the SCALING 400 study, for each of the five thermal inputs, detailed in-flame measurements of both an unstaged (high-$NO_x$) and a staged (low-$NO_x$) flame have been performed for extensive validation of mathematical models. This paper presents the main results of the very first predictions of the 4 MW and 12 MW unstaged swirling natural gas flames. The applied mathematical models for main flame properties and nitric oxide predictions, as they will be discussed, have been tested and optimized using detailed experimental data of an unstaged, swirling natural gas flame issued from a 2.25 MW burner different to the one used in this study.

## THE NATURAL GAS FLAMES CONSIDERED IN THIS PAPER

The burner input conditions for the 4 MW and 12 MW unstaged swirling natural gas flames and the compositions and properties of the natural gases fired are given in table 1 and table 2 respectively. The generic natural gas burner is shown in figure 1.

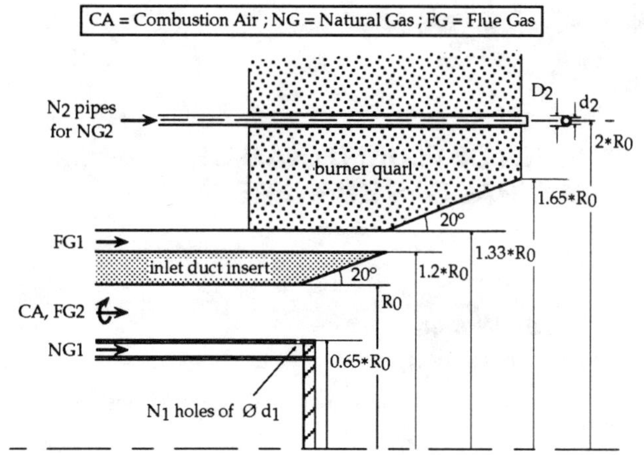


FIGURE 1: GENERIC NATURAL GAS BURNER (SEE ALSO TABLE 1).

At the applied excess air level of 15 %, this burner generated blue, non-sooty flames. The 4 MW flame was generated in a refractory-brick-lined furnace with a square cross-sectional area of $2*2\ m^2$ and a length of 6.44 m. The 12 MW flame was generated in a cylindrical furnace with a diameter of 3.4 m, a length of 5.9 m, a for the greater part water-cooled cylinder wall and insulated front and back walls. Both furnaces were equipped with a number of cooling loops. It is noted that the furnace sizes (diameters) are scaled to guarantee similarity of in-furnace fluid dynamics. Detailed information on the scaled 4 MW and 12 MW versions of the generic burner, the corresponding furnaces and the calibrations of the movable block swirlers in question is available (Bertolo et al., 1993, and Weber et al., 1993b).

In-flame temperatures and chemical species concentrations were measured at various traverses by means of standard IFRF sampling probes. The measured chemical species are $O_2$, $CO_2$, UHC, CO, NO and $NO_x$. Velocity and turbulence measurements were performed by means of an LDV technique.

Both flames were axis-symmetric and have been modeled accordingly. In the case of the 4 MW flame, the diameter of the modeled cylindrical furnace has been set to 2.26 m so that the cross-sectional areas of the modeled furnace and the real furnace agree while the cooling loops have been modeled as cylindrical rings with proper physical area and positioned on the modeled cylindrical furnace wall. In the case of the 12 MW flame, the total cooling loop area has been added to the water-cooled jacket area, resulting in a somewhat smaller insulated surface on the front part of the cylindrical furnace wall.

## MATHEMATICAL MODEL FOR MAIN FLAME PROPERTIES PREDICTIONS

### Mass and Momentum Balance Equations

The stationary, time-mean, mass balance equation for a gaseous mixture reads

TABLE 1: BURNER INPUT CONDITIONS FOR THE 4 MW AND 12 MW UNSTAGED SWIRLING NATURAL GAS FLAMES (SEE ALSO FIGURE 1).

| burner input condition | | 4 MW unstaged | 12 MW unstaged |
|---|---|---|---|
| **combustion air** | **dim.** | | |
| mass flow rate | kg/hour | 5676 | 17059 |
| duct diameter ($D_0=2R_0$) | mm | 317 | 549 |
| temperature | K | 308.15 | 308.15 |
| average axial velocity | m/s | 30.21 | 30.27 |
| swirl number ($S_2$) | - | 0.57 | 0.56 |
| average tang. velocity | m/s | 20.57 | 20.24 |
| excess air level | % | 15 | 15 |
| **natural gas** | **dim.** | | |
| mass flow rate | kg/hour | 314 | 914 |
| number of holes ($N_1$) | - | 36 | 72 |
| hole diameter ($d_1$) | mm | 5.2 | 6.5 |
| temperature | K | 347.15 | 335.93 |
| radial velocity | m/s | 184.3 | 169.2 |

TABLE 2: COMPOSITION AND PROPERTIES OF THE NATURAL GASES FIRED.

| natural gas | | IJmuiden 4 MW | Tulsa 12 MW |
|---|---|---|---|
| **composition** | **dim.** | | |
| $CH_4$ | vol% | 90.9 | 90.0 |
| $C_2H_6$ | ,, | 3.9 | 5.5 |
| $C_3H_8$ | ,, | 0.7 | 1.0 |
| $C_4H_{10}$ & higher | ,, | 0.3 | 0.5 |
| $CO_2$ | ,, | 1.3 | - |
| $N_2$ | ,, | 2.9 | 3.0 |
| **properties** | **dim.** | | |
| density (at 0 °C) | kg/m$^3$ | 0.787 | 0.772 |
| LCV | MJ/kg | 46.1 | 47.0 |
| air requirement | kg/kg | 15.72 | 16.23 |

$$\underline{\nabla}\cdot(\overline{\rho}\,\underline{\overline{v}})=0\ , \tag{1}$$

where $\rho$ and $\underline{v}$ stand for mixture density and velocity vector while the overbars denote Reynolds- or time-averaging. In the case that the turbulence is modeled by means of the standard k-epsilon model (Launder and Spalding, 1974), the corresponding momentum balance equation may be written as

$$\underline{\nabla}\cdot(\overline{\rho}\,\underline{\overline{v}}\,\underline{\overline{v}})-\underline{\nabla}\cdot\left(2\mu_{eff}\,\underline{\underline{\overline{D}}}\right)+\underline{\nabla}(\overline{P})=\underline{0}\ . \tag{2}$$

In the above equation, $\underline{\underline{D}}$ is the symmetric part of the velocity gradient tensor $\underline{\nabla}\underline{v}$ while $P$ stands for static pressure. Furthermore, $\mu_{eff}$ is the effective dynamic viscosity which equals

$$\mu_{eff} = \mu + \mu_t \; ; \; \mu_t = c_\mu \bar{\rho} \frac{\bar{k}^2}{\bar{\varepsilon}} \; ; \; c_\mu = 0.09 \; . \tag{3}$$

Here, $\mu$ and $\mu_t$ stand for (laminar) fluid dynamic viscosity and turbulent dynamic viscosity respectively. The time-mean, turbulent kinetic energy, $\bar{k}$, and the time-mean, turbulent viscous dissipation rate of turbulent kinetic energy, $\bar{\varepsilon}$, are solved for by means of two balance equations.

## Mass Balance Equations for Chemical Species

Six major chemical species are considered: hydrocarbons ($C_xH_y$), oxygen, carbon monoxide, carbon dioxide, water vapor and nitrogen. The stationary, time-mean, mass balance equations for these chemical species read

$$\underline{\nabla} \cdot (\bar{\rho} \underline{\bar{v}} \bar{\omega}_J) - \underline{\nabla} \cdot \left( \frac{\mu_{eff}}{Sc} \underline{\nabla} \bar{\omega}_J \right) = S_J \; ; \quad J = C_xH_y, O_2, CO, CO_2, H_2O, N_2 \; , \tag{4}$$

where $\omega_J$ stands for mass fraction of chemical species $J$ while $Sc$ is the Schmidt number ($Sc$=0.7). The time-mean mass sources $S_J$ include formation and/or reduction of chemical species $J$ due to combustion. These sources will be discussed further below.

The time-mean density is obtained from the ideal gas law:

$$\bar{\rho} = \frac{\bar{P}}{R\bar{T} \sum_J \frac{\bar{\omega}_J}{M_J}} \; . \tag{5}$$

Here $T$ denotes temperature, $R$ is the universal gas constant and $M_J$ stands for molecular weight of chemical species $J$.

## Enthalpy Balance Equation

The stationary, time-mean, enthalpy balance equation reads

$$\underline{\nabla} \cdot (\bar{\rho} \underline{\bar{v}} \bar{h}) - \underline{\nabla} \cdot \left( \frac{\mu_{eff}}{Pr} \underline{\nabla} \bar{h} \right) = S_{comb} + S_{rad} \; , \tag{6}$$

where $h$ stands for physical enthalpy while $Pr$ is the Prandtl number ($Pr$=0.7). The two time-mean, thermal energy sources $S_{comb}$ and $S_{rad}$ account for energy release due to combustion and energy decrease or increase due to flame radiation. The time-mean enthalpy source $S_{comb}$ will be discussed further below.

The time-mean temperature is obtained from the following equation:

$$\bar{T} = \frac{\bar{h}}{\tilde{c}_p(\bar{T})} \; ; \; \tilde{c}_p(\bar{T}) = \sum_J \bar{\omega}_J \tilde{c}_{p,J}(\bar{T}) \; . \tag{7}$$

Here, $\tilde{c}_p(T)$ and $\tilde{c}_{p,J}(T)$ are the average specific heats at constant pressure, over the temperature interval (0,$T$), of the gaseous mixture and chemical species $J$ respectively. In modeling the average specific heats of $CO_2$, $H_2O$ and $O_2$, molecular dissociation has been accounted for. For the average specific heat of $C_xH_y$, the average specific heat of $CH_4$ has been assumed.

Radiative heat transfer is described by means of a four/six-flux model of Gosman and Lockwood (1972). The linear absorption coefficient $\kappa_a$, appearing in this model, is obtained from the Exponential Wide Band Model (Edwards, 1976, and Lallemant and Weber, 1993). With this model, $\kappa_a$ may be calculated as a function of temperature, path length and various species concentrations. In the modeling, only the radiative species solved for, i.e. $CO_2$, $H_2O$, CO and $C_xH_y$ (as $CH_4$), are accounted for.

## Global Combustion Chemistry

Assuming that $C_xH_y$ oxidizes to CO and $H_2O$ while intermediate CO oxidizes to $CO_2$, the following reaction scheme applies:

$$1 \text{ kg\_}C_xH_y + s_1^* \text{ kg\_}O_2 \xrightarrow{H_1^*} \frac{28}{12} f_C \text{ kg\_CO} + \frac{18}{2}(1 - f_C) \text{ kg\_}H_2O \; ; \tag{RS1}$$

$$1 \text{ kg\_CO} + s_2 \text{ kg\_}O_2 \xrightarrow{H_2} (1 + s_2) \text{ kg\_}CO_2 \; .$$

Here, the stoichiometric oxygen requirement for burning CO to $CO_2$ and the corresponding heat of reaction are denoted by $s_2$ and $H_2$ respectively. The carbon fraction in $C_xH_y$ is denoted by $f_C$. The stoichiometric oxygen requirement $s_1^*$ for burning $C_xH_y$ to CO and $H_2O$ and the corresponding heat of reaction $H_1^*$ are defined as follows:

$$s_1^* = s_1 - \frac{28}{12} f_C s_2 \; ; \; s_1 = \frac{32}{12} f_C + \frac{16}{2}(1 - f_C) \; ;$$

$$H_1^* = H_1 - \frac{28}{12} f_C H_2 \; ; \; H_1 = LCV_{C_xH_y} \; ; \tag{8}$$

$$s_2 = \frac{16}{28} \; ; \; H_2 = LCV_{CO} = 10.16 \text{ MJ / kg\_CO} \; .$$

In these expressions, $s_1$ is the stoichiometric oxygen requirement for burning $C_xH_y$ to $CO_2$ and $H_2O$ while $H_1$ is the corresponding heat of reaction. $LCV$ stands for lower calorific value.

### Turbulent Combustion

The various source and sink terms $S_J$ entering the chemical species mass balance equations (4) due to gaseous combustion, are modeled following the eddy-break-up concept according to Magnussen and Hjertager (1976). Rates of formation and reduction of chemical species are related to the dissipation rate of eddies and expressed by means of the density, the chemical species mass fractions, the turbulent kinetic energy and the turbulent viscous dissipation rate. Furthermore, the concept makes use of a proportionality constant, the mixing rate coefficient $A_{mix}$. For turbulent diffusion flames with an overall thermal input of about 2.0-2.5 MW, an appropriate value for $A_{mix}$ was found to be 0.6 (Visser et al., 1990).

The maximum possible reduction rates of $C_xH_y$ and CO read:

$$R_{C_xH_y} = A_{mix}\frac{\bar{\varepsilon}}{\bar{k}}\bar{\rho}\,\overline{\omega}_{C_xH_y} \;;\; R_{CO} = A_{mix}\frac{\bar{\varepsilon}}{\bar{k}}\bar{\rho}\,\overline{\omega}_{CO} \;. \tag{9}$$

In the case that the two combustibles burn according to reaction scheme RS1 and at the above maximum possible rates, the required oxygen reduction rate reads

$$R_{O_2,req} = s_1^* R_{C_xH_y} + s_2 R_{CO} \;. \tag{10}$$

The maximum possible oxygen reduction rate equals

$$R_{O_2} = A_{mix}\frac{\bar{\varepsilon}}{\bar{k}}\bar{\rho}\,\overline{\omega}_{O_2} \;. \tag{11}$$

Thus, the following inequality must be obeyed:

$$R_{O_2,req} \le R_{O_2} \;. \tag{12}$$

If this inequality is obeyed, then the actual oxygen reduction rate equals the required one and the two combustibles burn at maximum possible rates. If the inequality is not obeyed, then the reduction rates of the two combustibles have to be limited while oxygen burns at maximum possible rate. Thus, the actual oxygen reduction rate may be written as

$$R_{O_2,actual} = \min\left(s_1^* R_{C_xH_y} + s_2 R_{CO}, R_{O_2}\right) \;. \tag{13}$$

If inequality (12) is not obeyed, then the limited reduction rates of the two combustibles equal

$$R_{C_xH_y,lim} = \frac{s_1^* R_{C_xH_y}}{R_{O_2,req}}\frac{R_{O_2}}{s_1^*} \;;\; R_{CO,lim} = \frac{s_2 R_{CO}}{R_{O_2,req}}\frac{R_{O_2}}{s_2} \;, \tag{14}$$

where the limiting is solely based on the oxygen requirements of the individual combustibles.

Based on inequality (12), the following local stoichiometry may be obtained for reaction scheme RS1:

$$\lambda_1 = \frac{\overline{\omega}_{O_2}}{s_1^*\overline{\omega}_{C_xH_y} + s_2\overline{\omega}_{CO}} \;. \tag{15}$$

Thus, oxygen-rich and oxygen-lean regions for reaction scheme RS1 may be denoted as follows:

$$\begin{aligned} &\lambda_1 \ge 1 \;\Leftrightarrow\; \text{oxygen-rich (i.e. fuel-lean) for RS1} \;; \\ &\lambda_1 < 1 \;\Leftrightarrow\; \text{oxygen-lean (i.e. fuel-rich) for RS1} \;. \end{aligned} \tag{16}$$

For oxygen-rich regions, the sources $S_J$, $J=C_xH_y, O_2, CO, CO_2, H_2O, N_2$, as well as the source $S_{comb}$ are listed below:

$$\begin{aligned} &S_{C_xH_y} = -R_{C_xH_y} \;;\; S_{CO} = -R_{CO} + \frac{28}{12} f_C R_{C_xH_y} \;; \\ &S_{O_2} = -s_1^* R_{C_xH_y} - s_2 R_{CO} \;;\; S_{CO_2} = (1+s_2) R_{CO} \;; \\ &S_{H_2O} = \frac{18}{2}(1-f_C) R_{C_xH_y} \;;\; S_{N_2} = 0 \;; \\ &S_{comb} = R_{C_xH_y} H_1^* + R_{CO} H_2 \;. \end{aligned} \tag{17}$$

The corresponding sources for oxygen-lean regions may be obtained by replacing $R_{C_xH_y}$ and $R_{CO}$ by $R_{C_xH_y,lim}$ and $R_{CO,lim}$ respectively.

## MATHEMATICAL MODEL FOR NITRIC OXIDE PREDICTIONS; $NO_x$-POSTPROCESSOR

### Thermal-NO Formation Mechanism

In combustion of fuel-lean and near-stoichiometric fuel-air mixtures, the principal reactions governing the formation of NO from the oxidation of molecular nitrogen are those originally proposed by Zeldovich:

$$O + N_2 \xleftrightarrow{k_{1f}, k_{1b}} NO + N \;; \tag{R1}$$

$$N + O_2 \xleftrightarrow{k_{2f}, k_{2b}} NO + O \;. \tag{R2}$$

These two reactions are usually referred to as the thermal-NO formation mechanism or the Zeldovich mechanism. Lavoie et al. have suggested that the reaction

$$N + OH \xleftrightarrow{k_{3f}, k_{3b}} NO + H \tag{R3}$$

may also contribute to the formation of thermal-NO, especially in fuel-rich and near-stoichiometric fuel-air mixtures. The reactions (R1), (R2) and (R3) are usually referred to as the extended Zeldovich mechanism (e.g. Bowman, 1975).

In the present work, the role played by the OH-radicals is not taken into account. Then, under the assumption that the N-radicals concentration may be calculated from a steady state approximation, the chemical reaction rate of thermal-NO formation reads

$$r_{t-NO} = 2k_{1f}[O]\frac{[O_2][N_2]-\left(K_{e,NO}\right)^{-2}[NO]^2}{[O_2]+\left(k_{1b}/k_{2f}\right)[NO]} \; ; \qquad K_{e,NO} = \left(\frac{k_{1f}k_{2f}}{k_{1b}k_{2b}}\right)^{\frac{1}{2}} \; , \tag{18}$$

where $K_{e,NO}$ denotes the equilibrium constant of the overall reaction

$$\frac{1}{2}O_2 + \frac{1}{2}N_2 \overset{K_{e,NO}}{\longleftrightarrow} NO \; . \tag{R4}$$

For the forward and backward rate constants $k_{1f}$, $k_{1b}$, $k_{2f}$ and $k_{2b}$, the expressions as proposed by Bowman (1975) have been applied.

The O-radicals concentration is assumed to be equal to the equilibrium O-radicals concentration as in the case of molecular oxygen dissociation, i.e.

$$[O] = \frac{K_{e,O}}{\sqrt{RT}}[O_2]^{\frac{1}{2}} \; , \tag{19}$$

where $K_{e,O}$ is the equilibrium constant for the reaction

$$\frac{1}{2}O_2 \overset{K_{e,O}}{\longleftrightarrow} O \; . \tag{R5}$$

For $K_{e,O}$, the expression as proposed by Westenberg (1975) has been applied.

Finally, note that, due to the backward reactions, the rate $r_{t\text{-}NO}$ also contains an NO reduction term. This reduction term is of importance in the case that the reactions (R1) and (R2) reach equilibrium very fast, i.e. in the case that in-flame temperatures are extremely high and/or in the case that NO concentrations are very high.

## Prompt-NO Formation Mechanism

In combustion of hydrocarbon fuels, the actual chemical reaction rates of NO formation can exceed those attributable to the direct oxidation of molecular nitrogen by the (extended) Zeldovich mechanism. This is especially true under fuel-rich conditions. This rapidly formed NO was termed prompt-NO by Fenimore since the rapid NO formation he observed, occurred very early in the flame front (e.g. Miller and Bowman, 1989). Reaction mechanisms involving hydrocarbon-radicals, like the reactions

$$CH + N_2 \longleftrightarrow HCN + N \tag{R6}$$

and

$$C + N_2 \longleftrightarrow CN + N \; , \tag{R7}$$

together with the subsequent reaction of N-radicals via the reactions (R2) and (R3) to form NO, play an important role in the formation of prompt-NO (e.g. Bowman, 1975).

For hydrocarbon fuels , De Soete (1975) proposes a roughly estimated chemical reaction rate appropriate for this prompt-NO formation mechanism. This rate reads

$$r_{p-NO} = C\frac{M^{1+b}}{\rho^{1+b}}[O_2]^b[N_2][C_xH_y]\exp\left(-\frac{E_a}{RT}\right) \; , \tag{20}$$

where $M$ is the molecular weight of the gaseous mixture. In natural gas combustion, for $C_xH_y$ the constants $C$ and $E_a$ are set to values corresponding with $CH_4$:

$$C = 6.4*10^6 \;\; s^{-1} \;\; ; \;\; E_a = 72.5*10^3 \;\; cal/g\_mole \; . \tag{21}$$

For the natural gas flames in question (with in-flame temperatures below 2000 K), an appropriate value for the power $b$ is 0.5.

## Turbulence/Chemistry-Interaction

The chemical reaction rates of thermal-NO and prompt-NO formation discussed above, cannot be used directly as source terms in the time-mean mass balance equation for chemical species NO. This is due to the fact that these chemical reaction rates are obtained from laminar premixed flame experiments or shock tube studies, and thus not directly applicable in the case that there is a strong interaction between chemistry and turbulence. To be of use, they must be time-averaged first.

The stationary, time-mean, mass balance equation for chemical species NO reads

$$\underline{\nabla}\cdot\left(\bar{\rho}\,\underline{\bar{v}}\,\bar{\omega}_{NO}\right) - \underline{\nabla}\cdot\left(\frac{\mu_{eff}}{Sc}\underline{\nabla}\,\bar{\omega}_{NO}\right) = 10^3 M_{NO}\left(\overline{r_{t-NO}} + \overline{r_{p-NO}}\right) \; . \tag{22}$$

Since the chemical reaction rates of thermal-NO and prompt-NO, show the stronger dependence on the in-flame temperature and the lesser dependence on the in-flame chemical species concentrations, the presumed, single-variable probability density function (pdf-) approach as proposed by Hand et al. (1989) has been chosen to model the time-mean rates $\overline{r_{t-NO}}$ and $\overline{r_{p-NO}}$. The original Hand et al. (1989) concept has been enhanced in our work by including expressions for the so-called burnt and unburned temperatures. Therefore, the complete pdf-submodel used is described below.

The starting-point of the pdf-submodel used is a presumed joint-pdf of the instantaneous chemical species mass fractions and the instantaneous density and temperature. All these quantities are assumed to be statistically independent, so that the joint-pdf can be written as a product of single-variable pdf's. Furthermore, to the chemical species mass fractions and to the density, $\delta$-Dirac type pdf's are assigned, such that, the chemical species mass fractions and the density have their time-means as expectation. For the instantaneous temperature, the Beta-pdf $B_{pdf}(T;a,b)$ has been chosen. This probability density function is defined as

$$B_{pdf}(T;a,b)=\frac{\left(T_b-T_u\right)^{2-a-b}}{B(a,b)}\left(T-T_u\right)^{a-1}\left(T_b-T\right)^{b-1} , \quad (23)$$

where the Beta-function $B(a,b)$ is given by

$$B(a,b)=\left(T_b-T_u\right)^{2-a-b}\int_{T_u}^{T_b}\left(T-T_u\right)^{a-1}\left(T_b-T\right)^{b-1}\frac{dT}{T_b-T_u} ;$$

$$a>0 \ ; \ b>0 . \quad (24)$$

Here, $T_u$ and $T_b$, or the unburned and the burnt temperature, are respectively the lowest-possible value and the highest-possible value of the instantaneous (read: fluctuating) in-flame temperature $T$. These temperature limits will be discussed further below. The expectation Exp($T$) of the fluctuating temperature $T$ is set to the time-mean temperature $\overline{T}$, i.e.

$$\mathrm{Exp}(T)=\int_{T_u}^{T_b}TB_{pdf}(T;a,b)\frac{dT}{T_b-T_u}=T_u+\frac{a}{a+b}\left(T_b-T_u\right) \quad (25)$$
$$=\overline{T} ,$$

while the variance Var($T$) of the fluctuating temperature $T$ is assumed to be approximately equal to a fraction of the maximum possible variance, i.e.

$$\mathrm{Var}(T)=\mathrm{Exp}\left(T^2\right)-\mathrm{Exp}^2(T)=\frac{ab}{(a+b)^2(a+b+1)}\left(T_b-T_u\right)^2$$
$$\cong s\left(\overline{T}-T_u\right)\left(T_b-\overline{T}\right) \ ; \ \frac{1}{2}\le s<1 . \quad (26)$$

Here, the fraction $s$ is called the variance coefficient. By means of (25) and (26) the following expressions for the Beta-pdf parameters $a$ and $b$ may be found:

$$a=\frac{1-s}{s}\frac{\overline{T}-T_u}{T_b-T_u} \ ; \ b=\frac{1-s}{s}\frac{T_b-\overline{T}}{T_b-T_u} . \quad (27)$$

From a statistical point of view, the variance coefficient $s$ can vary between 0.0 and 1.0. In (26) however, $s$ can vary only from 0.5 to 1.0. There are two reasons for this choice. Firstly, Hand et al. (1989) take for $s$ a value of 0.6, based on a wide range of experimental and theoretical results obtained by Missaghi (1987). Secondly, if $s$ can vary only from 0.5 to 1.0, then both the parameters $a$ and $b$ can vary only between 0.0 and 1.0. Consequently, all the possible Beta-pdf's have a shape similar to the shape depicted in figure 2 and thus give the largest fluctuations possible (those near the unburned and the burnt temperatures), the largest probability to occur. Note that the hatched areas $A_u$, $A_i$ and $A_b$ in figure 2 represent probabilities for the case $s$=0.5.

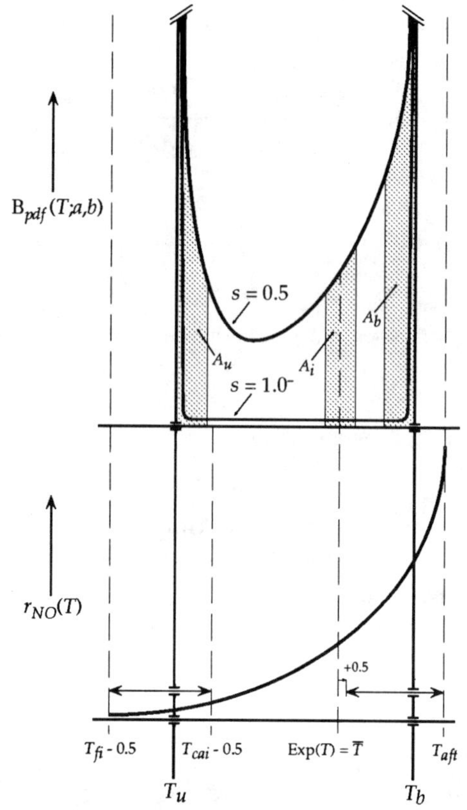


FIGURE 2: GRAPHICAL SUPPORT OF THE PROBABILITY DENSITY FUNCTION SUBMODEL.

To complete the pdf-submodel, expressions have to be given for the unburned and the burnt temperature. The choice of the unburned temperature $T_u$ is rather straightforward; it is nearly everywhere equal to the combustion air inlet temperature (see further below). Unfortunately, there is no such an obvious choice for the burnt temperature $T_b$. The burnt temperature is typically assumed to be either equal to the maximum, time-mean, in-flame temperature or equal to the adiabatic flame temperature. In the authors' opinion both alternatives are inappropriate in the case of turbulent diffusion flames. Firstly, there is an overwhelming experimental evidence that the instantaneous in-flame temperatures may exceed the maximum time-mean in-flame temperature. Secondly, in large-scale turbulent diffusion flames, in which radiation losses are significant, the maximum fluctuating in-flame temperature may be substantially lower than the adiabatic flame temperature. In the present pdf-submodel, the burnt temperature is calculated throughout the whole flame/furnace. For this purpose, use has been made of the local stoichiometry $\lambda_1$ defined by (15).

An imaginary enthalpy increase $\Delta h_{ima}$ may be defined as

$$\Delta h_{ima} = \begin{cases} \overline{\omega}_{C_xH_y} H_1^* + \overline{\omega}_{CO} H_2 & \text{for} \quad \lambda_1 \geq 1 \\ \lambda_1 \left( \overline{\omega}_{C_xH_y} H_1^* + \overline{\omega}_{CO} H_2 \right) & \text{for} \quad \lambda_1 < 1 \end{cases} \tag{28}$$

Thus, assuming that in oxygen rich regions of the flame the two combustibles might burn instantaneously while in oxygen lean regions all the oxygen might be used to burn a portion of the two combustibles instantaneously. The corresponding imaginary temperature increase $\Delta T_{ima}$ is the solution of the algebraic equation

$$\Delta T_{ima} = \frac{\overline{h} + \Delta h_{ima}}{\tilde{c}_p \left( \overline{T} + \Delta T_{ima} \right)} - \overline{T} \ . \tag{29}$$

The unburned and the burnt temperature, $T_u$ and $T_b$, are set to

$$T_u = \min\left(T_{cai}, \overline{T}\right) - 0.5 \ ; \qquad T_b = \overline{T} + \max\left(\Delta T_{ima}, 0.5\right) \ . \tag{30}$$

Here, $T_{cai}$ denotes the combustion air inlet temperature. Let $T_{aft}$ denote the adiabatic flame temperature and $T_{fi}$ the inlet temperature of the natural gas. Then, $T_b$ can vary, dependent on both the level of the local radiative heat losses and the local stoichiometry $\lambda_1$, between $\overline{T}+0.5$ and $T_{aft}$, while $T_u$ can vary between $T_{fi}$ - 0.5 and $T_{cai}$ - 0.5. The value of 0.5 is somewhat arbitrary. This "small" value is needed to make sure that $T_u$ is always less than $\overline{T}$ and that $T_b$ is always larger than $\overline{T}$.

The time-mean rates for formation of thermal-NO and prompt-NO are modeled as follows:

$$\overline{r_{k-NO}} = \mathrm{Exp}\left(r_{k-NO}\right) = \int_{T_u}^{T_b} r_{k-NO}(T) \mathrm{B}_{pdf}(T;a,b) \frac{dT}{T_b - T_u} \ ; \qquad k = t, p \ . \tag{31}$$

The only means to manually control this statistical averaging procedure, is the variance coefficient $s$ (see figure 2). For $s$-values near to 1.0, the largest fluctuations have a very a high probability to occur while the intermediate fluctuations have a negligibly small probability to occur. Consequently, for $s$-values near to 1.0, the pdf-submodel will predict relatively high statistical averages. On the other hand, for $s$-values near to 0.5, all fluctuations have a significant probability to occur. Consequently, for $s$-values near to 0.5, the pdf-submodel will predict relatively low statistical averages.

TABLE 3: INFORMATION ON THE APPLIED COMPUTATIONAL GRIDS FOR THE 12 MW UNSTAGED SWIRLING NATURAL GAS FLAME.

| 12 MW unstaged | 2D fine grid | 3D coarse grid |
|---|---|---|
| **number of live cells in** | (ax.*rad.) | (ax.*rad.*tang.) |
| inlet duct CA&FG2 | 30*68=2040 | 10*17*18=3060 |
| inlet duct FG1 | 88*9=792 | 27*3*18=1458 |
| the quarl | 82*112=9184 | 25*32*18=14400 |
| the furnace | 44*155=6820 | 28*52*18=26208 |
| the chimney | 4*132=528 | 3*37*18=1998 |
| **total number of live cells** | 19364 (76 %) | 47124 (64 %) |
| **total number of cells** (live, dead, boundary) | 162*157=25434 (100 %) | 68*54*20=73440 (100 %) |
| **size of** | (ax.*rad.) | (ax.*rad.) |
| smallest live cell (mm$^2$) | 1.07*0.76 | 2.88*2.07 |
| largest live cell (mm$^2$) | 509.58*224.31 | 627.88*255.99 |
| **largest aspect ratio in** | | |
| the quarl | 22 | 17 |
| the furnace | 671 | 303 |
| **smallest growth factor** | 0.84 | 0.73 |
| **largest growth factor** | 1.19 | 1.48 |

## NUMERICAL ASPECTS AND APPLIED COMPUTATIONAL GRIDS

All presented solutions have been obtained using staggered, polar cylindrical computational grids. As to minimize numerical diffusion, in the discretization of the balance equations the quadratic upstream differencing scheme bounded-QUICK of Leonard (1987) has been applied. Pressure-velocity coupling is accomplished by means of the PISO-method of Issa (1986). The applied under-relaxation factors and numbers of sweeps-per-iteration for pressure (0.975; 100), velocities (0.5-0.6; 1), turbulence quantities (0.7-0.8; 1), and enthalpy and chemical species (0.9-1.0; 20-25) guarantee non-frozen solutions.

Information on the applied 2D and 3D computational grids for the 12 MW unstaged swirling natural gas flame is given in table 3. Note that the 2D grid is very fine in the near-burner zone while the 3D grid is coarse there. In the 2D grid, the natural gas injection is modeled as a cylindrical ring. The 3D grid covers an angle of 15° and thus comprises three of the 72 natural gas injection holes. Additionally, a third grid has been applied for the 12 MW flame, namely, a 2D grid with the same coarseness as the 3D grid. For the 4 MW unstaged swirling natural gas flame a 2D fine grid, similar to the one for the 12 MW flame, has been applied. Obtaining full convergence on the 2D fine grid requires about four to six weeks of continuous computing on a Sun SparcStation 10 equipped with a 41 processor. Reasonable convergence on the 3D coarse grid is obtained after three to four months of continuous computing on the same workstation. At full [reasonable] convergence, the normalized residuals of pressure, velocities, turbulence quantities, enthalpy and chemical species are less than $3*10^{-3}$ [$5*10^{-3}$], $2.5*10^{-5}$ [$6*10^{-5}$], $2*10^{-6}$ [$2*10^{-5}$], $7.5*10^{-8}$ [$1*10^{-7}$] and $7.5*10^{-8}$ [$2*10^{-6}$] respectively. Full convergence is reached only on fine grids.

## MAIN FLAME PROPERTIES AND NITRIC OXIDE PREDICTIONS

### Fluid Dynamics

A comparison between the 2D-fine-grid predictions and the measurements of axial velocity of both the 4 MW flame and the 12 MW flame is presented in figure 3. This comparison shows that full, near-burner zone fluid dynamics similarity is obtained in the predictions whereas in the measurements some differences occur. The measured internal recirculation zone (IRZ) of the 4 MW flame is, relatively, longer and slightly thicker than the one of the 12 MW flame. This is quantified in table 4, showing scaled lengths ($X/D_0$) and diameters ($D/D_0$) of the predicted and measured IRZ's. Here the lengths of the IRZ's are measured from the quarl outlets.

TABLE 4: SCALING CHARACTERISTICS OF THE INTERNAL RECIRCULATION ZONE, IRZ (X=0 : QUARL OUTLET). PREDICTIONS CORRESPOND WITH THE 2D-FINE COMPUTATIONAL GRIDS.

| Flame | $X_{IRZ}/D_0$ | | $D_{IRZ}/D_0$ | |
|---|---|---|---|---|
| | pred. | meas. | pred. | meas. |
| **12 MW unstaged** | 1.665 | 0.965 | 1.611 | 1.730 |
| (tolerance) | - - | (0.783;1.148) | - - | (1.639;1.821) |
| **4 MW unstaged** | 1.659 | 1.987 | 1.593 | 1.893 |
| (tolerance) | - - | (1.672;2.303) | - - | (1.735;2.019) |

For the 12 MW flame a 3D-coarse-grid prediction has been performed as well. The comparison between axial and tangential velocities obtained by means of this prediction and those obtained by means of the 2D-fine-grid prediction is given in figure 4. This comparison shows the significant improvement to the quarl zone flow field resulting from modeling individual natural gas jets. That the improvement does not stem from the coarseness of the 3D grid becomes obvious when comparing the 3D-coarse-grid predictions with the 2D-coarse-grid predictions that are shown in figure 4 as well. Further improvement to the prediction of the near-burner zone flow field, as obtained by means of the 3D coarse grid, is believed to be achievable through applying 3D fine grids, eventually, together with a more advanced turbulence model, e.g. the Reynolds Stress Model. A reasonable 3D fine grid that incorporates just one natural gas injection hole contains about 80000 computational cells. Modeling three injection holes (15° angle) requires about 200000 cells. Modeling nine injection holes as well as one fuel staging pipe (45° angle for staged combustion) requires about 560000 cells.

### Temperature Distribution

The 2D-fine-grid predictions and the measurements of temperature of both the 4 MW flame and the 12 MW flame are compared in figure 5. In both the predictions and the measurements, the non-similarity of the thermal boundary conditions of the furnaces (see table 5) results in higher temperatures in the external recirculation zone (ERZ) of the 4 MW flame. In the measurements, the temperatures in the ERZ's of the 4 MW flame and the 12 MW flame are 1300-1400 K and 1150-1350 K, respectively. In the predictions, the temperatures in the ERZ's of the 4 MW flame and the 12 MW flame are about 1400 K and 1300 K, respectively. The measured peak temperatures in the 4 MW flame and the 12 MW flame are 1949 K and 1908 K, respectively. The corresponding predicted values are 1823 K and 1751 K.

TABLE 5: INDICATION OF THE DEVIATION FROM THERMAL SIMILARITY DUE TO DIFFERENCES IN THE FURNACES. PREDICTIONS CORRESPOND WITH THE 2D-FINE COMPUTATIONAL GRIDS.

| Flame | Heat extraction of cooling loops and/or water-cooled jacket | Percentage of thermal input | Flue gas temperature |
|---|---|---|---|
| **12 MW unstaged** | | | |
| predicted | 4595 kW | 38.5% | 1412 K |
| measured | 4560 kW | 38.2% | 1423 K |
| | *5 loops and jacket between* | *of 11933 kW* | |
| *$X/D_0$* | *1.18 and 10.93* | | |
| *cooled area* | *92.0%* | | |
| **4 MW unstaged** | | | |
| predicted | 1182 kW | 29.4% | 1394 K |
| measured | 1276 kW | 31.7% | 1390 K |
| | *6 loops between* | *of 4021 kW* | |
| *$X/D_0$* | *4.62 and 13.85* | | |
| *cooled area* | *16.9%* | | |

The improvement in the predictions resulting from 3D modeling, i.e. from modeling individual natural gas jets, is shown in figure 6.

### Major Chemical Species Concentrations

The figures 7 and 8 show predicted and measured scaling characteristics of oxygen and carbon monoxide concentrations. The similarity in the predictions is the result of the modeled turbulent combustion rates being solely mixing-dependent. As the measurements show a significant deviation from similarity, it would indicate that turbulent combustion is not governed by

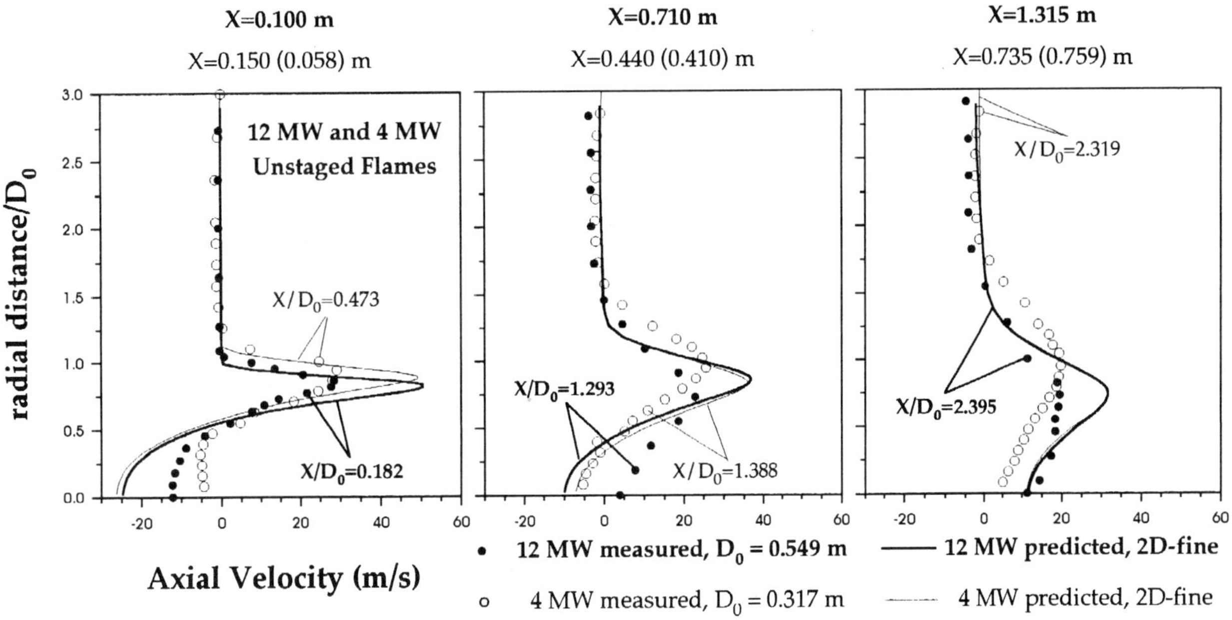


FIGURE 3: SCALING CHARACTERISTICS OF AXIAL VELOCITY (X=0 : QUARL OUTLET).

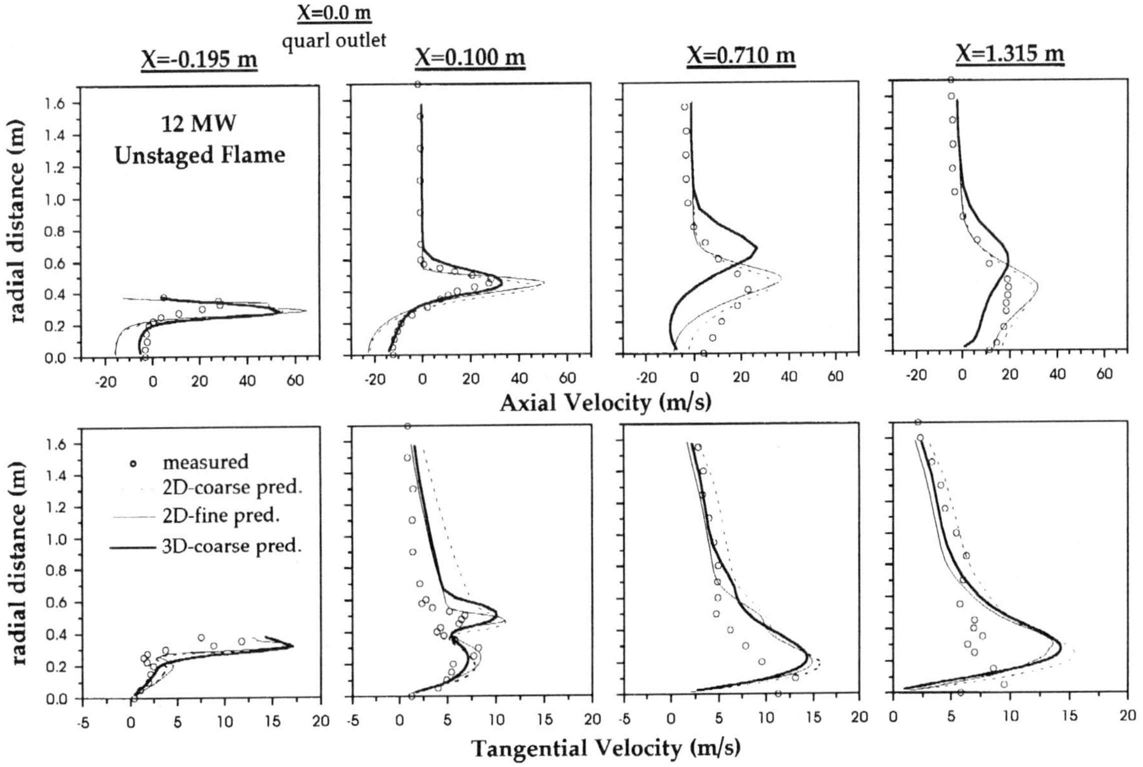


FIGURE 4: COMPARISON BETWEEN THE PREDICTIONS OF AXIAL AND TANGENTIAL VELOCITY OBTAINED BY MEANS OF THE 2D-COARSE, THE 2D-FINE AND THE 3D-COARSE COMPUTATIONAL GRID.

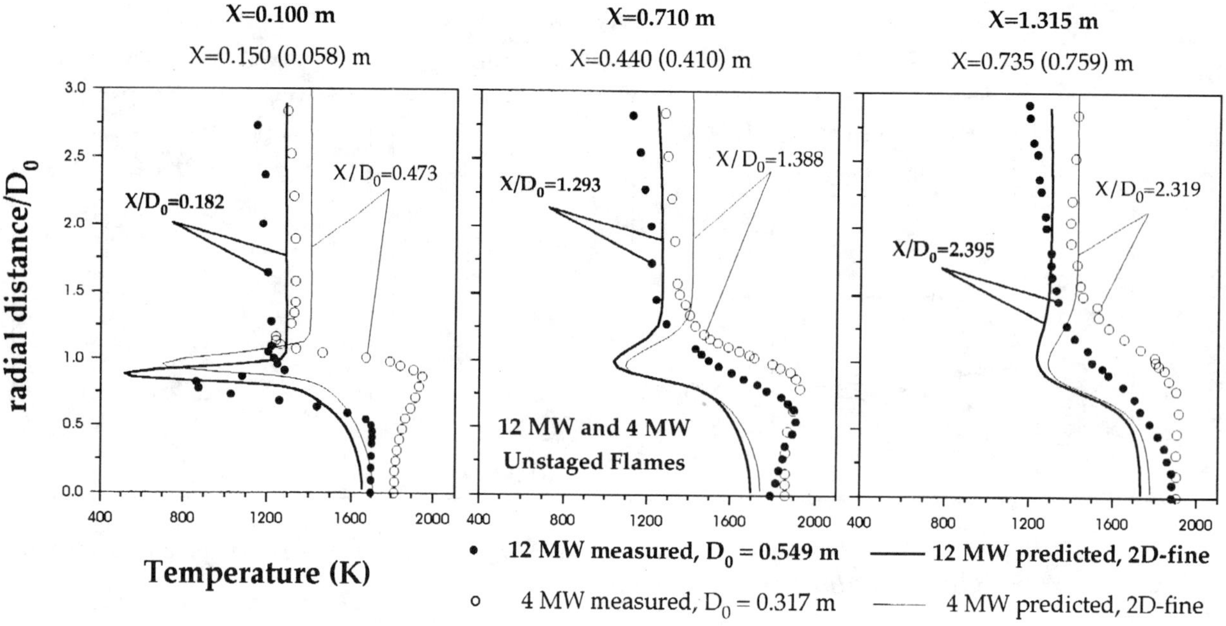


FIGURE 5: SCALING CHARACTERISTICS OF TEMPERATURE (X=0 : QUARL OUTLET).

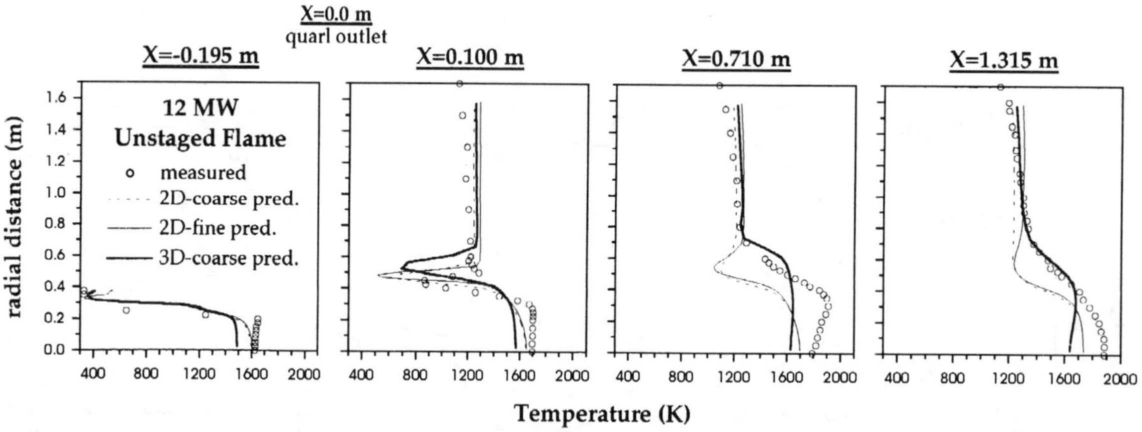


FIGURE 6: COMPARISON BETWEEN THE PREDICTIONS OF TEMPERATURE OBTAINED BY MEANS OF THE 2D-COARSE, THE 2D-FINE AND THE 3D-COARSE COMPUTATIONAL GRID.

mixing only. The non-similarity of the combustion characteristics in the shear layers of the flames can be nicely observed in the oxygen concentration distributions. On the traverse X=0.710 m, the 12 MW flame has a peak oxygen concentration of 5.30 vol%, dry (see middle plot in figure 7). This traverse is positioned 0.392+0.710=1.102 m downstream of the natural gas injection holes. On the traverse X=0.735 m, the 4 MW flame has a peak oxygen concentration of 5.66 vol%, dry (see left plot in figure 7). This traverse is positioned 0.226+0.735=0.961 m downstream of the natural gas injection holes. As the scaling criterion is constant-velocity-scaling, it can be concluded that the degree of combustion in the shear layer depends on the residence time of fluid pockets in this layer. This actually implies that incorporation of chemical reaction rates into the turbulent combustion model is desired. However, incorporating chemical reaction rates requires that turbulent combustion is modeled in a more complex manner, for example by means of a pdf-approach similar to the one applied in the $NO_x$-postprocessor. This implies a drastic increase of the computational efforts. Working with simplified models, like the model described in this paper, is therefore attractive, particularly for industrial and engineering applications. It is suggested that the eddy-break-up submodel can be expanded by adopting more reactions to the present reaction scheme, including overall chemical reaction rates for each individual reaction and/or altering the mixing rate coefficient $A_{mix}$.

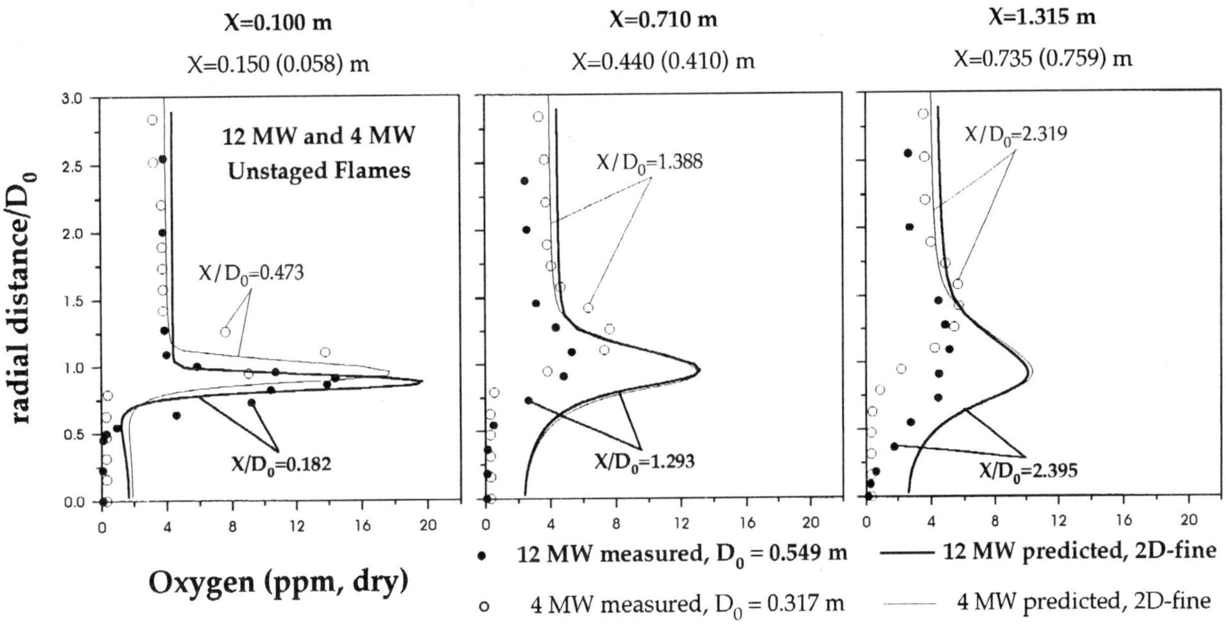


FIGURE 7: SCALING CHARACTERISTICS OF OXYGEN (X=0 : QUARL OUTLET).

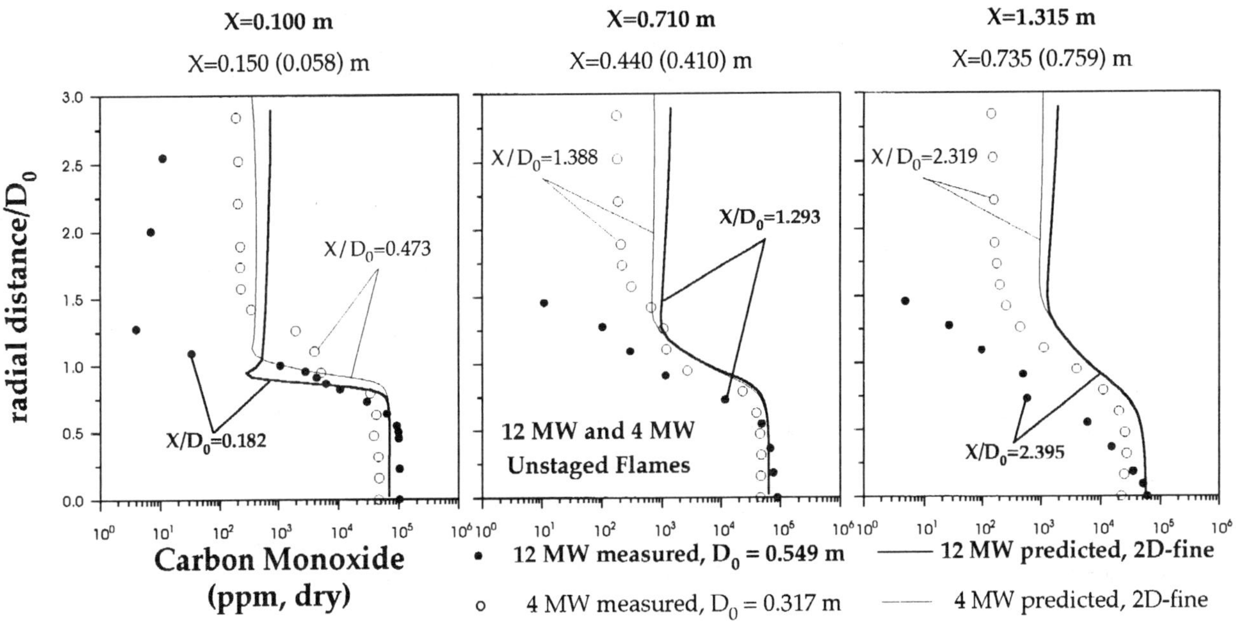


FIGURE 8: SCALING CHARACTERISTICS OF CARBON MONOXIDE (X=0 : QUARL OUTLET).

Predicted and measured scaled flame lengths ($X/D_0$) are presented in table 6. Here the flame length is based on the summed amounts of carbon monoxide and unburned hydrocarbons. From table 6 and the figures 7 and 8, the conclusion can be drawn that, in the measurements, the 4 MW flame is, relatively, slightly shorter and somewhat thicker than the 12 MW flame. Furthermore from table 6, it can be concluded that at scaled distances $X/D_0$ larger than 2.5, predicted carbon monoxide and unburned hydrocarbons are, firstly, present in too high amounts and, secondly, do not decay fast enough, as compared to the measurements. A possible explanation for this discrepancy in predictions and measurements is believed to lie in the fact that the reaction

$$CO + OH \longleftrightarrow CO_2 + H$$

is not represented in reaction scheme RS1. The reaction of CO with OH-radicals is said to be of significant importance in the combustion of hydrocarbon fuels and even to be preferred over the reaction of CO with $O_2$, especially at temperatures higher than 1000 K (Westenberg, 1975, and Miller and Bowman, 1989). Actually, in the region where the carbon monoxide concentration should be decreasing rapidly, the predicted water concentration reaches levels up to 16 vol%, dry. This amount of water is definitely sufficient to reduce, through the above reaction, a portion of the carbon monoxide present. Consequently, more oxygen will be available for the unburned hydrocarbons.

The predicted and measured furnace outlet values are given in table 7.

## Nitric Oxide Concentration

Figure 9 shows predicted and measured nitric oxide concentrations. In the predictions, the nitric oxide concentrations in the ERZ of the 4 MW flame are higher than those in the ERZ of the 12 MW flame merely because of the non-similarity in the thermal boundary conditions. In the measurements, there is an additional reason for the nitric oxide concentrations in the 4 MW flame being higher than those in the 12 MW flame, namely the non-similarity of the combustion characteristics in the shear layers of the flames that results in higher temperatures in the shear layer of the 4 MW flame. This is clearly visible in figure 9 as the nitric oxide concentrations in the near-burner zone of the 4 MW flame are much higher than those in the near-burner zone of the 12 MW flame.

The predicted and measured NO and $NO_x$ emissions of the two flames are given in table 7. The increase in the predicted NO due to the non-similarity of the thermal boundary conditions is about 20% (46/37=1.24 and 83/71=1.17). Note that this increase stems from the thermal-NO formation mechanism only and not from the prompt-NO formation mechanism. The measured $NO_x$ for the 4 MW flame is 40% higher than that for the 12 MW flame (73/52=1.40). So, roughly, it can be put that half of this 40%, i.e. about 10 ppm, is due to the non-similarity of the thermal boundary conditions while the other half is attributable to the non-similarity of the combustion characteristics in the shear layers of the flames. The predictions show that prompt-NO amounts to 26-28% of the total NO emission of the 4 MW flame while in the 12 MW flame the prompt-NO contribution is 34-38%.

## CONCLUSIONS

Predictions of two unstaged (high-$NO_x$) swirling natural gas flames of 4 MW and 12 MW thermal input have been performed and compared with the measurements. These predictions and measurements form part of a large project concerned with scaling of a generic natural gas burner over a thermal input range that stretches from 30 kW up to 12 MW. In scaling the burners, the constant-velocity-scaling criterion has been applied together with geometric similarity.

TABLE 6: SCALING CHARACTERISTICS OF THE FLAME LENGTH (X=0 : QUARL OUTLET). PREDICTIONS CORRESPOND WITH THE 2D-FINE COMPUTATIONAL GRIDS.

| Flame | $X_{flame}/D_0$ | | $X_{flame}/D_0$ | |
|---|---|---|---|---|
| | pred. **P1** | meas. **M1** | pred. **P2** | meas. **M2** |
| **12 MW unstaged** | 4.367 | 4.155 | 6.933 | 5.123 |
| **4 MW unstaged** | 4.533 | 3.785 | 7.467 | 4.890 |

**P1**: based on *0.05*-contour-line of summed wet mass fraction CO&UHC
**M1** : based on *0.005*-contour-line of summed dry volume fraction CO&UHC
**P2**: based on *0.01*-contour-line of summed wet mass fraction CO&UHC
**M2** : based on *0.001*-contour-line of summed dry volume fraction CO&UHC

TABLE 7: PREDICTED AND MEASURED FLUE GAS COMPOSITIONS. PREDICTIONS CORRESPOND WITH THE 2D-FINE COMPUTATIONAL GRIDS. SENSITIVITY TO THE VARIANCE COEFFICIENT *s* IS DEPICTED AS WELL: [*s*=0.5]#[*s*=1.0`].

| **chemical species** | **dim.** | **12 MW unstaged** | | **4 MW unstaged** | |
|---|---|---|---|---|---|
| | | pred. | meas. | pred. | meas. |
| $O_2$ | vol%, dry | 3.00 | 2.9 | 2.95 | 3.0 |
| CO | ppm, dry | 913 | 2 | 323 | 0 |
| UHC | ppm, dry | 62 | 8 | 18 | 0 |
| $CO_2$ | vol%, dry | 10.05 | 10.0 | 10.18 | 10.3 |
| $H_2O$ | vol% | 16.27 | -- | 16.36 | -- |
| $NO_x$ | ppm, dry | -- | 52 | -- | 73 |
| NO | ,, | 37#71 | -- | 46#83 | 71 |
| thermal-NO | ,, | 23#47 | -- | 33#61 | -- |
| prompt-NO | ,, | 14#24 | -- | 13#22 | -- |

The applied mathematical models for main flame properties and nitric oxide predictions have been discussed. The two-step-reaction turbulent combustion model is based on the eddy-break-up concept, i.e. mixing. In the $NO_x$-postprocessor, thermal-NO and prompt-NO formation rates are obtained by means of statistically averaging of the corresponding chemical reaction rates over the fluctuating temperature.

In the predictions, the 4 MW flame and the 12 MW flame scale ideally. This is with the exception of a slight non-similarity of both temperature distribution and nitric oxide concentration. The predicted non-similarity stems from the non-similarity of the

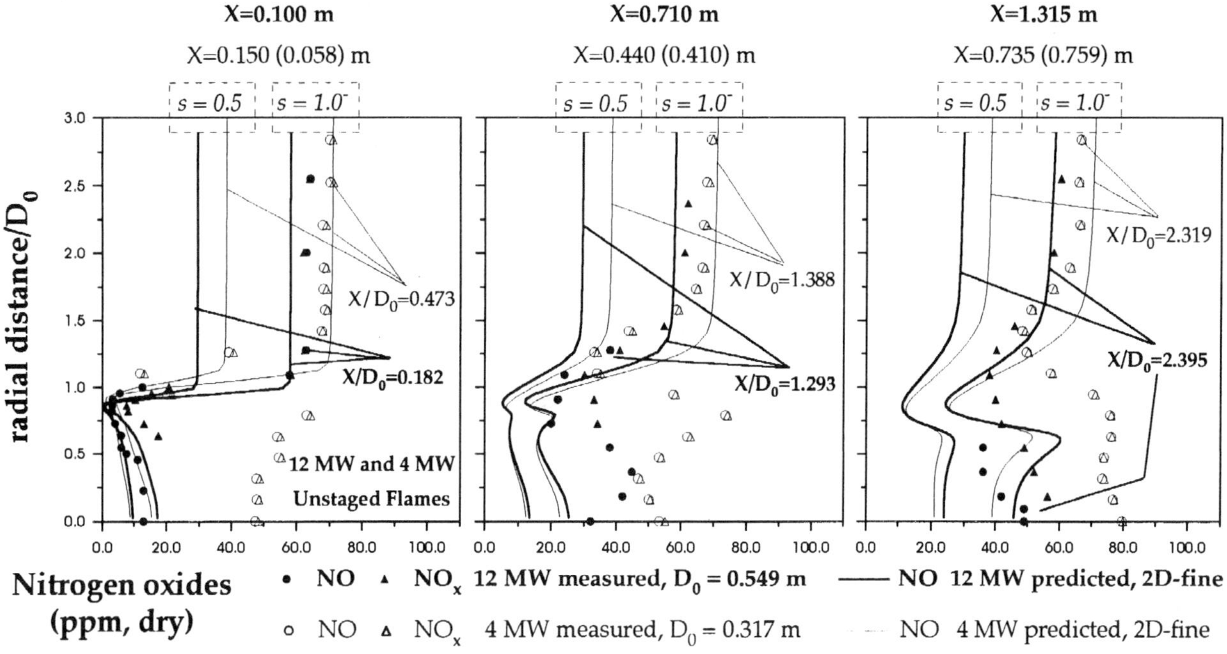


FIGURE 9: SCALING CHARACTERISTICS OF NITRIC OXIDE (X=0 : QUARL OUTLET). SENSITIVITY TO THE VARIANCE COEFFICIENT $s$ IS DEPICTED AS WELL.

thermal boundary conditions of the two furnaces. 3D-modeling, i.e. modeling of individual natural gas jets, shows a significant improvement to the predictions and is to be preferred over 2D-modeling.

In the measurements, the overall flame shape and near-burner zone fluid dynamics were similar for the 4 MW flame and the 12 MW flame. The near-burner zone temperatures of the 4 MW flame were higher than those of the 12 MW flame. This is partially attributed to the non-similarity of the thermal boundary conditions of the two furnaces. The temperatures in the external recirculation zone of the 4 MW flame varied between 1300 K and 1400 K whereas those in the 12 MW flame varied between 1150 K and 1350 K. The total heat extraction of cooling surfaces was 1276 kW for the 4 MW flame against 4560 kW for the 12 MW flame. Non-similarity of the combustion characteristics in the shear layers of the flames was observed and some residence time effects were identified.

As the modeled turbulent combustion rates are solely mixing-dependent, the predicted combustion characteristics in the shear layers of the flames are similar and thus yield near-burner zone similarity, both dynamically and thermally. From the measurements, the conclusion has been drawn that the degree of combustion in the shear layer depends on the residence time of fluid pockets in this layer. This actually implies that incorporation of chemical reaction rates into the turbulent combustion model is desired. However, incorporating chemical reaction rates requires that turbulent combustion is modeled in a more complex manner, for example by means of a statistical approach similar to the one applied in the $NO_x$-postprocessor. This implies a drastic increase of the computational efforts. Working with simplified models, like the model described in this paper, is therefore attractive, particularly for industrial and engineering applications. It is suggested that the eddy-break-up submodel can be expanded by adopting more reactions to the present reaction scheme, including overall chemical reaction rates for each individual reaction and/or altering the mixing rate coefficient $A_{mix}$.

Finally, it is concluded that, when applying the two-step-reaction scheme

$$C_xH_y + \left(\frac{x}{2}+\frac{y}{4}\right)O_2 \rightarrow xCO + \frac{y}{2}H_2O \ ;$$

$$CO + \frac{1}{2}O_2 \rightarrow CO_2 \ ,$$

the fast axial decreases in the carbon monoxide and unburned hydrocarbons concentrations as they occur in the measurements are not obtained. A possible explanation for this discrepancy between predictions and measurements is believed to lie in the fact that the reaction

$$CO + OH \longleftrightarrow CO_2 + H$$

is not represented in the above reaction scheme.

## ACKNOWLEDGEMENTS

The authors would like to thank the IFRF Joint Committee for the permission to publish this paper. The project was financed by the Gas Research Institute with the supervision of Dr. B.V. Gemmer.

## REFERENCES

Bertolo, M., Sayre, A.N., Dugué, J., and Weber, R., 1993, "Scaling Characteristics of Aerodynamics and Low-$NO_x$ Properties of Industrial Natural Gas Burners", The SCALING 400 Study, Part V: The 4 MW Test Results, IFRF Doc. No. F40/y/12.

Bowman, C.T., 1975, "Kinetics of Pollutant Formation and Destruction in Combustion", *Prog. Energy Combust. Sci.*, Vol. 1, pp. 33-47.

Edwards, D.K., 1976, "Molecular Gas Band Radiation", *Advances in Heat Transfer*, Ed.'s T.F. Irvine, Jr. and J.P. Harnett, Academic Press, New York, Vol. 12, pp. 115-193.

Gosman, A.D., and Lockwood, F.C., 1972, "Incorporation of a Flux Model for Radiation into a Finite-Difference Procedure for Furnace Calculations", *Fourteenth Symposium (Int.) on Combustion*, The Combustion Institute, pp. 661-672.

Hand, G., Missaghi, M., Pourkashanian, M., and Williams, A., 1989, "Experimental Studies and Computer Modeling of Nitrogen Oxides in a Cylindrical Natural Gas Fired Furnace", *The IFRF Ninth Members Conference*, Noordwijkerhout, The Netherlands.

Issa, R.I., 1986, "Solution of the Implicitly Discretized Fluid Flow Equations by Operator-Splitting", *J. of Computational Physics*, Vol. 62, No. 1, pp. 40-65.

Lallemant, N., and Weber, R., 1993, "Radiative Properties Models for Computing Non-Sooty Natural Gas Flames", IFRF Doc. No. G08/y/2.

Launder, B.E., and Spalding, D.B., 1974, "The Numerical Computation of Turbulent Flows", *Comp. Meth. in Appl. Mech. and Eng.*, Vol. 3, p. 269.

Leonard, B.P., 1987, "Locally Modified QUICK Scheme for Highly Convective 2D and 3D Flows", *Fifth International Conference on Numerical Methods in Laminar and Turbulent Flows,* Ed.'s C. Taylor, W.G. Habashi and M.M. Hafez, Vol. 5, Part 1, Montreal, Canada.

Magnussen, B.F., and Hjertager, B.H., 1976, "On Mathematical Modeling of Turbulent Combustion with Emphasis on Soot Formation and Combustion", *Seventeenth Symposium (Int.) on Combustion*, The Combustion Institute, pp. 719-729.

Miller, J.A., and Bowman, C.T., 1989, "Mechanism and Modeling of Nitrogen Chemistry in Combustion", *Prog. Energy Combust. Sci.*, Vol. 15, pp. 287-338.

Missaghi, M., 1987, "Mathematical Modeling of Chemical Sources in Turbulent Combustion", Ph.D. Thesis, University of Leeds.

De Soete, G.G., 1975, "Overall Reaction Rates of NO and $N_2$ Formation from Fuel Nitrogen", *Fifteenth Symposium (Int.) on Combustion*, The Combustion Institute, pp. 1093-1102.

Visser, B.M., Smart, J.P., van de Kamp, W.L., and Weber, R., 1990, "Measurements and Predictions of Quarl Zone Properties of Swirling Pulverized Coal Flames", *Twenty-third Symposium (Int.) on Combustion*, The Combustion Institute, pp. 949-955.

Weber, R., Driscoll, J.F., Dahm, W.J.A., and Waibel, R.T., 1993a, "Scaling Characteristics of Aerodynamics and Low-$NO_x$ Properties of Industrial Natural Gas Burners", The SCALING 400 Study, Part I: Test Plan, IFRF Doc. No. F40/y/08 or GRI Topical Report GRI-93/0227.

Weber, R., Sayre, A.N., Dugué, J., and Horsman, H., 1993b, "Scaling Characteristics of Aerodynamics and Low-$NO_x$ Properties of Industrial Natural Gas Burners", The SCALING 400 Study, Part II: The 12 MW Test Results, IFRF Doc. No. F40/y/09 or GRI Topical Report GRI-93/0079.

Westenberg, A.A., 1975, "Kinetics of NO and CO in Lean, Premixed Hydrocarbon-Air Flames", *Combust. Sci. and Techn.*, Vol. 4, pp. 59-64.

FACT-Vol. 18, Combustion Modeling, Scaling and Air Toxins
ASME 1994

# RESULTS FROM THE DEPARTMENT OF ENERGY'S ASSESSMENT OF AIR TOXIC EMISSIONS FROM COAL-FIRED POWER PLANTS

**Charles E. Schmidt and Thomas D. Brown**
Pittsburgh Energy Technology Center
U.S. Department of Energy
Pittsburgh, Pennsylvania

ABSTRACT

The Department of Energy has developed a program to assess the toxics emissions from coal-fired power plants. The program involved field testing eight coal-fired utility boilers for the hazardous air pollutants contained in Title III of the Clean Air Act Amendments of 1990. Data are presented on the concentrations of specific trace and minor species in all the major input and output streams of the power plants. Emission factors were determined for some of the hazardous air pollutants emanating from the power plant stacks.

INTRODUCTION

Coal has been, and will continue to be, the predominant fossil fuel used to generate electricity in the United States. In 1991, 777 million tons of coal (78% of the coal mined in the US) were combusted to produce steam for electricity generation (EIA, 1991). There are, however, undesirable products from coal combustion emitted into the atmosphere that must be controlled. The passage of the 1990 Clean Air Act Amendments ushered in a new era of environmental focus relating to the use of coal for energy production. Title III, Hazardous Air Pollutants (HAPS), of the Clean Air Act Amendments contains provisions that require the Environmental Protection Agency (EPA) to decide if air toxics (Hazardous Air Pollutants) emissions from electric steam generating units pose a threat to human health.

The EPA will conduct a risk assessment study that will use either actual or extrapolated air toxics emission data from all electric generating units larger than 25 MW(e) to determine if regulation of power plants is necessary for Title III HAPS. At present, there are insufficient data of good scientific quality on power plant emissions available to conduct a risk assessment that will properly evaluate the health hazards associated with electric power generating stations. Therefore, the U.S. Department of Energy (DOE) and other interested organizations developed programs to determine the air toxic emissions from utility boilers to augment the database for the EPA study.

In January, 1993, the DOE initiated a program to determine the concentrations of toxic emissions emanating from coal-fired electric utility boilers. Additional goals of the program were to: (1) determine the ability of conventional pollution control equipment to remove selected species from flue gas, (2) determine the concentrations of toxic substances associated with particulate matter as a function of particle size, (3) quantify toxic materials on the surfaces of particulate matter, (4) determine material balances for selected trace elements in all major input and output streams of a utility boiler, and (5) measure the concentrations of condensible toxic species in flue gas.

The DOE air toxics assessment program was structured in two phases that would provide emissions data from a total of sixteen power plants (eight power plants in each phase). Results from the field sampling and subsequent laboratory analyses of seven of the eight power plants will be presented in this report and discussed in terms of the program goals. The data from the air toxics assessment of the Bailly power plant were not available for this report. The main focus of this manuscript will be the results for the trace metals contained on the Title III HAPS list.

## EXPERIMENTAL

### Phase I Power Plants

The eight power plants are listed in Table 1 along with the DOE contractor responsible for obtaining the air toxics emissions data at each respective site. Also included in Table 1 are the pollution control devices, boiler size, and type of coal burned, for each respective power plant. The power plants chosen for Phase I have several pollution control technologies, and also include plants that burn bituminous, subbituminous, and lignite coals. The selection of power plants was coordinated with the Electric Power Research Institute, who is conducting a similar program, and with the Utility Air Regulatory Group and the EPA to acquire adequate emissions data from all the major coal-fired boilers types for the Title III study.

Three of the utilities in Phase I are also sponsoring demonstration projects in the Innovative Clean Coal Technology Program (ICCT) of DOE. They are; (1) the Bailly station of Northern Indiana Public Service Company which is hosting the Pure Air wet limestone scrubber demonstration, (2) Georgia Power's plant Yates with the Chioda CT-121 Jet Bubbling Reactor wet limestone scrubber, and (3) the Niles plant of Ohio Edison where the SNOX process is being demonstrated on a 35 MW(e) slipstream. At the Niles plant, air toxics emissions, material balances, and pollutant removal device efficiencies were determined for both the SNOX process (35MW(e) slipstream) and the Niles plant at full load (100MW(e)) with the SNOX process not in operation.

The ICCT plants were included in the air toxics emissions assessment program to provide data on pollutant removal capabilities of advanced pollution control technologies under development. Furthermore, the data will assist the EPA in complying with the Title III requirement that the risk assessment reflect the boiler population after all the Clean Air Act Amendments titles are enacted, i.e., in the year 2000.

The suite of power plants shown in Table 1 have a wide variety of boiler types including cyclones at Bailly, Baldwin, and Niles, tangential-fired boilers at Yates and Coal Creek, front-fired and drum-type at Springerville and Clay Boswell, respectively, and a cell burner at Cardinal. Sulfur control technologies include a spray dryer absorber at Springerville, limestone scrubbers at Yates and Bailly, a lime scrubber at Coal Creek, and a catalytic system that converts $SO_2$ to $SO_3$ which is then removed as sulfuric acid at Niles (SNOX process). Nitrogen oxides emissions are abated at Springerville and Coal Creek by using over-fire air combustion techniques, and by selected catalytic reduction as part of the SNOX process. Particulate control is effected by baghouses at Springerville, Clay Boswell, and in the SNOX process, with the rest of the power plants employing electrostatic precipitators.

### Field Sampling

The field sampling activities were designed to provide for a total of seven days of sampling for the solid, liquid, and gas streams. Three inorganic compound and three organic compound sampling days, plus one day for field blank determinations, were scheduled for flue gas sampling at each site. A typical sampling schedule for both inorganic and organic sampling days is shown in Figure 1 and Figure 2, respectively. Note that solid and liquid samples were collected each day. Each contractor provided for on-site QA/QC oversight during the field sampling to ensure the integrity of the samples. Additionally, the EPA provided a third-party auditor who monitored the QA/QC programs of the DOE contractors and also conducted field sampling QA/QC activities at each site.

Extensive power plant operating data were collected during sampling to provide for determination of process stream flows and calculation of component material balances. Samples were collected from all power plant input and output streams including process water streams and coal pile runoff (if available). Simultaneous sampling at the inlet and outlet of the pollution control devices was conducted to provide removal efficiency data for selected trace materials. The power plant was operated in as near steady-state manner as possible to provide representative samples of air toxics emissions. Also, when possible, soot blowing activities were suspended during sampling to promote consistent sampling for air toxics.

Table 2 contains the referenced flue gas sampling methods employed in this study, and the targeted analytes. For the most part, the same sampling methods developed by the EPA for municipal solid waste combustors, or modifications thereof, were used to sample the major power plant process streams.

### Sample Analysis

About 1000 analytical determinations were necessary to characterize each power plant for air toxics emissions. These included analyzing samples from all the site process streams, blank and QA/QC samples, as well as laboratory calibration standards and sample blanks. The analytical methods used to measure the concentrations of trace species in the field samples are listed in Table 3. In some cases, specific methods were used by individual contractors for selected trace elements and compounds. These are not listed in Table 3, but are contained in the respective DOE reports for each power plant (Dismukes and Bush, 1994; England, et al., 1994; Jackson, et al., 1994a; Jackson, et al., 1994b; Flora, et al., 1994; Sverdrup, et al., 1994a; Sverdrup, et al., 1994b; Sverdrup, et al., 1994c). All of the analytes listed in Table 3 with the exception of barium, boron, copper, molybdenum, vanadium, and ammonia are contained in the HAPS list in Title III of the Clean Air Act Amendments of 1990.

## RESULTS AND DISCUSSION

### Material Balances

Calculation of material balances for selected trace and minor elements requires accurate measurements of the concentrations of the respective elements in all the major process streams (input and output) and concise determinations of process stream flows. The results can be used to measure data quality, determine removal efficiencies of the pollution control devices, and compute the partitioning of the trace elements throughout the entire power plant. With regard to data quality, a performance goal of material balances closures (average values) from 70 to 130 percent was established for the sixteen trace elements listed in Table 3. Also, material balance closures for a few minor elements (titanium, potassium, silicon, and iron) were used as a means to assess mass flow measurements and assumptions, and overall data quality.

Material balance closures for minor elements potassium and titanium, and trace elements arsenic, cadmium, lead, and mercury are given in Table 4 for the seven plants and the SNOX process. For the minor element titanium, as expected, the material balance closure met the program performance goal. The closure for potassium exceeded the program goal at sites 3 and 6, and was not determined at site 1. At site 6, the non-closure for potassium was attributed to possible analytical bias in the analysis of the coal ash (Flora, et al., 1994). Material balance closures were generally poor for all the elements at site 3. This could be the result of the method used by England, et al. (1994) to ash the coal prior to measuring the trace and minor elements in the coal samples. Closure for a specific trace element was quite variable among the power plants. On the other hand, the material balance closures for all the trace and minor elements at sites 4 and 5 were fairly consistent. For the most part, the variation in material balance closures could be attributed to difficulties in determining mass flows and to the concentrations of some of the trace species being at or below the method detection limit.

<u>Distribution of HAPS as a Function of Particle Size</u>

When coal is combusted, trace elements associated with the mineral matter are released in both solid and vapor forms. Those trace elements not vaporized during combustion will become part of the bottom ash stream or exit the combustor entrained in the flue gas as flyash matter. The trace elements that are vaporized during combustion will exit the combustor as gases, and subsequently condense either as submicron particles, or on the surface of particulate matter in the flue gas stream. The condensation of trace elements on particles results in an enrichment in concentration of those specific elements (Coles, et al., 1979) on very fine particles (less than 5μm in diameter) due to their higher surface area.

Inhalation is the main pathway by which air toxics released by coal-fired power plants can enter the human body and affect human health. Electrostatic precipitators, the predominant particulate control technology used by the utility industry, are very effective at removing particles greater than 10 μm in size. However, this efficiency drops off considerably for particles less than 5 μm. The respirable size fraction of particles, nominally the 0.1 to 5 μm fraction, is most likely to impact human health. Therefore, an important goal of this air toxics assessment program was to determine the distribution of HAPS as a function of particle size.

At each power plant, the concentrations of all the trace metal species associated with particulate matter in three size ranges (> 10 μm, 10 - 5 μm, and < 5 μm) were measured. Enrichment for a specific element can be estimated by dividing the weight percent of the respective trace element in the smallest particulate size fraction by the analogous data for the largest size fraction. Trace elements exhibiting the most enrichment in fine particles were cadmium, molybdenum, arsenic, lead, chromium, and antimony. Fine particle enrichment was not significant for beryllium, nickel, and manganese.

<u>Comparison of Methods to Determine Mercury</u>

There is considerable interest in determining the amount of mercury released from the combustion of coal to generate electricity. Title III of the Clean Air Act of 1990 requires the EPA to conduct three separate investigations dealing with determining the sources and estimating the amounts of mercury emissions, as well as assessing the environmental and human health impacts of mercury emissions. Specifically, the EPA is responsible for reporting to Congress on the mercury emissions from electric utility steam generating units, municipal waste combustion units, and other sources. A second report involving mercury focuses on the atmospheric deposition of HAPS (including mercury) in the Great Lakes, Chesapeake Bay, Lake Champlain, and the coastal waters. The third report to Congress concerned with mercury emissions is the aforementioned study of the hazards to public health posed by emissions of HAPS (including mercury) from electric steam generating units. The DOE air toxics assessment program focused on measuring mercury emissions from coal-fired utilities as accurately as possible.

Two independent methods to measure mercury in flue gas were employed at each power plant. The EPA Draft Method 29 multi-metals train was used at all the utility sites to quantify mercury (concurrently with the major, minor, and trace elements) at stack locations. The Bloom method (Bloom, et al., 1993) for both speciated and total mercury measurements was used as the comparative method at Sites 1, 4, 5, and 6. Mercury concentrations in flue gas were also measured by the Hazardous Element Sampling Train (HEST) method developed by Cooper (1994) at Sites 7, 8, and 9. A carbon sorption tube method was used at Site 3.

Comparison of the mercury emission data from the different methods for each site is given in Table 5. The mercury emission data in Table 5 were derived from measurements made on the stack flue gas at each site. Reasonably good agreement was seen for all the methods when compared to the EPA Draft Method 29 except for Site 3, where the carbon sorption tube was used as the alternative mercury measurement method.

Also included in Table 5 are the concentrations of mercury in each respective feed coal. Generally, the mercury concentrations in the flue gas, based on the EPA Draft Method 29 data, corresponded somewhat to the amounts in the feed coals, except for Site 5. The material balance data (Jackson, et al., 1994a) for mercury at Site 5 indicated about half of the mercury in the flue gas exiting the boiler was removed along with the flyash in the baghouse. At Site 1, only about 17 percent of the mercury was removed from the flue gas by the baghouse (Dismukes and Bush, 1994). This could explain the difference in the values for mercury in the flue gas between Site 1 and Site 5, where the mercury concentrations in the feed coals were essentially the same.

The low concentration of mercury determined by the EPA method for Site 3, and disparity in the flue gas data between Sites 6 and 9, where the feed coal concentrations were about the same, underscores the difficulity in measuring mercury in combustion gases, and the need for additional research.

Emission Factor Data

All of the stack emission data for each power plant were converted to emission factors that relate the concentrations of HAPS released to the atmosphere to the amount of fuel consumed in the boiler. Emission factor ranges for the eleven trace elements in the Title III list of HAPS and selected organic compounds are given in Table 6. The ranges represent a composite of the emission data for selected HAPS from each site. For some of the organic compounds, the emission factors are quite small, thus the values are given in scientific notation.

Also included in Table 6 are literature data for some of the trace element and organic HAPS (Brooks, 1989). Two striking observations can be made from the comparison of the DOE emissions data with the literature data. Generally, the levels of emissions for the DOE data are considerably less than those for the literature data. Also, the range of emissions for each specific compound are less for the DOE data. It is interesting to note that the concentrations of chromium and nickel determined in the DOE study are several orders of magnitude lower than the previously reported data. The most recent flue gas sampling methods for trace metals use glass probes to draw flue gas from the duct through the impinger traps in the sample train, as opposed to the older methods that used metal probes that contained chromium and nickel. This explains the dramatic decrease in the concentrations of those trace metals measured in the DOE study.

## SUMMARY

Some of the results from the DOE study on assessing the HAPS emissions from coal-fired electric utilities have been presented and discussed. The final reports for each site contain all the sampling, analytical, and quality assurance/quality control data obtained for each respective project. The data shown in this report signify the difficulties in effecting material balances for pollution control subsystems and the entire power plant, the variability in the data from several methods used to measure mercury in flue gas, and the low levels of emissions for selected HAPS.

## REFERENCES


Bloom, N. S., Lupsina, V., and Prestbo, E., "Flue Gas Mercury Emissions and Speciation from Fossil Fuel Combustion", presented at the Second International Conference on Managing Hazardous Air Pollutants, Washington, D.C., July 13-15, 1993.

Brooks, G., "Estimating Air Toxic Emissions from Coal and Oil Combustion Sources", FY89 Final Report, EPA-450/2-89-001, April 1989.

Coles, D.G., R.C. Rahaini, J.M. Ondov, G.L. Fisher, D. Silberman, and B.A. Prentice, "Chemical Studies of Stack Fly Ash from Coal-Fired Power Plant", *Environmental Science and Technology*, Vol 13 (4) April 1979.

Cooper, J. A., "Recent Advances in Sampling and Analysis of Coal-Fired Power Plant Emissions for Air toxic Compounds", *Fuels Processing Technology*, 1994, To be published.

Energy Information Administration (EIA) ,"Inventory of Power Plants in the US 1991", DOE/EIA-0095(91).

Dismukes, E., and Bush, P. V, "Characterizing Toxic Emissions from a Coal-Fired Power Plant Demonstrating the AFGD ICCT Project and a Plant Utilizing a Dry Scrubber/Baghouse System", FY94 Final Report, DOE Contract No. DE-AC22-93PC93254, April 1994.

England, G. C., McGrath, T. P., and Hansell, D., "Assessment of Toxic Emissions from a Coal-Fired Power Plant Utilizing an ESP", FY94 Final Report, DOE Contract No. DE-AC22-93PC93252, April 1994.

Flora, H. B., Williams, A. and Maxwell, D. P., "A Study of Toxic Emissions from a Coal-Fired Power Plant Utilizing and ESP while Demonstrating the ICCT CT-121 FGD Project", FY94 Final Report, DOE Contract No. DE-AC22-93PC93253, April 1994.

Jackson, B. L., O' Neill, J. D., and DeVito, M. S., "Toxics Assessment Report: Illinois Power Company Baldwin Station - Unit 2", FY94 Final Report, DOE Contract No. DE-AC22-93PC93255, April 1994.

Jackson, B. L., O' Neill, J. D., and DeVito, M. S., "Toxics Assessment Report: Minnesota Power Company Boswell Energy Center - Unit 2", FY94 Final Report, DOE Contract No. DE-AC22-93PC93255, April 1994.

Sverdrup, G. M., Riggs, K, and Cooper, J. A., "A Study of Toxic Emissions from a Coal-Fired Power Plant - Niles Station Boiler No. 2", FY94 Final Report, DOE Contract No. DE-AC22-93PC93251, April 1994.

Sverdrup, G. M., Riggs, K, and Cooper, J. A., "A Study of Toxic Emissions from a Coal-Fired Power Plant Utilizing the SNOX Innovative Clean Coal Technology Demonstration", FY94 Final Report, DOE Contract No. DE-AC22-93PC93251, April 1994.

Sverdrup, G. M., Riggs, K, and Cooper, J. A., "A Study of Toxic Emissions from a Coal-Fired Power Plant Utilizing an ESP/Wet FGD System", FY94 Final Report, DOE Contract No. DE-AC22-93PC93251, April 1994.

Table 1. Phase I power plant descriptions.

| Site | Contractor | Power Plant | Operator | Size(MW) | Coal | $SO_2$ | $NO_x$ | Part |
|---|---|---|---|---|---|---|---|---|
| 1 | SRI | Springerville | Tucson Elec Company | 360 | Sub | SDA | OFA | BH |
| 2 | SRI | Bailly | NIPSCO | 528 | Bit | Limestone Scrubber | None | ESP |
| 3 | EER | Cardinal | Ohio Power Company | 615 | Bit | None | None | ESP |
| 4 | Weston | Baldwin | Illinois Power | 570 | Bit | None | None | ESP |
| 5 | Weston | Clay Boswell | Minnesota Power | 69 | Sub | None | None | BH |
| 6 | Radian | Yates | Georgia Power | 100 | Bit | Limestone Scrubber | None | ESP |
| 7 | Battelle | Niles | Ohio Edison | 100 | Bit | None | None | ESP |
| 8 | Battelle | SNOX | ABB | 35 | Bit | WSA | SCR | BH |
| 9 | Battelle | Coal Creek | Cooperative Power | 550 | Lig | Lime Scrubber | OFA | ESP |

Contractors
SRI - Southern Research Institute
EER - Energy and Environmental Research Corporation

Operators
ABB - Asea Bavaria Brown
NIPSCO - Northern Indiana Public Service Company

Pollution Control Technologies
OFA - Over-Fire Air
BH - Baghouse
SDA - Spray Dryer Absorber
ESP - Electrostatic Precipitator
SNOX - Wet Gas Sulfuric Acid-Selective Reduction of $NO_x$
WSA - Wet Sulfuric Acid Selective Catalytic Converter
SCR - Selective Catalytic Reduction

Table 2. Referenced flue gas sampling methods used in the DOE air toxics assessment program.

| Method | Target Analytes |
|---|---|
| 0010 (SW-846/23) | SVOCs, PCDD/PCDF,POM/PAH |
| 0010 (SW-846) | Formaldehyde, Ketones, Aldehydes |
| M26 | Acid Gases (HF, HCl, HBr, HCN) |
| M29 | Trace Metals, Total Particulate Emissions |
| BIF 11 | Hexavalent Chromium |
| M5 | Radionuclides, Chloride, Fluoride, Carbon, Sulfates, Phosphates, Macro Elements |
| VOST | VOCs |

SVOCs - Semivolatile Organic Compounds
VOST - Volatile Organic Sampling Train
PCDD - Polychlorinated dibenzo-p-dioxins
PCDF - Polychlorinated dibenzofurans
POM/PAH - Polycyclic Organic Material/Polycyclic Aromatic Hydrocarbons
VOCs - Volatile Organic Compounds

Table 3. Analytical methods used in DOE air toxics assessment.

| Analytical Method | Analyte(s) |
|---|---|
| ICP - AES | B, Ba, Be, Cr, Co, Cu, Mn, Mo, Ni, V |
| GF - AAS | As, Cd, Pb, Sb, Se |
| CV - AAS | Hg |
| ISE | $CN^-$, $NH_3$ |
| IC | $Cl^-$, $F^-$, $PO_4^{3-}$, $SO_4^{2-}$ |
| GC/MS | Benzene, Toluene, PAH |
| HPLC | Formaldehyde |
| Gamma Radiation | Radionuclides |
| GC/HRMS | Dioxins/Furans |

ICP - AES =Inductively Coupled Plasma - Atomic Emission Spectroscopy
GF - AAS =Graphite Furnace - Atomic Absorption Spectrometry
CV - AAS =Cold Vapor - Atomic Absorption Spectrometry
GC/MS =Gas Chromatography/Mass Spectrometry
GC/HRMS =Gas Chromatography/ High Resolution Mass Spectrometry
HPLC =High Performance Liquid Chromatography
ISE =Ion Selective Electrode
IC =Ion Chromatography

Table 4. Power plant material balances (percent) summary for trace and minor elements.

| Site | Titanium | Potassium | Arsenic | Cadmium | Lead | Mercury |
|---|---|---|---|---|---|---|
| 1 | 104 | ... | 122 | 44 | 476 | 46 |
| 3 | 117 | 23 | 25 | 172 | -7 | 52 |
| 4 | 107 | 101 | 82 | 79 | 94 | 79 |
| 5 | 119 | 108 | 74 | 90 | 97 | 99 |
| 6 | 78 | 62 | 134 | 136 | 113 | 101 |
| 7 | 87 | 90 | 53 | 82 | 53 | 90 |
| 8 | 110 | 129 | 80 | 62 | 143 | 118 |
| 9 | 120 | 109 | 73 | 34 | 42 | 119 |

Table 5. Comparison of methods to measure mercury in flue gas (mg/Nm$^3$).

| Site | EPA Draft Method 29 | Bloom Method | Carbon Sorption Tube | HEST | Feed Coal (mg/g) |
|---|---|---|---|---|---|
| 1 | 9.64 | 5.94 | ... | ... | 0.05 |
| 3 | 0.45 | ... | 9.21 | ... | 0.02 |
| 4 | 5.22 | 7.19 | ... | ... | 0.07 |
| 5 | 2.57 | 3.15 | ... | ... | 0.05 |
| 6 | 3.0 | 3.3 | ... | ... | 0.08 |
| 7 | 18.2 | ... | ... | 16.4 | 0.21 |
| 8 | 20.5 | ... | ... | 22.8 | 0.21 |
| 9 | 8.52 | ... | ... | 5.4 | 0.08 |

Table 6. Emission factor ranges for trace metal and selected organic HAPS.

| Trace Metals (lbs/$10^{12}$ Btu) | | | Trace Organics (lbs/$10^{12}$ Btu) | | |
|---|---|---|---|---|---|
| Metal | DOE | Ref. 13 | Organic | DOE | Ref. 13 |
| Antimony | <0.1- 2.4 | NA | Benzene | 3.4 - 170 | NA |
| Arsenic | 0.1 - 42 | <1- 860 | Toluene | 2.0 - 24 | NA |
| Beryllium | <0.1- 1.4 | <1- 32 | Naphthalene | <0.1- 1.9 | a/ |
| Cadmium | <0.1- 3.0 | 1 - 490 | Anthracene | (3.0 - 20) $10^{-3}$ | a/ |
| Chromium | <0.1- 51 | 10 - 5000 | Phenanthrene | (2.0 - 31) $10^{-2}$ | a/ |
| Cobalt | <0.1- 6.8 | NA | Pyrene | (3.0 - 40) $10^{-3}$ | a/ |
| Lead | 0.6 - 29 | NA | Benzo(a)pyrene | (2.0 - 12) $10^{-4}$ | a/ |
| Manganese | 1.1 - 22 | 30 - 2400 | Formaldehyde | 1.4 - 57 | 63 - 2100 |
| Mercury | 0.5 - 14 | 1 - 22 | 2-Butanone | 3.7 - 16 | NA |
| Nickel | 0.3 - 40 | 1 - 2500 | 2,3,7,8-TCDD | (8.1 - 25) $10^{-7}$ | NA |
| Selenium | <0.1- 130 | NA | 2,3,7,8-TCDF | (6.6 - 9.9) $10^{-7}$ | NA |

a/ The literature data are classified as Polycyclic Organic Matter and range from 0.03 to 565 lbs/$10^{12}$ Btu.

Figure 1. Sampling schedule for inorganic flue gas constituents, solids, and liquids.

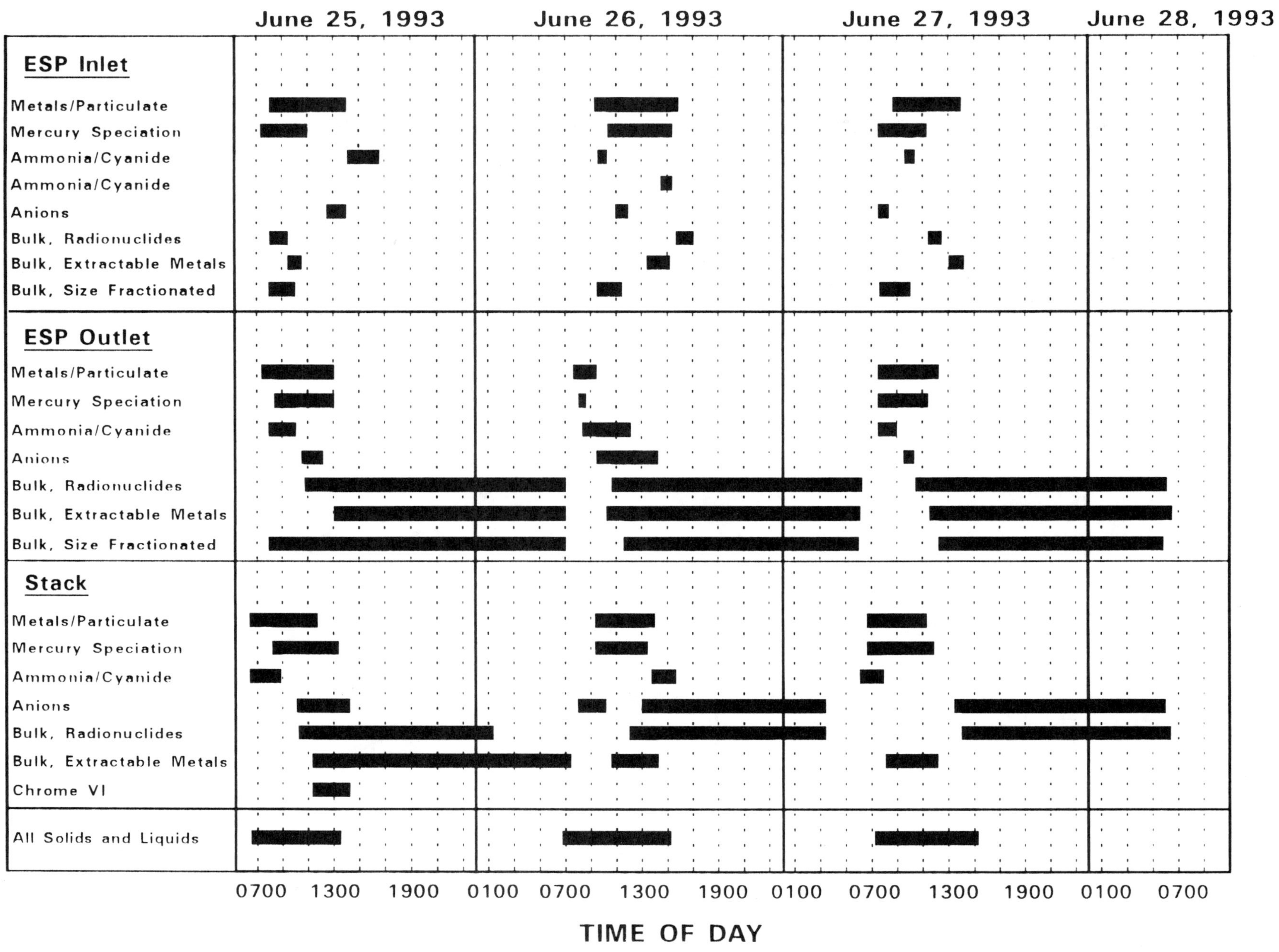

Figure 2. Sampling schedule for organic flue gas constituents, solids, and liquids.

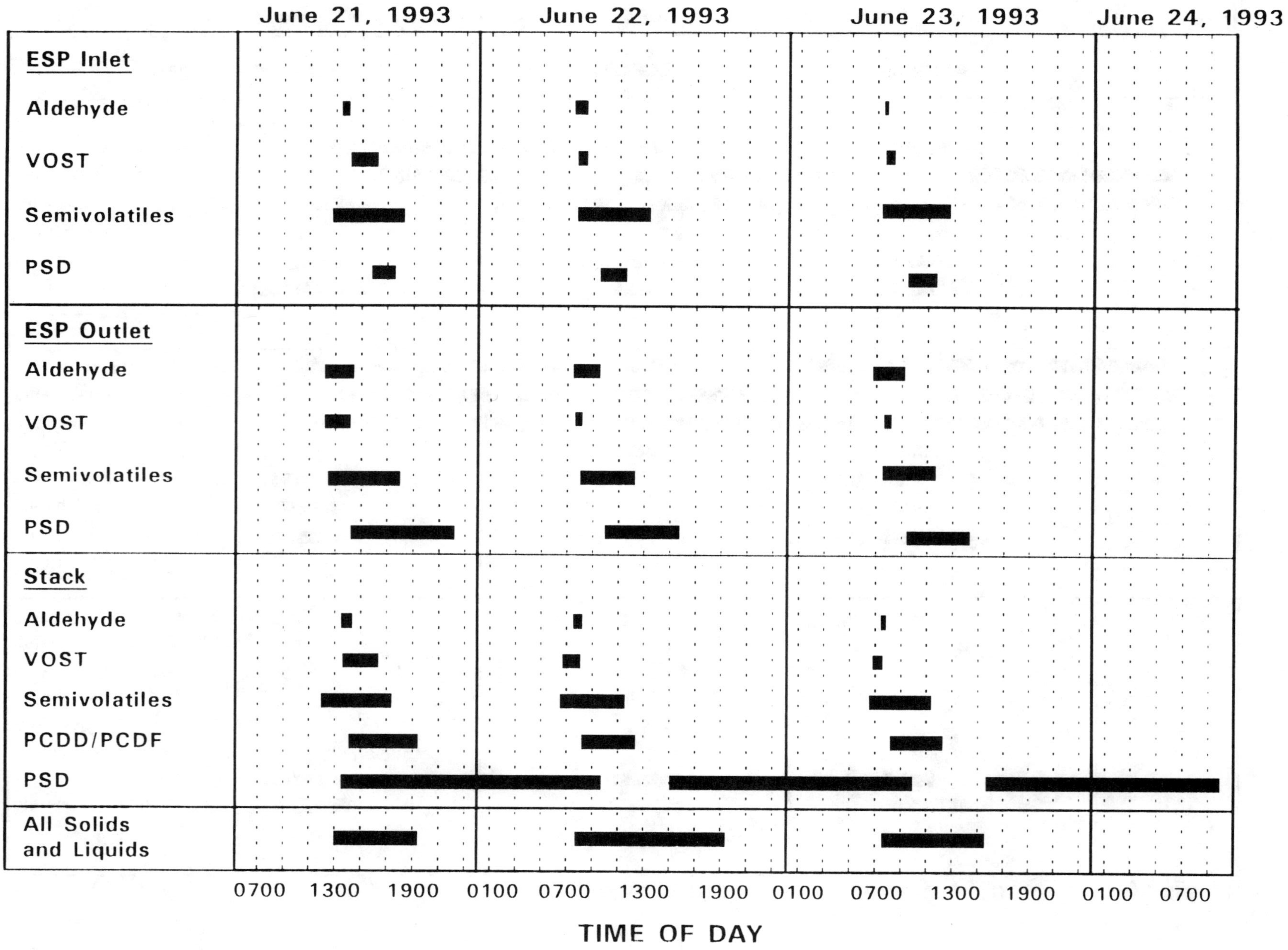

FACT-Vol. 18, Combustion Modeling, Scaling and Air Toxins
ASME 1994

# INTEGRATED FLUE GAS TREATMENT CONDENSING HEAT EXCHANGER FOR POLLUTION CONTROL

**D. W. Johnson**
Environmental Equipment Div.
Babcock & Wilcox
Barberton, Ohio

**J. J. Warchol and K. H. Schulze**
Research and Development Div.
Babcock & Wilcox
Alliance, Ohio

**J. F. Carrigan**
Condensing Heat Exchanger Corporation
Warnerville, New York

## ABSTRACT

Condensing heat exchangers recover both sensible and latent heat from flue gases. Using Teflon® to cover the heat exchanger tubes and inside surfaces that are exposed to the flue gas ensures adequate material lifetime in the corrosive environment encountered when the flue gas temperature drops below the acid dew point. A recent design improvement, called the integrated flue gas treatment (IFGT) concept, offers the ability to remove pollutants from the flue gas, as well as recover waste heat. It has been shown to remove $SO_2$, $SO_3$, particulates, and trace emissions. Babcock & Wilcox (B&W) is undertaking an extensive program to optimize this technology for a variety of flue gas applications. This paper summarizes the current status of IFGT technology and the development activities that are in progress.

## INTRODUCTION

Condensing heat exchangers are widely used to recover waste heat from flue gases. The most common applications to date are boilers firing oil or natural gas. Material corrosion problems normally encountered, when the flue gas temperature drops below the acid dew point are prevented by using Teflon® covered heat exchanger parts. Both the heat exchanger tubes and the inside of the tubesheets are covered with Teflon®. Single-stage commercial condensing heat exchangers have demonstrated satisfactory performance and lifetimes for over 100 industrial installations. A recent innovation to the commercial condensing heat exchanger design, called the integrated flue gas treatment (IFGT) concept, exhibits improved pollutant removal from the flue gas while recovering waste heat. It has been shown to remove $SO_2$, $SO_3$, particulates, and some trace element emissions soon to be regulated under the 1990 Clean Air Act Amendment, Title III. The IFGT condensing heat exchanger has been described as "the scrubber that pays for itself."

Condensing Heat Exchanger (CHX) Corporation is the developer of the condensing heat exchanger using Teflon® covered components. CHX Corporation also developed the IFGT concept and built a number of prototype units for testing. Babcock & Wilcox (B&W) and CHX entered into an exclusive license agreement in 1993 involving the marketing, manufacture, and development of condensing heat exchangers using Teflon® covered tubes and internals. B&W now offers both single-stage heat recovery and IFGT units to industrial, commercial, and utility customers. CHX continues to manufacture the heat exchangers and provides technical consultation for a wide variety of applications.

Babcock & Wilcox is undertaking an extensive program to further develop and optimize the IFGT technology for alternate fuels, such as coal, Orimulsion, and waste fuels, and emission control applications.

## PRODUCT DESCRIPTION

### Single-Stage Condensing Heat Exchanger Design

Commercial condensing heat exchangers remove both sensible and latent heat from the flue gas in a single stage. A typical commercial system is shown in Figure 1. Flue gas passes down through the heat exchanger while the water passes upward in a serpentine path through the tubes. Condensation occurs within the heat exchanger as the flue gas temperature at the tube surface is brought below the dew point.

Since Teflon® is hydrophobic, condensation on the surface of a tube occurs in drops rather than in a film. The condensate falls as a constant rain over the tube array and is removed from the bottom of the heat exchanger. Some cleaning of the gas occurs within the heat exchanger as the particulates impact the Teflon® covered tubes and falling condensate drops and also as gas condensation occurs. Collected particles, along with condensible and soluble pollutants, are removed with the liquid effluent at the bottom of the heat exchanger.

### IFGT Condensing Heat Exchanger Design

The IFGT condensing heat exchanger, shown schematically in Figure 2, is designed to enhance the removal of pollutants from the flue gas stream. This design uses many of the same heat exchanger components found in the proven commercial designs, so unit lifetimes will be comparable to current commercial units.

There are four major sections of the IFGT unit — the first heat exchanger stage, the interstage transition region, the second heat exchanger stage, and the mist eliminator.

Most of the sensible heat is removed from the gas in the first heat exchanger stage of the IFGT unit. The transition region is equipped with a water or alkali spray system. The spray saturates the flue gas with moisture before it enters the second heat exchanger stage and also assists in removing sulfur pollutants from the gas. The transition piece is normally made of corrosion resistant fiberglass-reinforced plastic.

The second heat exchanger stage is operated in the condensing mode, removing latent heat from the gas along with pollutants. The flue gas in this stage is flowing upward while the droplets in the gas fall downward. This provides a scrubbing mechanism that enhances particulate and pollutant capture. The dimensions and spacing of the heat exchanger tubes ensure that the larger particulates impact the wet tubes where droplet condensation is taking place. Sub-micron size particles act as condensation sites in the gas and are collected in the condensate stream. The top of the second heat exchanger stage can be equipped with an alkali reagent spray system to enhance $SO_2$ removal. The condensed gases, particulates, and reacted alkali reagent are collected at the bottom of the transition section. A portion of the condensate/alkali reagent solution is recirculated to the spray system to improve the efficiency of the process.

The flue gas outlet of the IFGT system is equipped with a mist eliminator to reduce the chance of moisture carryover and ensure that local environmental requirements are satisfied.

### Heat Exchanger Modules

Both the single-stage and the IFGT condensing heat exchangers are made up of heat exchanger modules that can be stacked in series in the gas stream. This modular design allows the size of the unit to be optimized for each application at minimum cost. Each module consists of a tubesheet and heat exchanger tubes. The tubes are made of Alloy 706 (10% nickel and

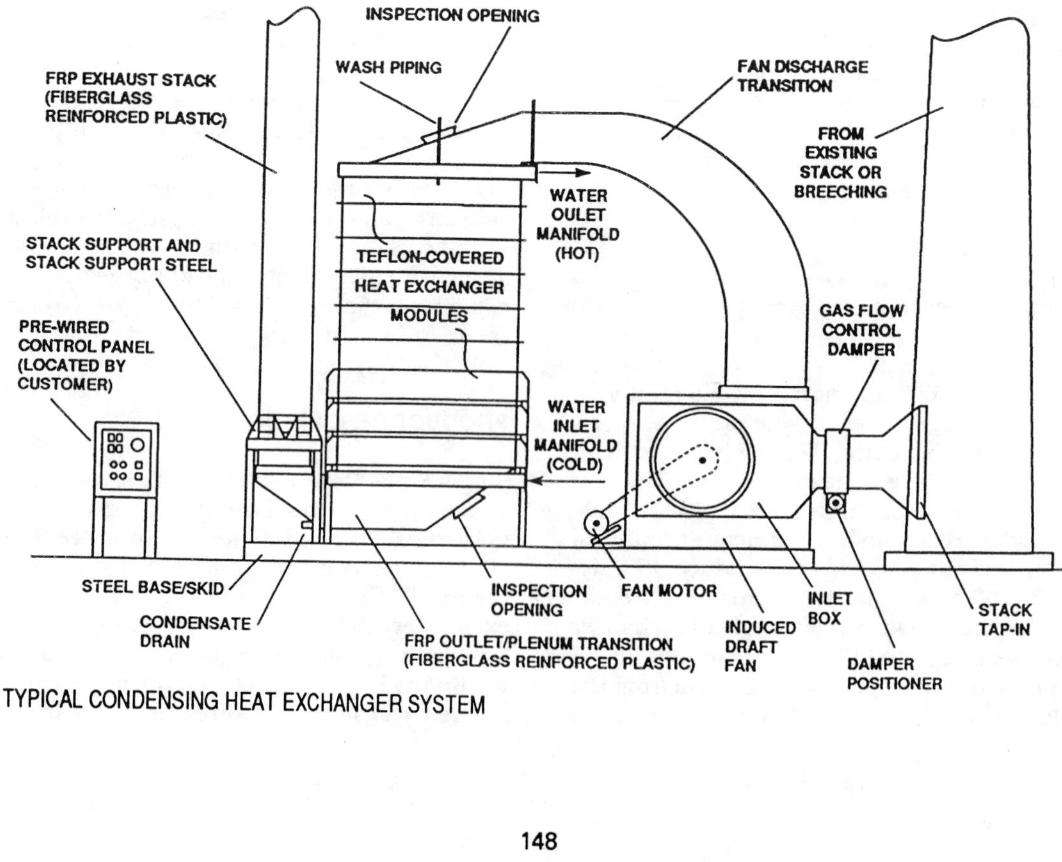


FIGURE 1. TYPICAL CONDENSING HEAT EXCHANGER SYSTEM

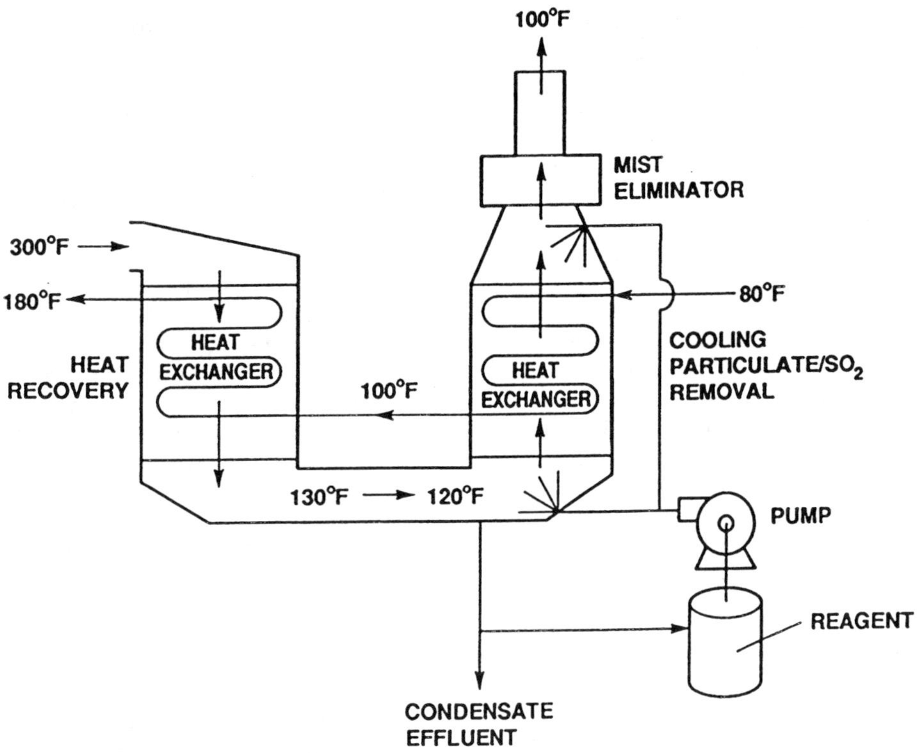


FIGURE 2. TYPICAL INTEGRATED FLUE GAS TREATMENT (IFGT) SYSTEM

90% copper), a material commonly used in boiler water applications. Each tube is covered with a 0.015-inch thick Teflon® covering that is extruded over the outside of the tube. The inside surfaces of the heat exchanger shell are covered with 0.060-inch thick Teflon® sheets. During fabrication, the Teflon® covered tubes are pushed through holes in the Teflon® tubesheet lining to form a Teflon®/Teflon® seal, ensuring that all heat exchanger surfaces exposed to the flue gas are protected against acid corrosion. This is shown in Figure 3. Interconnections between the heat exchanger tubes are made outside the tubesheet and are not exposed to the corrosive flue gas stream. To ensure the lifetime of the Teflon® covering on the tubes, the inlet gas temperature is kept below 500°F, and the flue gas velocity is maintained under 45 feet/second — conditions easily satisfied in most flue gas waste heat recovery applications.

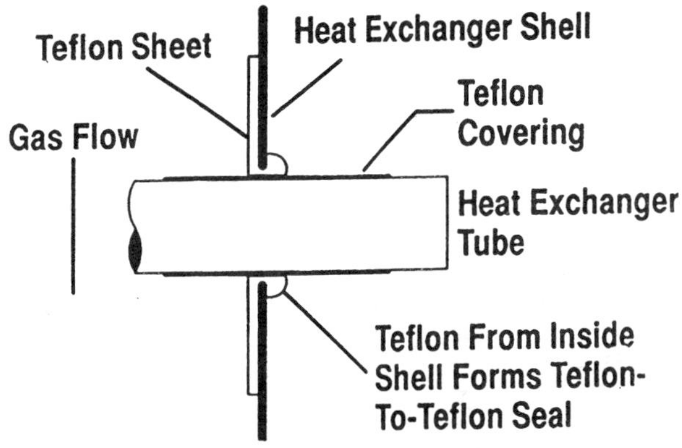


FIGURE 3. CONDENSING HEAT EXCHANGER TUBESHEET DESIGN

**IFGT System Pollutant Removal Mechanisms**

The temperature of the flue gas decreases as it passes through the IFGT condensing heat exchanger. The particulates and gaseous pollutants in the flue gas interact with the cooler tubes and the water droplets that are formed during condensation. Additional droplets and chemicals can be introduced into the gas stream by using the reagent spray nozzles in the transition region or at the top of the second stage of the IFGT system.

A number of removal mechanisms are occurring inside an IFGT condensing heat exchanger system:

- ***Inertial impaction*** occurs when the momentum of the particle causes it to collide with the heat exchanger tube or falling condensate drop instead of following the flow streamline around the tube. After losing its momentum, the captured particle is drained away from the bottom of the heat exchanger. This is the primary removal mechanism for fly ash particles.

- ***Interception*** occurs when the particle is small enough to follow the flue gas streamline around a heat exchanger tube, but its surface comes into contact with the tube diameter or droplets on the tube. Further, the local reduction of gas volume as flue gas cools near the tube surface can contribute to particle interception.
- ***Diffusion*** occurs for very small particles when there is a concentration gradient in the flue gas stream. Under stable conditions, the Brownian motion (which occurs when small particles randomly collide with gaseous molecules) does not impart a net movement to a particle. If the symmetry is disturbed, however, by removing material at the tube/gas interface (due to condensation), a gradient is introduced and there will be diffusion in the direction toward the heat exchanger tube.
- ***Thermophoresis*** is a diffusion process caused by a temperature gradient in the gas. The collisions between a particle and molecules are more energetic on the higher temperature side of the particle, causing a net movement of the particle in the direction of the lower temperature (toward the cooler heat exchanger tube).
- ***Direct condensation*** occurs when the gas becomes locally supersaturated and water droplets can form. Particles act as nucleation sites. Direct condensation results in increase in mass and size of the particle/water droplet. Removal by impaction or momentum forces then dominates. This effect is also dependent on the affinity of the particle for water.

Brookhaven National Laboratory (BNL) has undertaken an extensive study of each of the particle collection mechanisms described above and evaluated their contribution to the collection process in a condensing heat exchanger[1].

There are also a number of collection mechanisms occurring in an IFGT condensing heat exchanger that control its ability to remove gaseous pollutants. Some gases, such as HCl and $SO_3$, will interact with the water droplets and be removed with the condensate as an acid. $SO_3$ can also be removed by direct condensation on the cold tube surfaces. Other gases, such as $SO_2$, because of their higher concentration must be removed by having the gas interact with an alkali reagent, such as sodium bicarbonate or carbonate. This is accomplished by providing an alkali spray system above the second stage of the IFGT condensing heat exchanger. The turbulent action caused by the reagent falling over the tubes while the gas is flowing upward around the tubes increases the reagent/gas interaction and improves the $SO_2$ removal efficiency of the chemical process.

The collection mechanisms occurring in the second stage of an IFGT system will also assist with the removal of hazardous pollutants from the flue gas. This includes heavy metals, such as mercury, volatile organic compounds, polycyclic aromatic hydrocarbons, and other trace species of concern. Condensible heavy metals will be removed by condensation as the local flue gas temperature is lowered when it comes into contact with the cooler heat exchanger tubes or drops of water. The condensed elements and compounds will be collected as solids with the particles or collected in solution, if they are soluble. At the temperatures encountered in the IFGT unit, some of the high molecular weight polycyclic aromatic hydrocarbons may also be associated with the particulates and be removed with them. The scrubbing action provided by spraying reagent over the heat exchanger tubes is also expected to aid in hazardous pollutant removal (HCl, ionic mercury, etc.).

## INTEGRATED FLUE GAS TREATMENT (IFGT) SYSTEM PERFORMANCE

The IFGT condensing heat exchanger design has been evaluated at a number of test site locations. All tests have been performed on pilot units that extract a portion of a flue gas stream.

The first tests performed on an IFGT pilot unit were in 1991 using a simulated dirty flue gas (clean flue gas with particulates and pollutant gases added). This unit demonstrated a particulate removal efficiency of 90 to 98%. Using sodium bicarbonate as the reagent, removal efficiencies of 90 to 95% were realized for sulfur dioxide and about 90% for HCl removal[2].

A more extensive series of tests were performed on the same pilot unit at Morgan Linen Company in Menands, New York, in 1992. The IFGT system used a slipstream of the flue gas from their 14,000 lb/hr boiler which fired a 1.5% sulfur residual No. 6 fuel oil[3]. The particulate removal efficiency averaged 89.3%. Using sodium bicarbonate as the reagent, sulfur dioxide and sulfur trioxide removal efficiencies averaged over 99%. Measurements for trace metals yielded the following removal efficiencies: nickel (20%), chromium (11%), lead (36%), mercury (84%), cadmium (14%), and zinc (25%).

A performance test program was also conducted on an IFGT unit at the Ravenswood plant of Consolidated Edison of New York, Inc. (Con Ed) in 1993. The IFGT unit treated a 25,000 lb/hr flue gas slipstream of the residual oil-fired unit. $SO_2$ removal efficiencies were measured to be over 98%; the removal efficiency for $SO_3$ averaged 73 - 91%[4]. Other measurements performed as part of the test program demonstrated that stack particulate loading was limited to 0.005 lb/million Btu (approximately one-half of this material was residual sulfuric acid mist). The removal efficiency of the IFGT unit for mercury was 50%[5].

Shakedown tests were performed in early 1994 on Babcock & Wilcox's pilot IFGT unit installed at the B&W

Alliance Research Center. A Pittsburgh #8 coal having a sulfur level of 1.4 % was burned. Using sodium bicarbonate as the alkali reagent, $SO_2$ removal efficiencies of 97% were measured.

## POTENTIAL APPLICATIONS

Industrial condensing heat exchangers are economically justified on the basis of the heat recovered. Condensing heat exchangers are widely used for heating boiler makeup water and other industrial water heating applications. They can also be used in chemical processing, petroleum, paper, and other industrial applications for heat recovery, pollution control, and product recovery (such as a distillation process).

The IFGT design is attractive for applications where lower grade fuels, such as high-sulfur oil or coal, are fired. The justification for using an IFGT system is satisfying environmental regulations while increasing the plant capacity or thermal efficiency.

Because of its modular construction, the IFGT system can be designed for different size applications. It is uniquely attractive for smaller applications that can not justify many of the conventional pollutant removal processes currently available. For large applications, the IFGT system is also compatible with other gas cleaning methodologies currently available or under development.

An additional benefit realized by using a condensing heat exchanger for utility applications is the proportional reduction in carbon dioxide/MW for the plant because of its decreased heat rate and thus lower fuel requirements. A condensing heat exchanger could also enhance the performance of future $CO_2$ removal processes, since $CO_2$ removal is more efficient at lower temperatures.

Of increasing importance is the potential for the IFGT system design to remove air toxics from the flue gas stream. This is becoming more important in view of the Title III requirements of the 1990 Clean Air Act Amendment which established a list of 189 hazardous air pollutants and charged the EPA with the responsibility for regulating emissions of these substances into the atmosphere.

## IFGT TECHNOLOGY DEVELOPMENT

IFGT technology is being developed for industrial and utility applications through the joint efforts of B&W, CHX, BNL, and Con Ed.

### B&W — Research and Development Division

In 1993, an IFGT pilot unit was installed at B&W's Research and Development Division in Alliance, Ohio. The unit is currently connected to the division's 6-million Btu/hr small boiler simulator (SBS) test facility. The SBS is a combustion and fuel handling facility for evaluating various fossil fuels, combustion processes, and pollution abatement processes. The facility can burn liquid and gaseous fuels, as well as pulverized coal. The SBS models the geometry and provides a representative gas time-temperature profile of commercial boilers. The exhaust gas from the SBS is directed through the pilot IFGT system during testing. The pilot unit can handle the full gas flow capacity of the SBS.

A photograph of the pilot IFGT facility is shown in Figure 4. It consists of two heat exchanger sections separated by a fiberglass-reinforced plastic transition section. The arrangement models the design of commercial units. Each heat exchanger module contains 336 Teflon® covered tubes, each tube having a length of 3 feet and a diameter of 1.125 inches. The IFGT system also includes a storage tank for heated water, an induced draft fan and a damper to control the inlet flue gas volume, and an alkali reagent scrubbing system. The unit is also equipped with both operational and safety controls.

The near-term goal of the test program at the Alliance Research Center is to demonstrate the applicability of IFGT technology for alternate fuels. Shakedown tests were performed early in 1994 using a Pittsburgh #8 coal. This was the first time an IFGT unit was tested using coal as the fuel. To minimize fly ash erosion problems on the Teflon® covered tubes, the flue gas was passed through a baghouse before it entered the IFGT system. In the coming year, tests will be performed on additional coals and Orimulsion. Both thermal heat recovery and pollution removal performance measurements will be made. Of primary interest are: $SO_2/SO_3$ reduction, particulate removal, and the reduction of hazardous air pollutants (air toxics). Performance optimization studies will also be conducted.

The IFGT pilot unit can also accept flue gas from B&W's new 100-million Btu/hr clean environment development facility (CEDF). The CEDF is currently under construction and will be a state-of-the-art test facility that will integrate combustion and post-combustion air toxic testing capabilities. The facility is designed to allow full heat input on a wide range of pulverized coals, #2 and #6 oil, or natural gas. The unique feature of the CEDF is a prototypical convection pass that simulates the gas time-temperature history and tube metal temperatures of a commercial boiler convention pass[6]. This will allow for studies of volatile organic compounds (VOCs) and air toxics that are not possible in the SBS due to cold tube surfaces in the SBS convection pass. A flue gas slipstream from the CEDF will be connected to the inlet of the IFGT pilot.

### Brookhaven National Laboratory

Brookhaven National Laboratory (BNL) has been conducting tests on a small, single-stage condensing heat exchanger under a U.S. Department of Energy

(U.S. DOE) funded program. Much of BNL's work has been directed at identifying and measuring the particulate removal mechanisms taking place in the condensing heat exchanger.

BNL is currently performing research under a DOE-sponsored cooperative research and development agreement (CRADA). This effort is focused on improving the performance of the IFGT system design. Con Ed and B&W are providing in-kind financial contribution to the program. The technical issues being addressed in this program include optimizing particulate removal and characterizing the air toxics reduction capability of the technology. The major goal of this effort is to develop design procedures to predict particulate removal, sulfur oxides capture, and air toxics removal.

The participants in the CRADA — BNL, B&W, CHX, and Con Ed — will coordinate their research efforts and share test results. This will help ensure that the data gathered on different size facilities are integrated into the knowledge base and incorporated in commercial designs.

**Consolidated Edison of New York, Inc.**

Consolidated Edison of New York, Inc. (Con Ed) has been actively working with CHX for many years. This has included the operation of a large condensing heat exchanger at its 74th Street Station in New York City[5]. This heat recovery unit was installed as a first step of an R&D program designed demonstrate the technology at utility-scale. As previously mentioned, Con Ed has also operated a pilot-scale (25,000 lb/hr flue gas flow) IFGT unit at the utility's oil-fired Ravenswood Steam Station. This test program confirmed the pollutant removal capability of the IFGT system at five times the size of the

FIGURE 4. PILOT-SCALE INTEGRATED FLUE GAS TREATMENT (IFGT) FACILITY

initial testing at Morgan Linen. Recently, Con Ed purchased a 250,000 lb/hr IFGT unit from B&W to be installed at the Ravenswood Station. This unit will be used to demonstrate IFGT technology, including pollutant removal capability, at commercial scale.

## INNOVATIONS/NEW MARKETS

### Orimulsion

The discovery of large reserves of extra-heavy hydrocarbon and bitumen in the Orinoco belt area of Venezuela has led to a considerable effort to exploit these reserves for commercial applications. The result is a water-based emulsion of bitumen called Orimulsion. The utilization of Orimulsion as a replacement for higher grades of fuel oil is being considered by utilities primarily on the east coast of the United States as well as other locations worldwide. Because of its relatively high sulfur content, burning Orimulsion in existing oil-fired boilers will require the installation of pollution control equipment to meet environmental regulations.

The IFGT condensing heat exchanger is ideally suited for applications burning Orimulsion because:

- The heat lost by evaporating the water in the fuel during combustion will be recovered as latent heat in the heat exchanger.
- The higher water content of the flue gas will produce additional condensation droplets in the heat exchanger which will enhance the pollution removal mechanisms.
- The removal mechanisms in the condensing heat exchanger include condensing (for water and trace metal removal), scrubbing (for particulate and $SO_2$ removal), and dissolving (for $SO_3$ removal). No additional flue gas cleanup equipment should be required.
- The gas leaving has less water and, therefore, reduces the visible vapor plume.

### Coal-Fired Utility Applications

A limited number of tests have been performed on condensing heat exchangers relative to coal-fired applications. Union Carbide performed tests on a small single-stage condensing heat exchanger unit in 1983 at their coal-fired No. 2 steam plant in South Charleston, West Virginia. The four-week test confirmed the heat recovery predictions and no plugging or lifetime problems were encountered with the heat exchanger during the tests[7].

Southern Company Services exposed a Teflon® covered heat pipe to a coal-fired flue gas at Gulf Power Company's coal-fired Scholz generating station located in Sneads, Florida. The coal contained 2 to 3% sulfur. No problems were encountered during the 8000-hour test[8]. A broader Southern Company Services study that evaluated many candidate heat exchanger tube materials concluded that Teflon® clad heat exchanger tubes would give satisfactory performance in the coal flue gas[9].

B&W's Research and Development Division performed $SO_2$ removal tests on their IFGT pilot unit in 1994 using a Pittsburgh #8 coal as the fuel. Removal efficiencies of 97% were measured during the shakedown tests. Additional tests are planned for this unit using different types of coal.

## FUTURE DEVELOPMENT

Babcock & Wilcox and CHX have established plans to expand the use of Teflon® covered condensing heat exchanger technology. Development programs are planned for both heat recovery and IFGT applications. Important issues include cycle integration and further developing the product to address utility, industrial, government, and commercial concerns.

In the near term, the B&W development program will focus on expanding the application of IFGT technology for coal and other fuels to evaluate heat recovery, pollutant removal, reliability, by-product, and lifetime issues. Our continued involvement in external research and development programs, such as those currently underway at BNL and Con Ed, is critical to increasing our collective knowledge concerning the removal mechanisms taking place in the IFGT condensing heat exchanger. This will be especially important as we address the area of air toxics removal. The long-term goal of these programs is to be able to optimize both the design and operation of an IFGT system to address the specific pollutant removal needs of each industrial and utility customer.

## REFERENCES

1. Butcher, T. A., and Litzke, W., "Condensing Economizers for Small Coal-Fired Boilers and Furnaces," Brookhaven National Laboratory, January 1994.

2. "Performance Test Results on the Integrated Flue Gas Treatment System Developed by Condensing Heat Exchanger Corporation," Prepared by Arthur D. Little Inc., February 1992.

3. Source Test Report, "Source Emission Testing of a Pilot Scale Integrated Flue Gas Treatment System at Condensing Heat Exchanger Corporation," Warnerville, New York", Galson Project No. GE-234, October 1992, Galson Corporation, East Syracuse, New York.

4. Heaphy, J. P., Carbonara, J., Litzke, W., and Butcher, T. A., "Condensing Heat Economizers for Thermal Efficiency Improvements and Emissions Control," Tenth International Pittsburgh Coal Conference, September 20 - 24, 1993.

5. "Utility Seeks to Integrate Heat Recovery, Flue Gas Treatment," *Power*, May 1993.

6. Flynn, T. J., LaRue, A. D., and Nolan, P. S., "Introduction to Babcock and Wilcox's 100 MBtu/hr Clean Environment Development Facility," 1994 Annual Meeting of the American Power Conference, April 25 - 27, 1994.

7. Roy, J. G., "Evaluation of CHX Condensing Economizer," Union Carbide Corporation Test Report, March 4, 1983.

8. Mills, K. J., Southern Company Services Report, April 2, 1992.

9. "Material and Cleaning Options for Cyclic Reheat Systems: Final Report," *EPRI Report CS-6738.*

FACT-Vol. 18, Combustion Modeling, Scaling and Air Toxins
ASME 1994

# ABB'S INVESTIGATIONS INTO AIR TOXIC EMISSIONS FROM FOSSIL FUEL AND MSW COMBUSTION

**James D. Wesnor**
ABB Environmental Systems
Birmingham, Alabama

## ABSTRACT

Since passage of the Clean Air Act, Asea Brown Boveri (ABB) has been actively developing a knowledge base on the Title III hazardous air pollutants, more commonly called air toxics. As ABB is a multinational company, US operating companies are able to call upon work performed by our European counterparts, who have faced similar legislation several years ago. In addition to the design experience and database acquired in Europe, ABB Inc. has been pursuing several other avenues to expand its air toxics knowledge. ABB Combustion Engineering (ABB CE) is presently studying the formation of organic pollutants within the combustion furnace and partitioning of trace metals among the furnace outlet streams. ABB Environmental Systems (ABBES) has reviewed available and near-term control technologies and methods. Also, both ABB CE and ABBES have conducted source sampling and analysis at commercial installations for hazardous air pollutants to determine the emission rates and removal performance of various types of equipment. Several different plants hosted these activities, allowing for variation in fuel type and composition, boiler configuration, and air pollution control equipment. This paper discusses the results of these investigations.

## INTRODUCTION

The possible effects of the Clean Air Act of 1990 could be similar to those seen in Europe, particular Germany and Sweden: ever-tightening control of gaseous pollutant emissions, both macropollutants (particulate, $SO_2$, and NOx), and micropollutants (hazardous air pollutants). As a multinational company with strong European experience, ABB is able to use its engineering and technical experience gained in developing systems to meet European pollution regulations in applications in the United States. Even though at this time regulation of utility hazardous air pollutant emissions is uncertain, ABB is committed to offering commercially viable solutions if regulations are promulgated. To do this, ABB initiated a three phase program:

The first phase consisted of a literature review on the present state of technology for hazardous air pollutants from sources around the world. This, combined with previous information gathered during the 1980's, should create possibly the largest hazardous air pollutants library.

The second phase consists of sampling both pilot- and full-scale units to determine actual concentrations, forms, and possible control efficiencies of Title III pollutants with present-day air pollution control technologies. Information of the final fate of the pollutants will also be obtained. Analytical data gathered will be combined with data gathered from other programs funded by other organizations and incorporated into the above mentioned database.

The third phase will consist of development of advanced control technologies and measurement systems. It is possible that regulations may require control levels beyond those available with present-day pollution control systems. If so, utilities may have to add dedicated Title III control systems. These systems should be designed towards minimizing the cost impact while achieving needed control. Also, it is certain that once any pollutant is regulated, its emissions must be monitored, both continuously in real time and accurately.

## PHASE I: LITERATURE SEARCH

During 1991-1993, ABB performed an extensive literature search, gathering information on hazardous air pollutants in the following subjects:

- Form of the pollutant (metal) in the fuel, flue gas, and final fate
- Concentrations of the pollutant (metal) in fuel, flue gas, and final fate
- Formation of organic pollutants through combustion of fuel
- Concentration of organic pollutants in flue gas
- Control of pollutants with existing control systems (ESP, FF, DFGD, WFGD, etc.)
- Existing dedicated pollutant control technologies (activated carbon, etc.)
- Novel and advanced pollutant control technologies
- Sampling and analytical methodologies

An abundance of data was available for the trace metals, particularly mercury. However, organic emissions had apparently not been studied in great detail.

As a result, a library of over 400 literature sources was obtained. This library has been distilled down to a technical database and has been made available to ABB offices worldwide. The findings of this literature search were discussed in greater detail, although limited to coal combustion, in a previous paper.

## PHASE II: PILOT- AND FULL-SCALE SAMPLING

During the past year, ABB has sampled four full-size facilities in fashions similar to those of the EPRI PISCES program. It was intended to gather data of similar quality as the PISCES data to allow incorporation of the PISCES data into our database and allow direct comparison for a quality check.

The plants sampled were as follows:

**Coal-A**: A midwestern utility boiler burning a Kentucky subbituminous coal. Existing pollution control systems include a 1990's generation electrostatic precipitator and limestone wet flue gas desulfurization system.

**MSW-A**: A midwestern MSW fired boiler, permitted at greater than 250 tpd. Existing pollution control systems include a lime spray dry flue gas desulfurization system and a fabric filter.

**MSW-B**: A eastern MSW fired boiler, permitted at greater than 250 tpd. Existing pollution control systems include a lime spray dry flue gas desulfurization system and a fabric filter.

Additionally, three other coal-fired utilities will be sampled in the latter half of 1994. Data was available for inclusion at time of publication, however, it may be included in a later revision.

Sampling was performed by an independent, recognized subcontractor at all plants. Sampling and analysis was performed according to the following methods:

| | |
|---|---|
| Gas Velocity | EPA Method 1 |
| Volumetric Flow Rate | EPA Method 2 |
| Gas Composition | EPA Method 3 |
| Gas Moisture Content | EPA Method 4 |
| $PM_{10}$ Emissions | EPA Method 201A |
| Gaseous hydrocarbon emissions | EPA Method 18 |
| Dioxins/Furans | EPA Method 23 |
| Volatile organic emissions | EPA Method 25A |
| Volatile metal emissions | SW846-0030 |
| Semi-volatile metal emissions | SW846-0010 |
| Metal emissions | EPA Draft Method 29 |

In most cases, Method 29 samples were analyzed for the following metals: total chromium, cadmium, arsenic, nickel, manganese, beryllium, copper, zinc, lead, selenium, phosphorous, thallium, silver, antimony, barium, and mercury. Title III currently lists only antimony, arsenic, beryllium, cadmium, chromium, cobalt, lead, manganese, mercury, nickel, and selenium as pollutants.

For purposes of this presentation, results are presented in emission factors, with units of pound of pollutant emitted per trillion Btu for utility furnaces and pound of pollutant emitted per ton waste per day fired for MSW systems, enabling direct comparison between the different plants and to other presentations.

Results from the field testing are presented in the following tables.

Analysis of the multi-metals sampling from Coal-A show that most metals emissions are significantly reduced by the particulate control system itself, and that additional control systems will add little benefit to particulate metal emissions if that control system does not yield significant particulate removal. As a Wet FGD system will not lower the particulate emissions beyond a certain point, it's use as an ultra-high efficiency control device would be limited. However, systems such as catalytic reactors or spray dryers/fabric filters would, as they tend to collect particulate, even at low inlet loadings. An interesting note is that in the literature review and subsequent field testing, not one individual metal removal efficiency met or exceeded the particulate control efficiency.

Mercury emissions are addressed in a separate table. In most cases, mercury removal efficiencies for the particulate control system varied greatly, often swinging to large negative values. For cases of significant positive removal efficiencies, mercury could not be detected in the fly ash, making it doubtful that any significant reduction of mercury emissions could be achieved by particulate control alone at typical cold-side temperatures (275-350°F). Also, since the control mechanism is driven by vapor pressure, it would also be highly unlikely that particulate control alone would be effective for organic compounds with similar or higher volatilities as mercury, unless concentrations were high enough to cause precipitation.

Polychlorinated dibenzo-p-dioxin (PCDD) and polychlorinated dibenzofuran (PCDF) emissions are shown in Table 3 for Coal-A. Entries noted with "(max)" indicate that the particulate congener was not detected and the detection limit is given for calculations. Removal efficiencies noted "(min)" indicate that the actual removal efficiency exceeds the value given, which is calculated using the detection limit. The more toxic compounds were not found in detectable quantities. Emission factors for PCDD/PCDF congeners are six to eight orders of magnitude below metal emissions. From the data obtained, it would be very difficult to state with certainty whether any significant removal was found and if so, what would be the most effective device for control.

Other organic emissions sampled at Coal-A were formaldehyde and total hydrocarbons (volatile and semivolatile organic compounds), reported as total methane equivalents. Formaldehyde emissions were approximately the same order of magnitude as metal emissions. Total hydrocarbon emissions were several orders of magnitude greater than metals emissions, with a significant increase in emissions noted across the Wet FGD system. Speculation for this increase in THC emissions include steam stripping of oils and solvents washed into the recirculated water from the pad.

As a rule, metals concentrations in MSW combustion gas were several orders of magnitude greater than from coal combustion gases. However, the spray dryer/fabric filter was quite capable of efficient collection of these metals from the gas stream. The metals is Table 5 are associated with the particulate phase, and should as a rule be collected with removal efficiencies approaching the particulate collection efficiency.

Mercury emissions are shown in Table 6, along with improving SDA/FF removal efficiencies due to increasing activated carbon injection. Carbon was added to and injected with the lime slurry. The injected carbon was specifically manufactured for gas phase applications and was not impregnated with any sulfur compounds. Depending of the carbon content of the fly ash, it is possible to achieved in excess of 90 % removal efficiency with little carbon injection.

PCDD/PCDF concentrations in MSW combustion gas were also higher than in coal combustion gas. However, due to the proximity to the detection limit and variability in analytical results, the data obtained make it very tenuous to state with certainty if any significant collection was obtained and what was the method of collection.

Polyaromatic compound analytical results are shown in Table 8. Most compounds could not be detected, however those that were did show significant collection by the spray dryer/fabric filter. It is believed that activated carbon injection would also be effective in collected the compounds in manner similar to mercury. Polychlorinated biphenyl compounds were not found in MSW combustion gas in levels that were detectable.

**TABLE 1. COAL-FIRED UTILITY METALS EMISSIONS AND APC REMOVAL EFFICIENCIES**

| Coal-A Metal Pollutant | ESP Inlet Emission Factor, lb./$10^{12}$ Btu | ESP Outlet Emission Factor, lb./$10^{12}$ Btu | ESP Removal Efficiency | WFGD Outlet Emission Factor, lb./$10^{12}$ Btu | WFGD Removal Efficiency |
|---|---|---|---|---|---|
| Aluminum (Al) | | | | | |
| Antimony (Sb) | 110 | 16 | 85.5 % | 16 | No Add 'l |
| Arsenic (As) | 720 | 22 | 96.9 % | 22 | No Add 'l |
| Barium (Ba) | 2100 | 11 | 99.5 % | 11 | No Add 'l |
| Beryllium (Be) | 45 | 1.1 | 97.6 % | 1.1 | No Add 'l |
| Cadmium (Cd) | 37 | 2.8 | 92.4 % | 2.8 | No Add 'l |
| Chromium (Cr) | 680 | 720 | Not Avail | 710 | No Add 'l |
| Cobalt (Co) | | | | | |
| Copper (Cu) | 430 | 26 | 94.0 % | 26 | No Add 'l |
| Lead (Pb) | 590 | 27 | 95.4 % | 27 | No Add 'l |
| Manganese (Mn) | 940 | 70 | 92.6 % | 70 | No Add 'l |
| Nickel (Ni) | 1200 | 330 | 72.5 % | 330 | No Add 'l |
| Phosphorous (P) | 3200 | 66 | 97.9 % | 66 | No Add 'l |
| Selenium (Se) | 590 | 610 | Not Avail | 610 | No Add 'l |
| Silver (Ag) | 12 | 2.2 | 81.7 % | 2.2 | No Add 'l |
| Thallium (Tl) | 140 | 22 | 84.3 % | 22 | No Add 'l |
| Tin (Sn) | | | | | |
| Vanadium (V) | | | | | |
| Zinc (Zn) | 1600 | 59 | 96.3 % | 60 | No Add 'l |

**TABLE 2. COAL-FIRED UTILITY MERCURY EMISSIONS AND APC REMOVAL EFFICIENCIES**

| Plant ID | Boiler Emission Factor, lb./$10^{12}$ Btu | ESP Emission Factor, lb./$10^{12}$ Btu | ESP Removal Efficiency | Stack Emission Factor, lb./$10^{12}$ Btu | WFGD Removal Efficiency |
|---|---|---|---|---|---|
| Coal-A | 8.2 | 7.3 | 10.9 % | 1.2 | 83.1 % |
| Coal-A | 6.4 | 10.7 | -67.6 % | 4.0 | 62.0 % |
| Coal-A | 4.8 | 6.9 | -46.0 % | 1.6 | 77.2 % |

TABLE 3. COAL-FIRED UTILITY ORGANIC EMISSIONS AND APC REMOVAL EFFICIENCIES

| Organic Pollutant | ESP Inlet Emission Factor, lb./$10^{12}$ Btu | ESP Outlet Emission Factor, lb./$10^{12}$ Btu | ESP Removal Efficiency | FGD Outlet Emission Factor, lb./$10^{12}$ Btu | FGD Removal Efficiency | APC Removal Efficiency |
|---|---|---|---|---|---|---|
| 2,3,7,8 TCDD | $2.3 * 10^{-6}$ (max) | $9.6 * 10^{-6}$ (max) | Not Avail | $3.2 * 10^{-6}$ (max) | Not Avail | Not Avail |
| 1,2,3,7,8 PeCDD | $1.2 * 10^{-5}$ (max) | $6.7 * 10^{-6}$ (max) | Not Avail | $1.9 * 10^{-6}$ (max) | Not Avail | Not Avail |
| 1,2,3,4,7,8 HxCDD | $1.5 * 10^{-5}$ (max) | $9.4 * 10^{-6}$ | Not Avail | $3.3 * 10^{-6}$ (max) | 64.4 % (min) | Not Avail |
| 1,2,3,6,7,8 HxCDD | $2.8 * 10^{-5}$ | $1.5 * 10^{-5}$ | 44.0 % | $2.2 * 10^{-6}$ (max) | 85.6 % (min) | 92.0 % (min) |
| 1,2,3,7,8,9 HxCDD | $3.1 * 10^{-5}$ | $2.1 * 10^{-5}$ | 32.3 % | $2.5 * 10^{-6}$ (max) | 88.0 % (min) | 91.9 % (min) |
| 1,2,3,4,6,7,8 HpCDD | $1.8 * 10^{-4}$ | $1.1 * 10^{-4}$ | 37.2 % | $8.5 * 10^{-6}$ | 92.4 % | 95.2 % |
| 1,2,3,4,6,7,8,9 OCDD | $2.9 * 10^{-4}$ | $2.7 * 10^{-4}$ | 7.5 % | $2.1 * 10^{-5}$ | 92.1 % | 92.7 % |
| Total PCDD | $5.6 * 10^{-4}$ (max) | $4.4 * 10^{-4}$ (max) | | $4.3 * 10^{-5}$ | | 92.3 % |
| 2,3,7,8 TCDD Eq PCDD | $3.0 * 10^{-4}$ (max) | $2.8 * 10^{-4}$ (max) | | $2.6 * 10^{-5}$ | | 91.5 % |
| 2,3,7,8 TCDF | $1.1 * 10^{-5}$ | $6.6 * 10^{-6}$ (max) | 38.9 % (min) | $7.9 * 10^{-6}$ (max) | Not Avail | 27.2 % (min) |
| 1,2,3,7,8 PeCDF | $1.4 * 10^{-5}$ (max) | $7.5 * 10^{-6}$ | Not Avail | $1.2 * 10^{-6}$ (max) | 84.3 % (min) | Not Avail |
| 2,3,4,7,8 PeCDF | $3.6 * 10^{-5}$ (max) | $1.6 * 10^{-5}$ | Not Avail | $1.2 * 10^{-6}$ (max) | 92.6 % (min) | Not Avail |
| 1,2,3,4,7,8 HxCDF | $1.4 * 10^{-4}$ | $5.7 * 10^{-5}$ | 59.5 % | $2.0 * 10^{-6}$ (max) | 96.5 % (min) | 98.6 % (min) |
| 1,2,3,6,7,8 HxCDF | $3.6 * 10^{-5}$ (max) | $1.6 * 10^{-5}$ | Not Avail | $1.3 * 10^{-6}$ (max) | 92.3 % (min) | Not Avail |
| 2,3,4,6,7,8 HxCDF | $5.0 * 10^{-6}$ (max) | $3.5 * 10^{-6}$ | Not Avail | $2.3 * 10^{-6}$ (max) | 35.1 % (min) | Not Avail |
| 1,2,3,7,8,9 HxCDF | $7.0 * 10^{-5}$ (max) | $2.6 * 10^{-5}$ | Not Avail | $1.6 * 10^{-6}$ (max) | 93.7 % (min) | Not Avail |
| 1,2,3,4,6,7,8 HpCDF | $2.2 * 10^{-4}$ | $8.1 * 10^{-5}$ | 62.4 % | $3.6 * 10^{-6}$ (max) | 95.5 % (min) | 98.3 % (min) |
| 1,2,3,4,7,8,9 HpCDF | $2.3 * 10^{-5}$ (max) | $1.8 * 10^{-5}$ | Not Avail | $2.2 * 10^{-6}$ (max) | 87.4 % (min) | Not Avail |
| OCDF | $7.1 * 10^{-5}$ | $5.9 * 10^{-5}$ | 17.3 % | $2.5 * 10^{-6}$ (max) | 95.7 % (min) | 96.5 % (min) |
| Total PCDF | $6.2 * 10^{-4}$ (max) | $2.9 * 10^{-4}$ (max) | | $2.6 * 10^{-5}$ (max) | | |
| 2,3,7,8 TCDD Eq PCDF | $8.9 * 10^{-6}$ (max) | $4.2 * 10^{-6}$ (max) | | $1.1 * 10^{-6}$ (max) | | |
| Total PCDD+PCDF | $1.2 * 10^{-3}$ (max) | $7.3 * 10^{-4}$ (max) | | $6.9 * 10^{-5}$ (max) | | |
| 2,3,7,8 Eq PCDD+PCDF | $3.1 * 10^{-4}$ (max) | $2.9 * 10^{-4}$ (max) | | $2.7 * 10^{-5}$ (max) | | |

**TABLE 4. COAL-FIRED UTILITY ORGANIC EMISSIONS AND APC REMOVAL EFFICIENCIES**

| Coal-A Organic Pollutant | ESP Inlet Emission Factor, lb./$10^{12}$ Btu | ESP Outlet Emission Factor, lb./$10^{12}$ Btu | ESP Removal Efficiency | FGD Outlet Emission Factor, lb./$10^{12}$ Btu | FGD Removal Efficiency |
|---|---|---|---|---|---|
| Formaldehyde | 40.9 | 3.7 | 91.0 % | 1.6 | 56.8 % |
| THC | 17500 | 7800 | 55.4 % | 17900 | -129.5 % |

**TABLE 5. MSW COMBUSTION METALS EMISSIONS AND APC REMOVAL EFFICIENCIES**

| MSW-B Metal Pollutant | Inlet Emission Factor, lb./hr per ton/day waste fired | Outlet Emission Factor, lb./hr per ton/day waste fired | Removal Efficiency, SDA/FF |
|---|---|---|---|
| Aluminum (Al) | $7.5 * 10^{-2}$ | $6.6 * 10^{-4}$ | 99.1 % |
| Antimony (Sb) | $5.2 * 10^{-4}$ | $8.2 * 10^{-6}$ | 98.4 % |
| Arsenic (As) | $1.2 * 10^{-4}$ | $2.1 * 10^{-7}$ (max) | 99.8 % (min) |
| Barium (Ba) | $5.7 * 10^{-4}$ | $5.0 * 10^{-6}$ | 99.1 % |
| Beryllium (Be) | $1.3 * 10^{-6}$ | $6.2 * 10^{-8}$ (max) | 95.3 % (min) |
| Cadmium (Cd) | $4.0 * 10^{-4}$ | $5.6 * 10^{-6}$ | 98.6 % |
| Chromium (Cr) | $1.8 * 10^{-3}$ | $2.0 * 10^{-6}$ | 99.9 % |
| Cobalt (Co) | $1.0 * 10^{-4}$ | $7.2 * 10^{-7}$ | 99.3 % |
| Copper (Cu) | $7.6 * 10^{-4}$ | $9.2 * 10^{-6}$ | 98.8 % |
| Lead (Pb) | $1.7 * 10^{-3}$ | $6.1 * 10^{-5}$ | 96.4 % |
| Manganese (Mn) | $9.6 * 10^{-4}$ | $1.2 * 10^{-5}$ | 98.7 % |
| Nickel (Ni) | $8.4 * 10^{-5}$ | $1.8 * 10^{-6}$ | 97.8 % |
| Phosphorous (P) | $5.9 * 10^{-3}$ | $2.8 * 10^{-5}$ | 99.5 % |
| Selenium (Se) | $5.4 * 10^{-6}$ | $4.0 * 10^{-7}$ | 92.7 % |
| Thallium (Tl) | $5.5 * 10^{-6}$ (max) | $3.8 * 10^{-6}$ (max) | Not Avail |
| Tin (Sn) | $7.2 * 10^{-4}$ | $1.6 * 10^{-5}$ | 97.7 % |
| Vanadium (V) | $7.4 * 10^{-5}$ | $7.2 * 10^{-7}$ (max) | 99.0 % (min) |
| Zinc (Zn) | $1.7 * 10^{-2}$ | $2.2 * 10^{-4}$ | 98.7 % |

**TABLE 6. MSW COMBUSTION MERCURY EMISSIONS AND APC REMOVAL EFFICIENCIES**

| Plant ID | Rel. AC Injection Rate | Inlet Emission Factor, lb./hr per ton/day fired | Outlet Emission Factor, lb./hr per ton/day fired | Removal Efficiency, SDA/FF |
|---|---|---|---|---|
| MSW-A | 0 | $1.6 * 10^{-4}$ | $6.9 * 10^{-5}$ | 56.1 % |
| MSW-A | 0 | $9.4 * 10^{-5}$ | $3.8 * 10^{-5}$ | 60.0 % |
| MSW-A | 0 | $1.2 * 10^{-4}$ | $4.7 * 10^{-5}$ | 60.7 % |
| MSW-B | 0 | $5.9 * 10^{-5}$ | $1.9 * 10^{-5}$ | 68.4 % |
| MSW-B | 0 | $1.7 * 10^{-4}$ | $2.0 * 10^{-5}$ | 88.4 % |
| MSW-A | 0.13 | $7.6 * 10^{-5}$ | $3.3 * 10^{-5}$ | 56.5 % |
| MSW-A | 0.26 | $1.1 * 10^{-4}$ | $1.2 * 10^{-5}$ | 88.9 % |
| MSW-A | 0.27 | $1.0 * 10^{-4}$ | $1.5 * 10^{-5}$ | 85.4 % |
| MSW-B | 0.35 | $9.0 * 10^{-5}$ | $2.9 * 10^{-6}$ | 96.8 % |
| MSW-B | 0.35 | $2.3 * 10^{-4}$ | $3.4 * 10^{-6}$ | 98.5 % |
| MSW-A | 0.43 | $1.3 * 10^{-4}$ | $1.8 * 10^{-5}$ | 86.4 % |
| MSW-A | 0.43 | $1.3 * 10^{-4}$ | $9.0 * 10^{-6}$ | 93.1 % |
| MSW-A | 0.49 | $1.1 * 10^{-4}$ | $1.5 * 10^{-5}$ | 86.8 % |
| MSW-A | 0.57 | $5.2 * 10^{-4}$ | $9.1 * 10^{-5}$ | 82.6 % |
| MSW-A | 0.59 | $1.4 * 10^{-4}$ | $7.7 * 10^{-6}$ | 94.7 % |
| MSW-B | 0.61 | $1.1 * 10^{-4}$ | $2.4 * 10^{-6}$ | 97.8 % |
| MSW-B | 0.61 | $1.8 * 10^{-4}$ | $3.1 * 10^{-6}$ | 98.3 % |
| MSW-B | 0.61 | $1.2 * 10^{-4}$ | $1.9 * 10^{-6}$ | 98.5 % |
| MSW-B | 0.81 | $7.5 * 10^{-5}$ | $1.4 * 10^{-6}$ | 98.1 % |
| MSW-B | 0.81 | $2.2 * 10^{-4}$ | $3.3 * 10^{-6}$ | 98.5 % |
| MSW-A | 1.00 | $7.7 * 10^{-5}$ | $5.4 * 10^{-6}$ | 92.9 % |

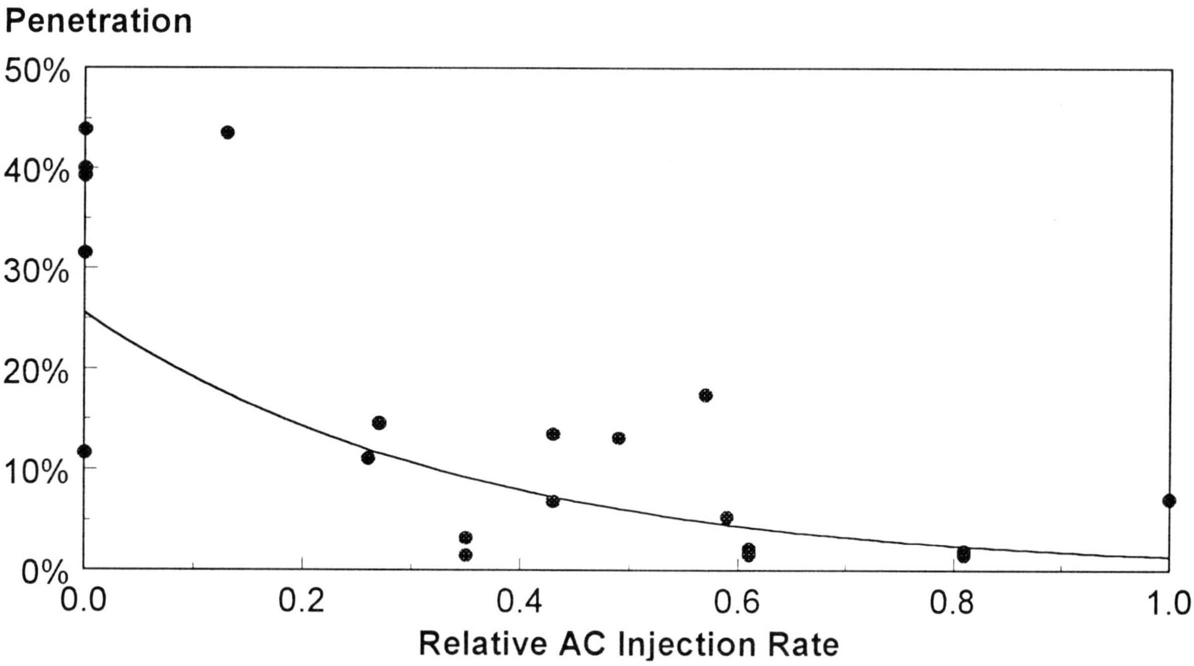


FIGURE 1. MERCURY PENETRATION AS A FUNCTION OF ACTIVATED CARBON INJECTION

## TABLE 7. MSW COMBUSTION ORGANIC EMISSIONS AND APC REMOVAL EFFICIENCIES

| MSW-B Organic Pollutant | Inlet Emission Factor, lb./hr per ton/day waste fired | Outlet Emission Factor, lb./hr per ton/day waste fired | Removal Efficiency, SDA/FF |
|---|---|---|---|
| 2,3,7,8 TCDD | $4.6 * 10^{-10}$ | $3.8 * 10^{-11}$ | 91.7 % |
| Other TCDD | $6.2 * 10^{-9}$ | $9.6 * 10^{-9}$ | -54.2 % |
| 1,2,3,7,8 PeCDD | $6.5 * 10^{-10}$ | $3.4 * 10^{-10}$ | 46.9 % |
| Other PeCDD | $5.3 * 10^{-9}$ | $1.4 * 10^{-8}$ | -165.1 % |
| 1,2,3,4,7,8 HxCDD | $2.7 * 10^{-10}$ | $3.2 * 10^{-10}$ | -19.9 % |
| 1,2,3,6,7,8 HxCDD | $3.6 * 10^{-10}$ | $6.3 * 10^{-10}$ | -75.7 % |
| 1,2,3,7,8,9 HxCDD | $5.3 * 10^{-10}$ | $9.6 * 10^{-10}$ | -79.6 % |
| Other HxCDD | $4.5 * 10^{-9}$ | $8.4 * 10^{-9}$ | -86.8 % |
| 1,2,3,4,6,7,8 HpCDD | $2.8 * 10^{-9}$ | $3.0 * 10^{-9}$ | -7.0 % |
| Other HpCDD | $2.8 * 10^{-9}$ | $3.0 * 10^{-9}$ | -7.4 % |
| 1,2,3,4,6,7,8,9 OCDD | $3.9 * 10^{-9}$ | $9.7 * 10^{-10}$ | 75.0 % |
| | | | |
| Total PCDD | $2.8 * 10^{-8}$ | $4.1 * 10^{-8}$ | -48.8 % |
| 2,3,7,8 TCDD Eq PCDD | $9.2 * 10^{-10}$ | $4.6 * 10^{-10}$ | 50.3 % |
| | | | |
| 2,3,7,8 TCDF | $8.8 * 10^{-9}$ | $2.5 * 10^{-9}$ | 71.8 % |
| Other TCDF | $7.0 * 10^{-8}$ | $1.5 * 10^{-8}$ | 79.0 % |
| 1,2,3,7,8 PeCDF | $2.9 * 10^{-9}$ | $7.6 * 10^{-10}$ | 74.3 % |
| 2,3,4,7,8 PeCDF | $2.9 * 10^{-9}$ | $1.2 * 10^{-9}$ | 56.9 % |
| Other PeCDF | $3.8 * 10^{-8}$ | $1.8 * 10^{-8}$ | 52.2 % |
| 1,2,3,4,7,8 HxCDF | $3.5 * 10^{-9}$ | $3.1 * 10^{-9}$ | 11.4 % |
| 1,2,3,6,7,8 HxCDF | $1.8 * 10^{-9}$ | $1.3 * 10^{-9}$ | 29.0 % |
| 2,3,4,6,7,8 HxCDF | $1.5 * 10^{-9}$ | $1.5 * 10^{-9}$ | -0.5 % |
| 1,2,3,7,8,9 HxCDF | $1.1 * 10^{-10}$ | $3.2 * 10^{-11}$ | 70.2 % |
| Other HxCDF | $1.1 * 10^{-8}$ | $9.8 * 10^{-9}$ | 13.0 % |
| 1,2,3,4,6,7,8 HpCDF | $3.4 * 10^{-9}$ | $3.7 * 10^{-9}$ | -7.9 % |
| 1,2,3,4,7,8,9 HpCDF | $4.0 * 10^{-10}$ | $5.5 * 10^{-10}$ | -37.7 % |
| Other HpCDF | $2.0 * 10^{-9}$ | $1.5 * 10^{-9}$ | 24.0 % |
| OCDF | $4.5 * 10^{-10}$ | $2.6 * 10^{-10}$ | 42.0 % |
| | | | |
| Total PCDF | $1.5 * 10^{-7}$ | $5.9 * 10^{-8}$ | 59.8 % |
| 2,3,7,8 TCDD Eq PCDF | $1.6 * 10^{-9}$ | $5.4 * 10^{-10}$ | 66.8 % |
| | | | |
| Total PCDD+PCDF | $1.8 * 10^{-7}$ | $1.0 * 10^{-7}$ | 42.6 % |
| 2,3,7,8 Eq PCDD+PCDF | $2.6 * 10^{-9}$ | $1.0 * 10^{-9}$ | 60.9 % |

**TABLE 8. MSW COMBUSTION ORGANIC EMISSIONS AND APC REMOVAL EFFICIENCIES**

| MSW-B Organic Pollutant | Inlet Emission Factor, lb./hr per ton/day waste fired | Outlet Emission Factor, lb./hr per ton/day waste fired | Removal Efficiency, SDA/FF |
|---|---|---|---|
| 2-Chloronaphthalene | $3.5 * 10^{-8}$ (max) | $5.5 * 10^{-9}$ (max) | Not Avail |
| Benzo(j)fluoranthene | $1.6 * 10^{-8}$ (max) | $1.3 * 10^{-8}$ (max) | Not Avail |
| Anthanthrene | $1.3 * 10^{-8}$ (max) | $1.1 * 10^{-8}$ (max) | Not Avail |
| Naphthalene | $7.8 * 10^{-6}$ | $1.8 * 10^{-6}$ | 76.4 % |
| 2-Methylnaphthalene | $2.7 * 10^{-7}$ | $3.0 * 10^{-8}$ | 88.7 % |
| Acenaphthylene | $6.0 * 10^{-8}$ (max) | $6.8 * 10^{-9}$ (max) | Not Avail |
| Acenaphthene | $1.2 * 10^{-7}$ | $1.0 * 10^{-8}$ (max) | 91.0 % (min) |
| Fluorene | $8.4 * 10^{-8}$ | $2.8 * 10^{-8}$ (max) | 67.4 % (min) |
| Phenanthrene | $6.0 * 10^{-7}$ | $2.6 * 10^{-7}$ | 57.4 % |
| Anthracene | $4.6 * 10^{-8}$ | $1.2 * 10^{-8}$ | 73.0 % |
| Fluoranthene | $1.4 * 10^{-7}$ | $2.7 * 10^{-8}$ | 80.8 % |
| Pyrene | $1.2 * 10^{-7}$ | $1.8 * 10^{-8}$ | 84.6 % |
| Benz(a)anthracene | $7.4 * 10^{-9}$ (max) | $5.5 * 10^{-9}$ (max) | Not Avail |
| Chrysene | $1.0 * 10^{-8}$ (max) | $5.5 * 10^{-9}$ (max) | Not Avail |
| Benzo(b)fluoranthene | $1.6 * 10^{-8}$ (max) | $1.3 * 10^{-8}$ (max) | Not Avail |
| Benzo(k)fluoranthene | $1.5 * 10^{-8}$ (max) | $1.3 * 10^{-8}$ (max) | Not Avail |
| Benzo(e)pyrene | $1.5 * 10^{-8}$ (max) | $1.3 * 10^{-8}$ (max) | Not Avail |
| Benzo(a)pyrene | $1.5 * 10^{-8}$ (max) | $1.3 * 10^{-8}$ (max) | Not Avail |
| Perylene | $1.5 * 10^{-8}$ (max) | $1.3 * 10^{-8}$ (max) | Not Avail |
| Indeno(1,2,3-c,d)pyrene | $1.0 * 10^{-8}$ (max) | $7.7 * 10^{-9}$ (max) | Not Avail |
| Dibenzo(a,h)anthracene | $1.2 * 10^{-8}$ (max) | $1.1 * 10^{-8}$ (max) | Not Avail |
| Benzo(g,h,i)perylene | $1.2 * 10^{-8}$ (max) | $7.7 * 10^{-9}$ (max) | Not Avail |
| Total Polyaromatic Hydrocarbons (PAHs) | $9.4 * 10^{-6}$ (max) | $2.4 * 10^{-6}$ (max) | |

**TABLE 9. MSW COMBUSTION ORGANIC EMISSIONS AND APC REMOVAL EFFICIENCIES**

| MSW-B Organic Pollutant | Inlet Emission Factor, lb./hr per ton/day waste fired | Outlet Emission Factor, lb./hr per ton/day waste fired | Removal Efficiency, SDA/FF |
|---|---|---|---|
| Chlorobiphenyls | $1.3 * 10^{-8}$ (max) | $2.8 * 10^{-9}$ (max) | Not Avail |
| Bichlorobiphenyls | $6.7 * 10^{-8}$ (max) | $5.8 * 10^{-9}$ (max) | Not Avail |
| Trichlorobiphenyls | $1.7 * 10^{-8}$ | $7.7 * 10^{-9}$ (max) | 54.4 % (min) |
| Tetrachlorobiphenyls | $8.4 * 10^{-9}$ | $2.1 * 10^{-9}$ (max) | 75.6 % (min) |
| Pentachlorobiphenyls | $1.9 * 10^{-8}$ (max) | $1.1 * 10^{-8}$ (max) | Not Avail |
| Hexachlorobiphenyls | $6.9 * 10^{-9}$ (max) | $6.3 * 10^{-9}$ (max) | Not Avail |
| Heptachlorobiphenyls | $3.9 * 10^{-9}$ (max) | $3.4 * 10^{-9}$ (max) | Not Avail |
| Octachlorobiphenyls | $1.3 * 10^{-8}$ (max) | $1.1 * 10^{-8}$ (max) | Not Avail |
| Nonachlorobiphenyls | $3.5 * 10^{-9}$ (max) | $2.9 * 10^{-9}$ (max) | Not Avail |
| Decachlorobiphenyls | $2.6 * 10^{-9}$ (max) | $2.3 * 10^{-9}$ (max) | Not Avail |
| Total Polychlorobiphenyls (PCBs) | $1.5 * 10^{-7}$ (max) | $5.6 * 10^{-8}$ (max) | |

FACT-Vol. 18, Combustion Modeling, Scaling and Air Toxins
ASME 1994

# MULTIPURPOSE ACTIVE COKE ADSORBERS: CLEAN EXHAUST GAS FROM WASTE INCINERATION

**Andreas Wecker**
RZR Herten
Herten, North-Rhine-Westfalia, Germany

## Abstract

Using mulitpurpose active coke adsorbers for additional gas cleaning of waste incineration is one possibility to reach the low emission limits required in Germany. The operating temperature of these adsorbers is about 120 - 160°C. Due to the self-inflaming coke temperature of over 200°C it is necessary to lead the gas equally through the coke layers. To indicate local temperature increases in the coke layers (so-called hot-spots), due to disturbances in equal gas flow, the CO-concentration is measured before and after the adsorber as ΔCO-value. Increase of ΔCO-value indicates a hot-spot due to the oxidation reaction with $O_2$ from exhaust gas and coke. A safety concept stipulates the preventive measures.

Three years in operating such multipurpose active coke adsorbers at RZR Herten show good results of this technology. The emission limits are not only observed, but the values are drastically below these limits.

## Introduction

The National Law for Air Pollution in Germany is based on the legislation for the European Union (EU). This legislation takes four forms (Bartaire, 1991):

- *Regulations* are binding to all Member States in their entirety.
- *Directives* are binding to Member States as to the result to be achieved, while leaving a degree of flexibility as to how the measures are implemented in national legislation.
- *Decisions* are binding in their entirety and may be addressed to Government, private enterprise, or individuals.
- *Recommendations* and *opinions* are not binding.

As to the reduction of pollutant emissions, the European Council has issued 3 Directives:

- 1. Council Directive 89/369 of June 8, 1989 on the prevention of air pollution from new municipal waste incineration plants (MWI).
- 2. Council Directive 89/429 of June 21, 1989 on the prevention of air pollution from existing municipal waste incineration plants (MWI).
- 3. Council Directive 94/3 (ENV) of June 28, 1993 on the prevention of air pollution from hazardous waste incineration plants (HWI).

At the end of 1990 the first two Council Directives were implemented in the National German Law with the 17th. Ordinance (the so-called 17. BImSchV) of the Federal Immission Control Act (FICA). The emission limits required in the third Council Directive arise partly from the 17th. Ordinance (Fig. 1).

FICA came into force on April 1, 1974 and forms the legal basis for almost all immission control measures. The law requires all industrial plants to be built and operated in such a way as not to cause any harmful or dangerous affects on the environment. Plant operators must employ "state of the art" precautions to prevent harmful environmental affects.

The Technical Instruction on Air Pollution Control (TI Air) forms the most important administrative regulation to implement the Federal Immission Control Act. It plays a central and decisive part in the enforcement of FICA by the 16 German States and constitutes an established instrument of licencing and supervising authorities.

In addition, the statutory regulations for immission control in Germany (laws, ordinances, etc.), technical standards, e.g.

| Units in $mg/m^3$ (i.N., dry, 11 Vol-% $O_2$)<br>* other classification than 17th.Ordinance<br>** to be observed from 12/01/1996 | Germany | | European Council Directives | | |
|---|---|---|---|---|---|
| | TI Air '86<br>Daily Average Values | 17th.Ordinance of FICA<br>Daily Average Values | 89/369 New MWI<br>Weekly Average Values | 89/429** Existing MWI<br>Weekly Average Values | 93/4 HWI<br>Daily Average Values |
| Dust | 30 | 10 | 30 | 30 | 10 |
| HCl | 50 | 10 | 50 | 50 | 10 |
| HF | 2 | 1 | 2 | 2 | 1 |
| $SO_2$ | 100 | 50 | 300 | 300 | 50 |
| $NO_2$ | 500 | 200 | - | - | - |
| Corg. | 20 | 10 | 20 | 20 | 10 |
| CO | 100 | 50 | 100 | 100 | 50 |
| | | sample period 0.5 - 2 h | | | sample period 0.5 - 8 h |
| Σ Cd, Tl | 0.2* | 0.05 | 0.2* | 0.2* | 0.05 |
| Hg | | 0.05 | | | 0.05 |
| Σ Sb, As, Pb, Cr, Co, Cu, Mn, Ni, V, Sn | 5* | 0.5 | 5 / 1* | 5 / 1* | 0.5 |
| | | sample period 6 - 16 h | | | sample period 6 - 8 h |
| PCDD, PCDF<br>calculated as toxicity equivalence "ΣTE" | – | 0.000 000 1 | – | – | 0.000 000 1 |

Figure 1 Emission Limits

guidelines of VDI (Association of German Engineers), and standards of DIN (German Institute of Standardization) serve as a decision making aid for specialists in this field, in the preparatory stages of legislation, and the elaboration of regulations and directives.

Starting in 1996, the emission limits of the 17th. Ordinance of the FICA which are drastically reduced against the presently existing values of TI Air (Fig. 1) must be observed. Therefore, all existing waste incineration plants in Germany must be equipped with additional gas cleaning systems. One possibility to reach the required emission limits is to install a multipurpose active coke adsorber in the existing process in order to clean the exhaust gas. This adsorber is equipped to reduce (Eicken et al., 1990):

- Organic compounds like polychlorinated dioxins/furans and others.
- Heavy metals.
- Acidic compounds like HCl and $SO_2$.
- Dust.

With a special sort of activated carbon or on the basis of ceramics as a catalyst, it is possible to reduce the $NO_x$-values by adding ammonia to the exhaust gas before admission.

In general, the active coke adsorbers are equipped to function as:

- fixed bed adsorbers or
- moving bed adsorbers.

To clean exhaust gas derived from waste incineration plants of compounds other than $NO_x$ the moving bed is used. When $NO_x$-reduction must be achieved the fixed bed is used.

There are two different types of moving bed adsorbers (Fig. 2):

- Crossflow adsorbers.
- Counterflow adsorbers.

Both types are used at RZR Herten waste incineration plant.

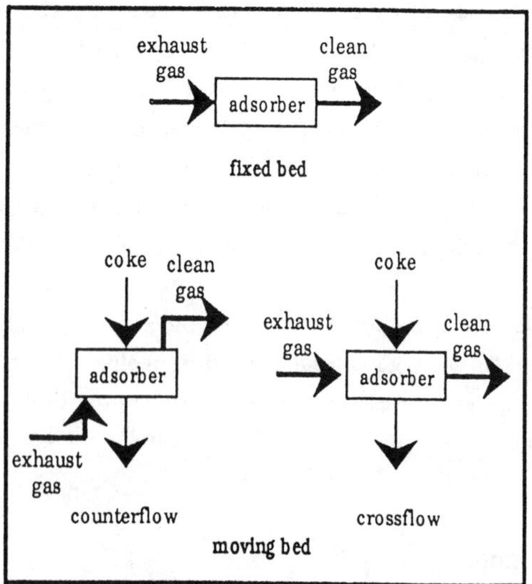


Figure 2 Types of Coke Adsorber Technologies

## The Facilities at RZR Herten

In 1978 AGR Waste Management Company of Essen, Germany, (AGR - Abfallentsorgungs-Gesellschaft Ruhrgebiet mbH) started in Herten, a town of the Ruhr region, the construction of facilities for:

- Production of fuel from domestic refuse (RDF).
- Incineration of municipal and bulky waste (MWI).
- Incineration of hazardous and hospital waste (HWI).

| | MWI | HWI |
|---|---|---|
| Waste input | 2 x 20 Mg/h | 2 x 6 Mg/h |
| Gas quantity (i.N., dry) | 2 x 113,000 m³/h | 2 x 56,300 m³/h |
| Steam production | | |
| Steam capacity | 2 x 57.5 Mg/h | 2 x 27 Mg/h |
| Pressure | 32 bar | |
| Temperature | 320 °C | |
| Steam utilization | | |
| Two turbo sets | 2 x 14.9 MW | |
| District heating system (3 heat exchangers) | 3 x 8 MW | |

Table 1: Total Steam Production/Utilization at RZR Herten

The operational management was assigned to STEAG, Germany, a power generating and plant engineering company. Operations started in 1981. Waste treatment capacity is:

| | |
|---|---|
| 395,000 Mg/a | municipal/bulky waste |
| 57,000 Mg/a | hazardous waste |
| 3,000 Mg/a | hospital waste |

for two lines of each plant. The values of production und utilization of steam are shown in Table 1.

## Waste Sorting and Treatment Plant

The process flow sheet for fuel production from domestic refuse is shown in Fig. 3. The domestic refuse is pre-screened in a rotary screen to pieces of > 40 mm and < 350 · 700 mm, separated from Fe-metal by magnetic separation, and crushed by a hammer mill. It is further separated and dried in a hot air stream and last briquetted to small pieces of 40 - 80 mm in length and 13 mm in diameter, the so-called "ECO-BRIQ" or "RDF".

Unfortunately the "RDF" facility failed to acquire any significance in the field of waste management, as due to its potential for air pollutants like HCl, $SO_2$ and heavy metals, the fuel produced finds no outlet even in the cement industry. Therefore, in the near future, the municipal bulky waste incineration will be expanded by two lines, whereas the waste sorting and treatment plant will be dismantled.

## Municipal Waste Incineration (MWI)

The two existing incinerator lines for municipal and bulky waste (capacity in 1993 was 250,000 Mg/a) are each equipped with an incinerator grate and the below specified exhaust gas cleaning technologies, based on a stipulation that does not allow to lead processed water into the canalisation system (Fig. 4):

- Cyclones for dust separation.
- Spray drier for evaporation of scrubbing water.
- Electrostatic precipitator for separation of fine dust and reaction products from scrubbing water.
- First and second step of the scrubbing system.

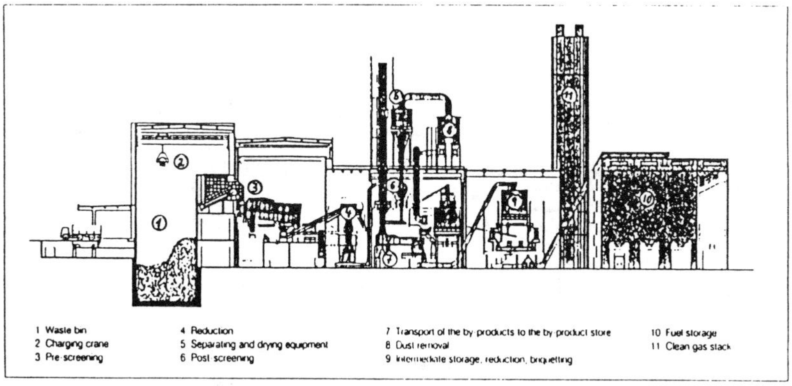


| Sorting and treatment | | | |
|---|---|---|---|
| Through-put capacity | t/h | pro line | 25 |
| Drying temperature | °C | | 200 |
| Moisture outlet | % | | 20 |
| ECO-BRIQ (BRAM, fuel from household waste) | t/h | pro line | 7,5 |
| FE metal separation | t/h | pro line | 1 |

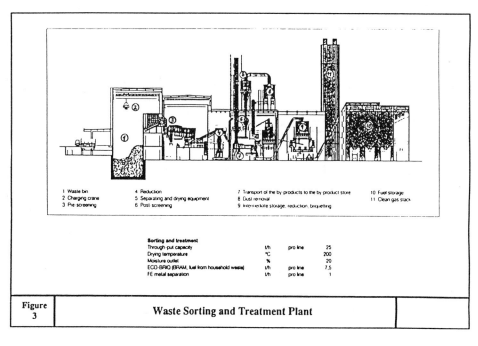

Figure 3 — Waste Sorting and Treatment Plant

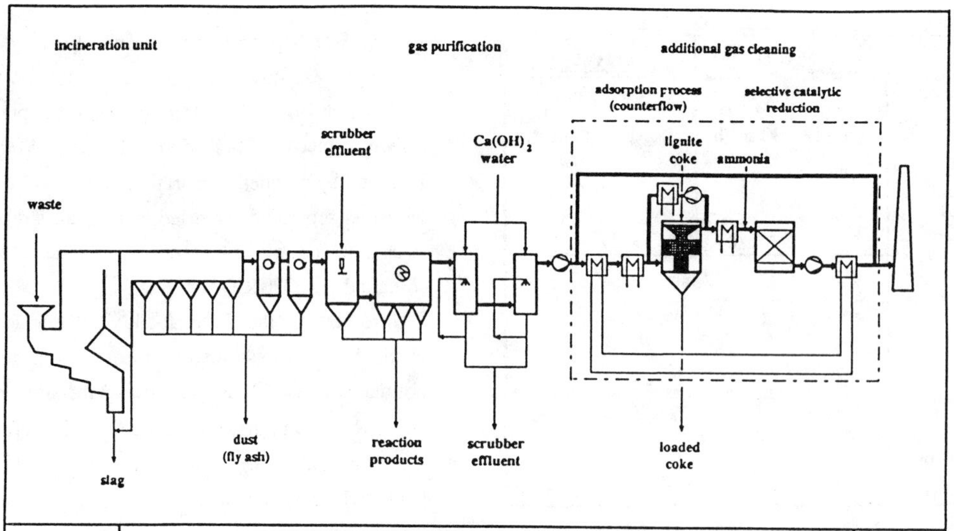


Figure 4 | Municipal Waste Incineration Facilities (MWI) at RZR Herten

First step of the wet cleaning system is to separate HCl and heavy metals, second step is to separate $SO_2$. Lime milk is used in both steps to achieve neutralization. As a result the reaction products consist of $CaCl_2$ and $CaSO_4$ as main ingredients.

## Hazardous/Hospital Waste Incineration (HWI)

The hazardous waste incineration facilities at RZR Herten are equipped as shown in Fig. 5 (both lines) with rotary kiln and reheating chamber, electrostatic precipitator for dust removal, and a two stage scrubbing system for neutralization using lime milk. The scrubbing water treated in an external evaporation plant, a so-called "waterhouse", is in contrast to the MWI.

In the "water house" both water streams from the two step scrubbing system are joined in a concentrator. The precipitation ($CaSO_4$) is separated by filtration using a vacuum tape filter.

The removal of heavy metal is achieved by adding $Na_2S$ to the salt solution ($CaCl_2$) after neutralization and filtration of precipitation. At last the solution is evaporated in two stages to crystallize the calcium chlorid for reuse in the chemical industry. The water derived from the evaporation as condensate is readded to the scrubbing system.

The above descriptions show that all processed water from the incineration facilities leaves the plant in a clean gas stream.

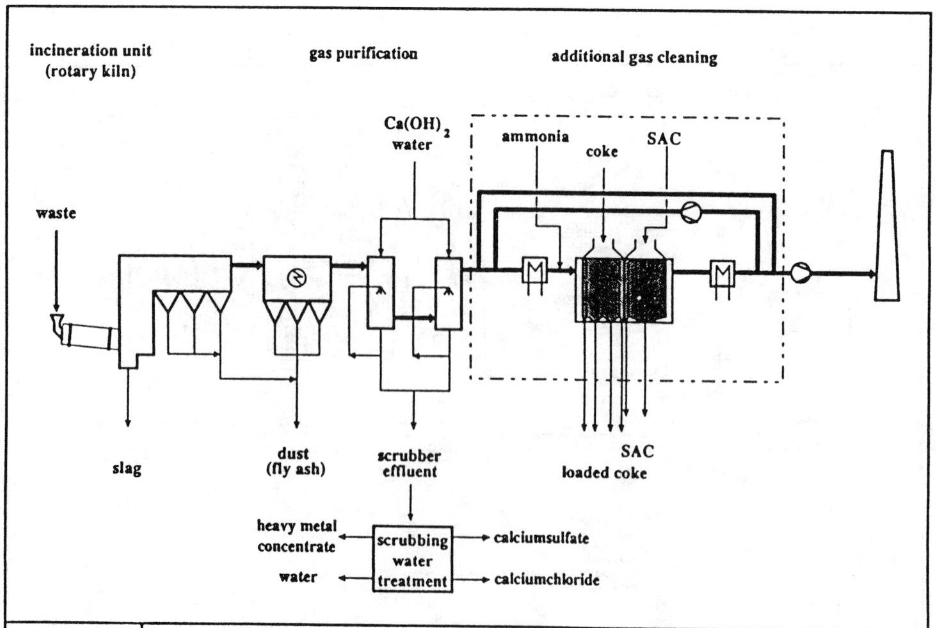


Figure 5 | Hazardous Waste Incineration Facilities (HWI 1) at RZR Herten

## Active Coke Adsorbers

At RZR Herten two different technologies are used:

- Crossflow adsorbers for hazardous/hospital waste incineration.
- Counterflow adsorbers for municipal and bulky waste incineration.

The same principle applies in both cases:

- $SO_2$ is adsorbed on the coke surface, catalytically oxidized to $SO_3$, and combined with water to sulphuric acid. This acid is neutralized with the basic ingredients of coke (CaO) to $CaSO_4$.
- HCl and HF are adsorbed on the coke surface and partly neutralized like sulphuric acid.
- Mercury ($Hg^0$) is adsorbed, oxidized, and combined to $HgSO_4$.
- Dust and heavy metals/organic compounds are filtered either bound to the dust surface or as particulate matter.
- Gaseous organic compounds are adsorbed.

The adsorbent is lignite coke.

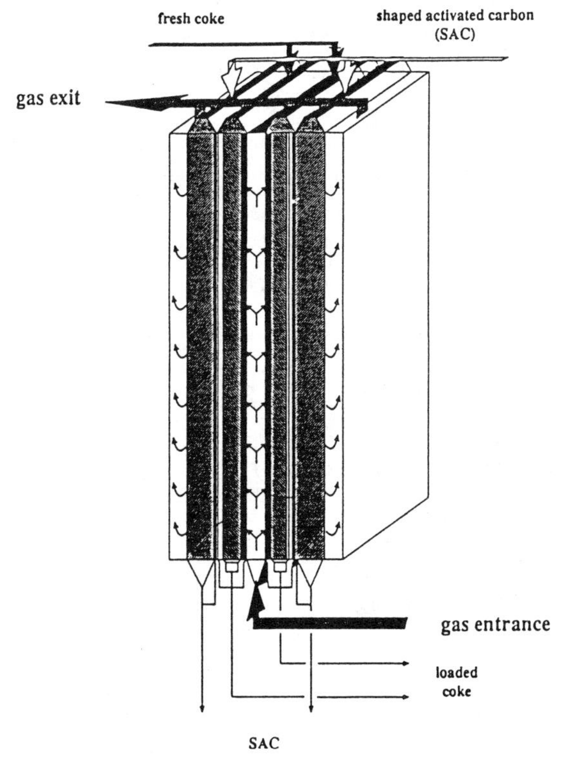


Figure 6 Crossflow Adsorber of HWI 1 at RZR Herten

## Crossflow Adsorber

The principle of the crossflow adsorber is to lead the exhaust gas horizontally across the vertically moving coke (Fig. 6). When design work began in 1986 with a view to obtain authorization,

| | |
|---|---|
| Adsorber unit dimensions (L x D x H ) | 9 m x 6 m x 22 m |
| Layer depth each side | |
| Coke | 1.20 m |
| Shaped activated carbon (SAC) | 1.75 m |
| Layer content each | |
| Coke | 160 m³ (80 Mg) |
| Shaped activated carbon (SAC) | 230 m³ (120 Mg) |
| Admission layer surface | 2 x 132 m² |
| Operation start-up | 08/08/1991 |

Table 2: Design Data of Crossflow Adsorber HWI 1

practical experience had already been gained with this kind of adsorbers for power plant application. Therefore, this technology was preferred when the first hazardous waste incineration line (HWI 1) was equipped with an additional gas cleaning system. The second line will be equipped by the end of 1995.

The exhaust gas is distributed left and right after entering the adsorber through the admission chamber. First the gas flows horizontally through the coke layers 1.20 m deep, symmetrically arranged on both sides of the admission chamber (Fig. 6). The fresh coke is added to the adsorber from above, while loaded coke is removed from below. Due to the filter action and the risk of pressure increase through dust loading, the admission layers (0.1 m) on both sides can be, separated from the rest, moved and transported to the outside.

In the second stage, exhaust gas flows through large layers (1.75 m) of shaped activated carbon (SAC) on both sides of the adsorber directly arranged next to the coke layers. The SAC is used for catalytic nitrogen oxide reduction with the aid of ammonia added to the exhaust gas before entering the adsorber (Fig. 7). The SAC-layers are operating as a quasi-fixed bed as only small amounts are transported to the outside where the fine dust is screened. The screened residue reenters the adsorber from the top.

Fig. 7 shows the process flow sheet of the adsorber facility. The exhaust gas arrives from the second scrubbing stage at a temperature of approx. 65°C. The gas is heated by a steam/gas exchanger (steam at 5 bar, 150 °C) to 120 °C, and, passing the adsorber by adding ammonia at the inlet, is blown through the stack by a clean gas fan. This figure shows that starting from incineration the flowing gas is below atmospheric pressure, its entire way including the adsorber. It is impossible, therefore, that the loaded coke can be blown to the outside even in cases of leakage. A

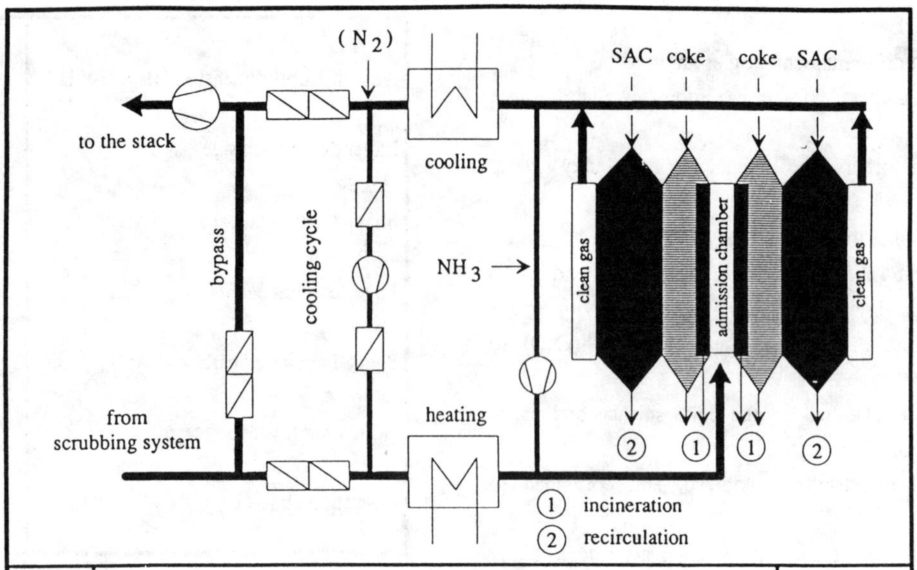


Figure 7 Process Flow Sheet of Additional Gas Cleaning System (HWI 1)

bypass and a cooling cycle (medium air) have been installed for safety. The design data are listed in Table 2.

## Counterflow Adsorber

Significant for the counterflow adsorber is that the gas reaches the coke layer via a special distribution floor from below, whereas the coke is moved in opposite direction (Fig. 8). The counterflow principle guaranties even loading covering the total coke surface at the inlet. Increase of pressure loss due to dust deposits in the coke layer can be readjusted to normal by moving the layer by only a few millimeters to the outside. Now, both lines of MWI are equipped with counterflow technology.

Before entering the adsorber units the clean gas, having been heated in two steps, is separated in two single streams of 50% each (Fig. 9) before the first heating step. The clean gas heats the exhaust gas up by means of plate heat exchangers (WT 1) increasing the temperature from approx. 65°C to approx. 135°C. Steam gas heat exchangers (WT 2) assist in raising the temperature by a further 20°C. The two gas streams are then led separately to the two adsorber units before being united. Before entering the $DeNO_x$-step it is possible to increase the gas temperature to 185°C (depending on the activity of the catalysator) by means of gas steam heat exchangers (WT 3). Denoxation takes place by adding ammonia to the gas before entering the ceramic catalysator. Behind the pressure increasing fan the clean gas stream is separated again into two streams, and after passing the heat exchangers (WT 1), reunited and led to the stack.

There are, however, other possibilities in addition to the afore described gas lead, i.e.:

- Safety bypass for start-up and shut-down of the entire additional gas cleaning system.
- $DeNO_x$-bypass to pass the catalysator.
- Preheating cycle of adsorber units and catalysator to reach approx. 100°C.
- Recirculation for maintaining a minimum stream capacity through the adsorber units during shut-down procedures or part-time operation, respectively.
- Cooling cycle (medium air) for rapid cooling of gas during plant shut-down periods or in case of hot-spot warning.

| | |
|---|---|
| Adsorber unit dimension (L x D x H ) | 12 m x 5.4 m x 13 m |
| Coke layer height each (4 per unit) | 1.5 m |
| Adsorber unit content | 90 Mg |
| Admission layer surface (per adsorber unit) | 4 x 32.4 m² |
| Catalysator dimension (L x D x H ) | 5 m x 4 m x 11.5 m |
| Number of catalysator stages | 3 (+1 for reserve) |
| Operation start-up | MWI 1 : 05/05/1994<br>MWI 2 : 02/07/1994 |

Table 3: Design Data of Counterflow Adsorber MWI

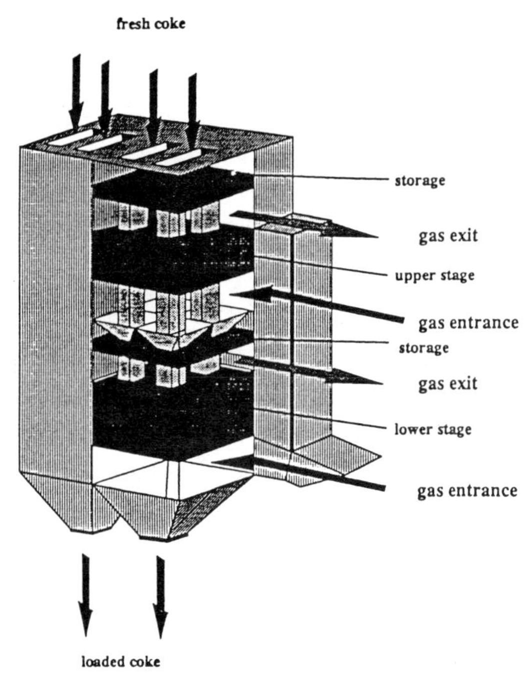

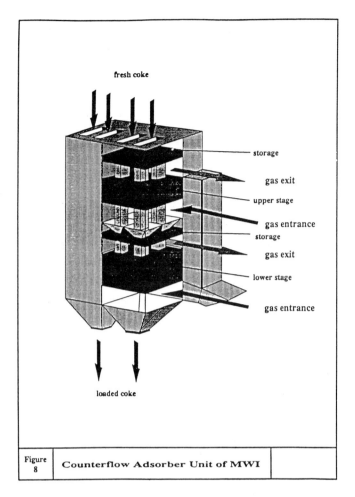

| Figure 8 | Counterflow Adsorber Unit of MWI | |
|---|---|---|

Via 2 adsorber units the gas lead offers the possibility to continue the operation on a part-load basis (50%) in case of break-down (e.g. hot-spot). One adsorber unit consists of two stages one on top of the other, which again are separated into two single beds (Fig. 8). The gas flows through a coke layer 1.5 m high. There are 2 feeding holes per single bed to fill the adsorbers with fresh coke. The design data are listed in Table 3.

The loaded coke is moved from the adsorber funnels (one per single bed) to a pressure container conveyer and finally, assisted by the carrier medium nitrogen, transported to a dosing device for dispatch to the hazardous waste incineration line 1.

## Safety Concept

As coke possesses a self-inflaming temperature of over 200°C and the coke adsorbers are operating at a temperature of 120°C to 155°C, special measures have to be taken to prevent or recognize so-called hot-spots.

Suitable construction of exhaust gas leads through the coke beds provide/guaranty a complete flow through the layers, so as not to form local temperature increases reaching critical limits, due to adsorption and reaction heat.

Should this occur, however, due to dust deposits or coke conglomerations, early recognition is obligatory in order to take respective preventive measures. As an inflammation or warm-up of the coke results in a strong CO-formation due to oxidation action taking place, a hot-spot can be detected by comparing the CO-concentration before and after the active coke plant with the ΔCO-value.

Suitable preventive measures to fight a hot-spot are:

- Moving the respective coke layer to reachieve a complete gas flow through the layer resulting in heat reduction.
- Adsorber shut-off from the exhaust gas stream through

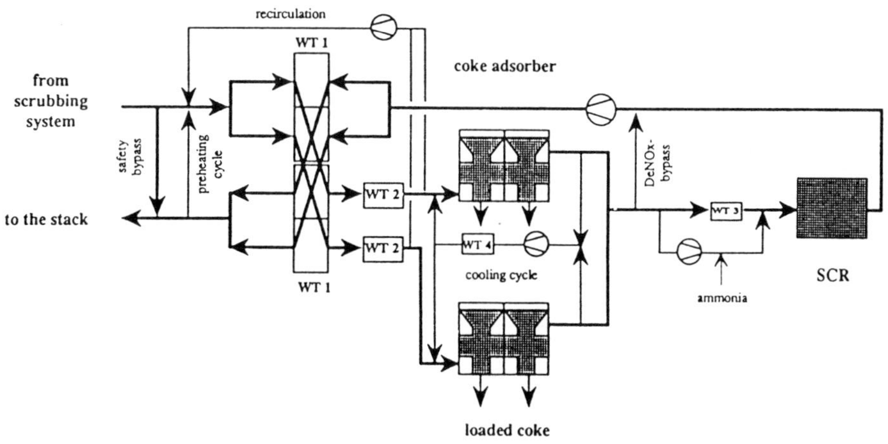


| Figure 9 | Process Flow Sheet of Additional Gas Cleaning System (MWI) | |
|---|---|---|

bypass operation and therefore self-inertisation due to the consumption of remaining oxygen by oxidation. Cooling with cooling cycle (medium air) under inert conditions by adding $N_2$.

## Handling of Loaded Coke

Due to the potential of hazardous ingredients the loaded coke is classified as hazardous waste. To eliminate this hazardous potential one possibility is to incinerate the loaded coke. Given the possibility of processing in the rotary kiln of the HWI at RZR Herten and thus the profit of steam production no fundamental problems arise for this elimination. Therefore, all loaded coke from the adsorbers at RZR Herten is transported to a dosing tank by pneumatic conveyor, with $N_2$ as a carrier gas, and then blown with an air stream into the rotary kiln.

To make sure that there is no upgrading of the mobilized ingredients from the coke along the exhaust gas cleaning system, it is necessary to have enough potential for destroying or separating it. Possibilities for destruction and separation are shown in Table 4.

| Destroying/Separating Position | Compound |
|---|---|
| Combustion | Organic Compounds |
| Electrostatic precipitator | Dust, heavy metals as particulate matter |
| First step of scrubbing system | Gaseous heavy metals (e.g. Hg), HCl and HF |
| Second step of scrubbing system | $SO_2$ |

Tab. 4: Destroying/Separating of Coke Ingredients

## Measuring Results

### Continuous Measuring

In Germany all waste incineration plants must be equipped with meter facilities to continuously measure the compounds listed in Fig. 1 in order to obtain daily average values. The results of these measurings for the HWI 1 are shown in Fig. 10. As can be seen in this figure, all data are far below the emission limits, or partly below detection limits and detection accuracy.

**Dust.** Continuous measurings in the stack have resulted in values of less than 3 mg/m³. The problem, however, is that the values depend on the test material used for the calibration of the meter facility. As the calibration was not carried out with coke dust the continous values are not correct.

**Chlorine Compounds.** The HCl values measured behind the coke adsorber are below 0.1 mg/m³. However, the permitted meter facility with a scale of 200 mg/m³ has an error margin of 5% i.e. 10 mg/m³.

**Fluorine Compounds.** The HF meter facility has a scale of 10 mg/m³. Here too, the error margin is indicated at 5%, i.e. 0.5 mg/m³. Readings obtained are in the zero region.

**Sulphur Compounds.** The values are less than 5 mg/m³ and therefore ranging in the error margin. Brief peak values in the adsorber gas inlet do not lead to any increase in the emission (Fig. 11).

| Units in mg/m³ (I.N., dry, 11 Vol-% $O_2$)<br>* Half-hourly Average Values<br>** Below detection limit or below detection acurracy | | 17th. Ordinance (1990) Emission Limits<br>Daily Average Values | Hazardous Waste Incineration at RZR Herten HWI 1<br>Daily Average Values 1993 | <br>Measuring Results of TÜV |
|---|---|---|---|---|
| | Staub | 10 | 2.7** | < 1 |
| Inorganic chlorine compounds calculated as | HCl | 10 | 0.11** | < 0.3** |
| Inorganic fluorine compounds calculated as | HF | 1 | 0.1** | < 0.03** |
| Sulphur compounds calculated as | $SO_2$ | 50 | 4.6** | < 1** |
| Nitrogen oxides calculated as | $NO_2$ | 200 | 92.1 | 35...105* |
| Organic carbon compounds | Corg. | 10 | 0.4** | < 1 |
| Carbon monoxide | CO | 50 | 9.9 | 4...7* |
| Σ Cd, Tl | | 0.05 | -- | < 0.000 1** |
| Hg | | 0.05 | -- | < 0.001** |
| Σ Sb, As, Pb, Cr, Co, Cu, Mn, Ni, V, Sn | | 0.5 | -- | 0.003 |
| PCDD, PCDF calculated as toxicity equivalence ΣTE | | 0.000 000 1 | -- | 0.000 000 004 |

Figure 10 Emission Values of HWI 1 at RZR Herten

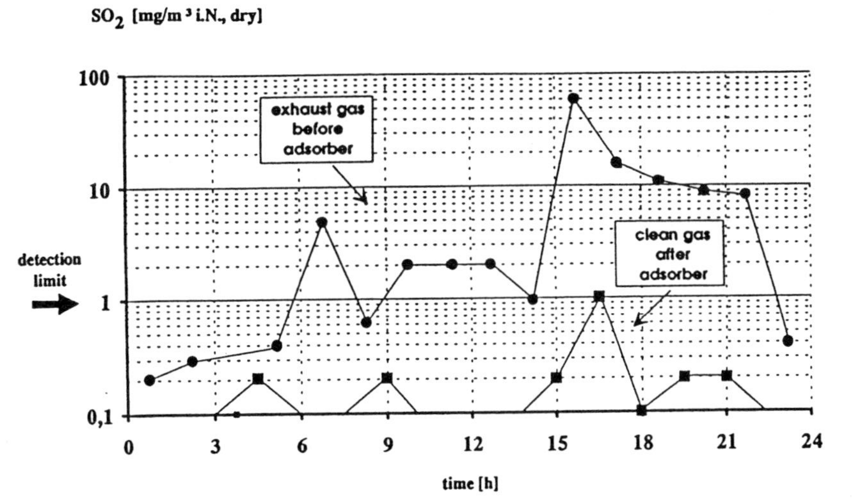


| Figure 11 | $SO_2$ - Concentration | |
|---|---|---|

**Nitrogen Oxides.** Concentrations on the admission side are generally between 170 and 250 mg/m³. About 50 to 70% of the nitrogen oxide is converted depending on the amount of ammonia added and the admission concentration of nitrogen oxide. Fig. 1 shows that there is an emission limit for $NO_x$ in the 17th. Ordinance of FICA and not in the Council Directives for municipal and hazardous waste incineration. The emission values are about half the limit.

**Total Organic Carbon.** The average values for total organic carbon are less than 0.4 mg/m³.

**Carbon Monoxide.** The average increase in CO-concentration as a result of the coke oxidation in the coke layers amounts to approx. 5 mg/m³. CO-peaks in excess of the average value of 5 mg/m³ are attributable to the waste incineration process.

As shown above the continuous measurings are within the error margin. This means that the meter facility technology must be improved to reach correct emission values.

## Discontinuous Measurings

The measurings of heavy metals and dioxins/furans are carried out in a research program of the Institute for Enviromental Technology and Analytic (IUTA Institut für Umwelttechnologie und Analytik, Duisburg, Germany) and during the demonstration of the specified performance of the additional gas cleaning system at the HWI 1 by TÜV (Technical Survey).

**Dust and Heavy Metals.** The dust concentration is below a detection limit of 1 mg/m³. Analysis have shown that the dust consists of coke only.

Heavy metal concentrations over detection limits are listed in Table 5 against the values of contents of the applied coke/carbon. This shows, that a significant emission of the 4 listed elements does exist.

| Element | Clean Gas [µg/m³] i.N., dry. | Coke [µg/mg] | SAC [µg/mg] |
|---|---|---|---|
| Ni | 0.1 - 0.2 | 0.002 | 0.025 |
| Pb | 0.1 - 2 | < 0.001 | 0.025 |
| Cu | 0.8 - 1 | 0.002 | 0.015 |
| Mn | 0.1 - 0.2 | 0.25 | 0.025 |

Tab. 5: Heavy Metal Concentrations

The mercury which is present in the exhaust gas in its elementary form, as a chemical compound or adsorbed on dust particles was measured constantly with a meter facility (HAGE HM 1400) partly used by IUTA in the research program. All forms of mercury are converted into the elementary form by automatically adding sodium hydroboron for the purpose of photometric detection. The smallest range of the meter facility is between 0 and 0.1 mg/m³. The daily average values were between 0.001 an 0.005 mg/m³. Discontinuous measurings show concentrations below detection limit of 0.001 mg/m³ (Fig. 10).

**Organic Compounds.** The concentration of polychlorinated dioxins and furans is 0.004 ng TE/m³ as an average value during three test periods of 6 hours each. Other organic compounds like polyaromatic hydrocarbons, polychlorinated biphenyls or benzene are not detectable.

### Auxiliaries

As shown in Fig. 10 the emission limits of the 17th. Ordinance of FICA are not only observed, but the values are drastically below these limits. This is reached as follows:

- 31 kg/h coke
- 3.5 - 4.5 m³/h ammonia
- 2.5 - 3 Mg/h steam.

### Summary

Based on adsorbent lignite coke, the multipurpose active coke adsorbers, used as gas cleaning system for incineration of waste offer a broad separation performance for many pollutants in the exhaust gas. Given the fact that fluctuations in the composition of waste are inevitable, and that no continuous monitoring of organic compounds in the exhaust gas like dioxins/furans will be possible in the near future, the application of this technology provides great certainty to the operator that low emission limits can be reached in all situations. Based on this knowledge, it is likely to enhance the public acceptability as to waste incineration facilities. We hope that in future these adsorbers will be able to separate or adsorbe compounds from the gas stream which are not yet known to date.

### References

Bartaires J.-G., 1991, " The Policy of the Commission of the European Communities in Respect of Atmospheric Pollution", *Clean Air Around the World*, L. Murley, International Union of Air Pollution Prevention Associations, Brighton (UK) pp. 11-29.

Eicken, M Esser-Schmittmann, W., Lambertz, J., Ritter, G., 1990, "Braunkohlenkoks zur Rauchgasreinigung und Reststoffminimierung von Abfallverbrennungsanlagen", *Brennstoff-Wärme-Kraft*, Vol. 10.

FACT-Vol. 18, Combustion Modeling, Scaling and Air Toxins
ASME 1994

# DEVELOPMENTS CONCERNING MODELS FOR THE CALCULATION OF ATMOSPHERIC DISPERSION OF AIR POLLUTANTS IN THE FEDERAL REPUBLIC OF GERMANY AND THE EU

**Stefanos E. Biniaris**
Department of Immission Protection
RWE Energie AG
Essen, Nordhein-Westfalen, Germany

## ABSTRACT

In the course of the licensing procedure of a power plant in Germany the following procedure, which is the same for any plant requiring a licence, is applied in order to examine the immissions:

Before building a plant the previous impact is measured (usually for one year) including any air pollution for which immission limits are set in the "Technical Instructions on Air Quality Control" (TA Luft, 1986). Furthermore, the additional impact caused by the plant to be built is calculated for the same substances by means of a dispersion model. The total environmental impact, the properties of which (annual mean average and 98-percentile) must be below the limit values set in the TA Luft is calculated from the sum of previous and additional impact.

The dispersion model of the TA Luft is a simple Gauss model which, although it does not consider several effects in its formula, is quite reliable in its picture of reality since it is calibrated for several cases.

In Europe, there are several initiatives to harmonise the dispersion models applied in the different European states. Harmonization does not mean that just one dispersion model should be applied by all states but that the models used for a set task should present results which agree within a certain limit.

In this, it is recommended to use an advanced Gauss dispersion model which takes into account the following effects:

- Dry deposition,
- wet deposition,
- chemical transformation,
- inversion layer,
- mixing layer,
- sedimentation.

The advanced Gauss model shall describe the atmosphere's turbulences not by dispersion categories but by boundary layer parameters such as Monin-Obukhov-length, friction velocity and mixing layer height.

## INTRODUCTION

A model for calculating the dispersion of air pollutants in the atmosphere is primarily required by a power plant operator when requesting a licence for a power plant. The total environmental impact of each air pollutant for which a permissible immission level is contained in the Technical Instructions on Air Quality Control (TA Luft, 1986) must be determined for each assessment area and compared with the permissible immission levels of the TA Luft. The total environmental impact is the sum of the measured previous impact (the impact without the plant to be licensed) and the additional impact (the impact resulting from the plant to be licensed) calculated with the help of a dispersion model. If the total environmental impact - characterised by the two properties: annual mean value and 98-percentile - for each air pollutant in all assessment areas is smaller than the corresponding permissible immission levels of the TA Luft IW1 and IW2 then the licence cannot be denied.

Table 1 indicates the measured immission burden of the three most important air pollutants $SO_2$, $NO_2$ and suspended dust (annual mean values of 1992) measured in those assessment areas in the surroundings of RWE Energie AG's brown coal-fired power plants which present the maximum immission burden. These values are compared with the corresponding permissible immission

Table 1. Immission concentration in the surroundings of RWE Energie AG's brown coal power plants (figures in μg/m³).

| Air-polluting material | maximum previous impact | permissible immission level IW1 | maximum additional impact of a 600 MW unit |
|---|---|---|---|
| $SO_2$ | 25 (18%) | 140 | 0.9 |
| $NO_2$ | 35 (44%) | 80 | 0.5 |
| suspended dust | 42 (28%) | 150 | 0.2 |

levels according to the TA Luft IW1. All concentrations are indicated in μg/m³. For an existing plant, this immission burden becomes the total environmental impact. However, if a new unit is to be built the measured immission burden becomes the previous impact (background). The figures in the brackets in the column for maximum previous impact refer to the shares of the maximum impact in the relevant permissible IW1 levels. Following the completion of the retrofit programme resulting from the 13th Ordinance for the Implementation of the Federal Immission Control Act (13. BImSchV, 1983) the background values usually measured in the vicinity of the RWE power plants and almost everywhere in the Federal Republic of Germany are considerably lower than the permissible levels. The right column contains, for reasons of comparison with the maximum previous impact and the relevant permissible IW1 level, the maximum additional impact (annual mean value) which is caused by a 600 MW unit complying with the emission limits of the 13th Ordinance for the Implementation of the Federal Immission Control Act (stack height 200 m, flue-gas volume rate under standard conditions (wet) $2.642 \times 10^6$ m³/h, temperature of flue gas at stack discharge 130°C). Table 1 reveals that the maximum additional impact of today's units

- hardly makes a contribution to the development of the total environmental impact from the previous impact and

- is negligible in relation to the permissible IW1 immission levels.

The following report on dispersion models which are used to calculate the above-mentioned additional impact is to be viewed against the background outlined above.

## THE DISPERSION MODEL

The process of atmospheric dispersion of air pollutants depends on a several variables. We will list the most important variables for the case of emissions from a power plant stack with the height H (fig. 1).

- The exhaust gas is discharged from the stack at point A with a vertical exit velocity and a temperature that is higher than the temperature of the ambient air. For this reason, two vertically upward-oriented forces act on the exhaust gas, i.e. the exit momentum and the thermal lift.
- At the point of discharge from the stack pressure and momentum forces of the wind deflect the exhaust gas and carry the stack plume downwind.

The consequence of these forces, is the curved path of the stack plume axis AB.

At point B the axis of the stack plume has achieved its maximum height of rise h. The effective source height h is equal to the stack-height H plus the plume rise $\Delta h$.

- Turbulent diffusion is the main cause for the dilution of the exhaust gas during its transport through the atmosphere and for the funnel-shaped development of the stack plume.

- When the stack plume reaches the ground, gases or aerosols may be deposited on the ground and on other receptors, i.e. humans, animals, plants, objects, water, etc. This process is called dry deposition.

# Fig. 1: Dispersion of a stack plume in the atmosphere

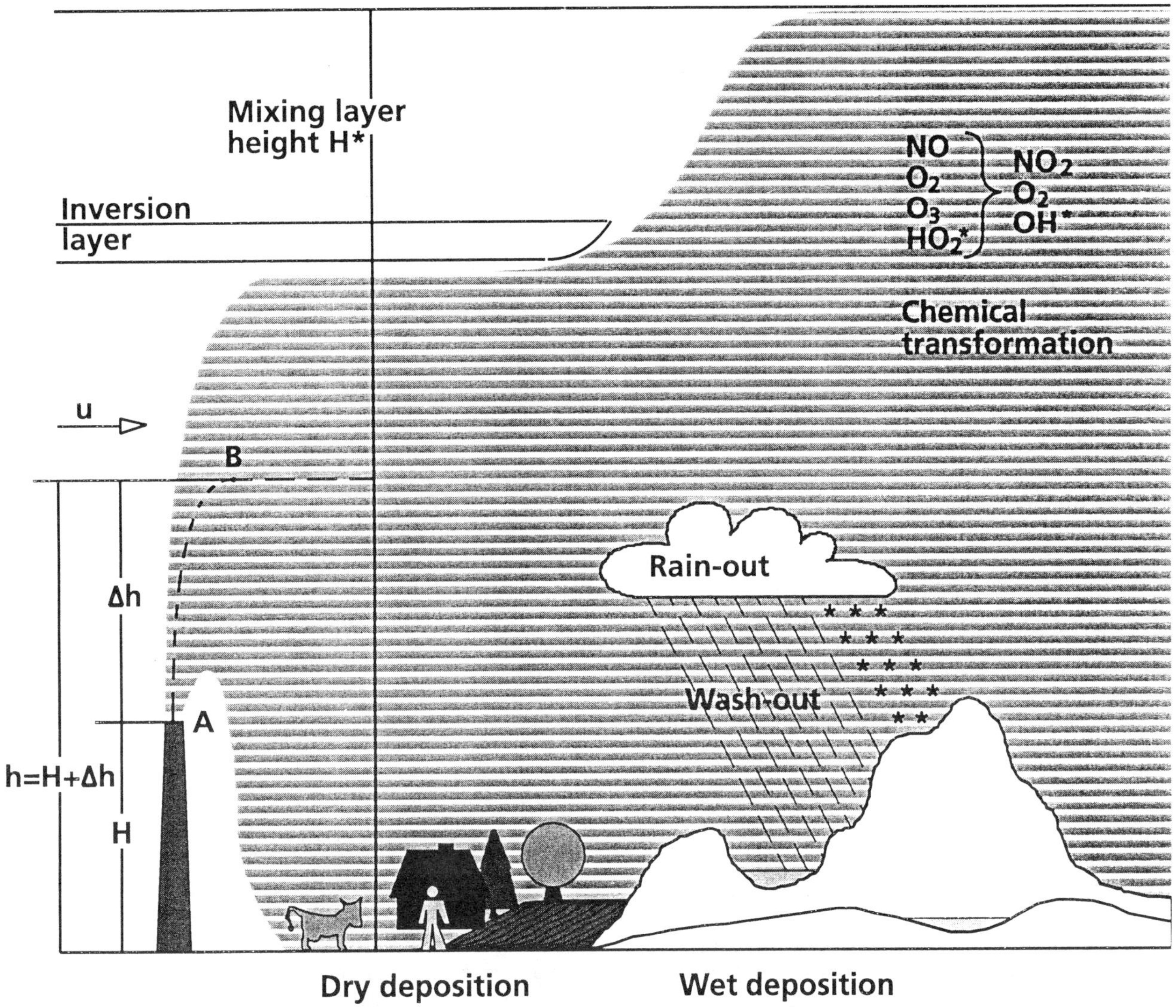


- Individual gases or aerosols contained in the stack gas can react with the moisture of the atmosphere. We distinguish between "rain-out" and "wash-out" depending on wether this process happens within or below a cloud. Together, rain-out and wash-out are called wet deposition.

- During dispersion, chemical reactions within the stack plume may transform gases or aerosols or create new ones. Fig. 1 notes the chemical transformation from NO to $NO_2$.

- It quite often occurs that the potential temperature within an air layer does not decrease with its height, but that it increases (inversion layer). Depending on its size, this layer may not be penetrated by a stack plume (barrier layer) so that if, for instance, the stack plume is beneath this layer (see fig. 1) the available room for dispersion is limited to the region below.

- However, even if there is no inversion or barrier layer room for vertical dispersion of the plume is limited by the mixing layer height of the atmosphere H*.

- For dust like air pollutants, sedimentation, i.e. gravitational settling of particles, is a factor involved. Sedimentation causes a decline of the stack plume axis which becomes steeper with a growing distance from the source.

- According to the type and distribution of buildings as well as vegetation, the ground surface presents different types of roughness and it may not be level due to orographic conditions.

## SOLVING THE DISPERSION MODEL

The mathematic model which describes the atmospheric dispersion of pollutants and which considers all processes which play a relevant role (see following chapter "The dispersion model") includes a complex system of non-linear partial differential equations plus corresponding initial and boundary conditions.

The individual differential equations give information about the changes of

- the concentration of the air-polluting material examined (advection diffusion equation),
- the wind field,
- the temperature,
- the humidity,

etc.

Due to the complex nature of the equations, an analytic solution, i.e. a solution in which the pollutant concentration is expressed as a function of space and time, is only possible in very specific cases. In all other cases, a solution of the dispersion model can only be of a numerical nature. Here, a three-dimensional grid is usually put over the dispersion area. At each grid point every single partial differential equation of this system is applied, the derivatives of the individual variables being replaced by the derivative of the differences of the variables in the neighbouring grid points. This means that an algebraic system of equations must be solved at discreet intervals, instead of solving a system of partial differential equations continuously.

## THE DISPERSION MODEL OF THE TA LUFT

The dispersion model of the TA Luft works in two steps. During the first step, the effective source height h is calculated from the stack height H and the plume rise $\Delta h$. The formulas for calculating the plume rise $\Delta h$ are mainly based on the semi-empiric equations published by Briggs (Briggs, 1971), which were developed from the evaluation of the plume rise measurements available at that time. However, the plume rise formula of the TA Luft also considers knowledge gained later (cp. VDI Guideline 3782, Part 3) (VDI guideline, 1985). During the second step, the concentration of the pollutant examined is calculated as a function of the location. Equation (1) calculates the concentration of a pollutant (no dust) according to the TA Luft.

$$C(x,y,z) = \frac{10^6}{3600} \frac{Q}{2\pi u_h \sigma_y \sigma_z} \exp\left[-\frac{y^2}{2\sigma_y^2}\right] \left[\exp\left[-\frac{(z-h)^2}{2\sigma_z^2}\right] + \exp\left[-\frac{(z+h)^2}{2\sigma_z^2}\right]\right] \quad (1)$$

| | |
|---|---|
| C | concentration of the pollutant to be examined, |
| x,y,z | cartesian co-ordinates of a point in space, |
| Q | mass flow of emitted substances, |
| $u_h$ | wind velocity at the effective source height, and |
| $\sigma_y$, $\sigma_z$ | horizontal and vertical dispersion parameters |

Equation (1) stipulates that the concentration C is a Gaussian distribution, i.e. normal distribution, perpendicular to the wind direction horizontally as well as vertically. Therefore, dispersion models with an equation for concentration similar to equation (1) are called Gaussian dispersion models.

Equation (1) is an analytic solution of the general advection diffusion equation, however, under a large number of simplifying assumptions which are listed in table 2. Since these assumptions are usually not fulfilled equation (1) is strictly valid only in very specific cases.

The advantage of equation (1) as an equation for determining the concentration in the dispersion model of the TA Luft is that it is calibrated, i.e. that the dispersion parameters $\sigma_y$ and $\sigma_z$ are specified such that they match mean concentrations measured in a number of tracer experiments.

The consequence is that for each dispersion case which is similar to the tracer experiments, equation (1) produces reasonable concentration values. For dispersion cases which differ strongly from the tracer experiments, concentrations calculated with equation (1) are uncertain. The probability that these concentrations are incorrect is the greater the more simplifying assumptions have been employed in the derivation of equation (1).

## DEVELOPMENTS IN THE FIELD OF DISPERSION MODELS IN THE FEDERAL REPUBLIC OF GERMANY

Considering the contribution of a power plant stack to the existing immission burden (see table 1) it is no surprise, that, from the point of view of a power plant operator, there is no reason for changing the dispersion model of the TA Luft. Also the fact that the present TA Luft model often overestimates ground-level concentrations (Biniaris and Wilhelm, 1988) does not provide sufficient motivation for a change. Of course, there are cases in which the dispersion model of the TA Luft is not applicable, or only with reservations. Examples are

- short-term and/or unsteady emissions,
- dispersion over complex terrain,
- dispersion at large distances from the source,
- complex chemical transformation of pollutants during dispersion, and
- episodic dispersion.

For such special cases, several advanced dispersion models have been developed. Examples are (Martens and Maßmeyer, 1990)

- Gauss-Puff models,
- flow models,
- Eulerian grid models, and
- Lagrangian models (particle simulation models).

The term "advanced dispersion models" should not obscure the fact that these models also present problems which means that there is no guarantee that the results achieved with these more sophisticated models are more accurate than those achieved with the simple TA Luft dispersion model. The fact that the advanced models require a much more input information which is often not easily available, reduces their attraction for power plant operators even more.

However, as described, the quality of a Gaussian dispersion model improves with a decreasing number of simplifying assumptions.

A step to reduce the number of simplifications has been made in the derivation of the dispersion model contained in the VDI guideline 3782, part 1 "Gaussian dispersion model for air quality management" which was issued in October 1992 (VDI guideline, 1992). Although this dispersion model is also of Gaussian type ( just like the model of the TA Luft). The equation for determining the concentration includes further terms which enable the inclusion of the following processes or situations (see table 2):

- dry deposition,
- wet deposition,
- chemical transformation,
- inversion layer,
- mixing layer, and
- sedimentation.

Table 2. Capacity of the dispersion models of the TA Luft and of the VDI guideline 3782, part 1.

| Dispersion model considers | TA Luft | VDI 3782 |
|---|---|---|
| - unsteady conditions | no | no |
| - wind shear | no | no |
| - variable eddy diffusivities | no | no |
| - vertical wind velocity | no | no |
| - very small wind velocity | no | no |
| - dry deposition | no | yes |
| - wet deposition | no | yes |
| - chemical transformation | no | yes |
| - inversion layer | no | yes |
| - mixing layer | no | yes |
| - sedimentation | no | yes |
| - various types of ground roughness | no | no |
| - uneven terrain | no | no |
| calibrated | yes | no |

These advantages of the new dispersion model as compared to the TA Luft must be contrasted with the disadvantage that the dispersion model of the VDI guideline 3782 is not calibrated . The dispersion parameters $\sigma_y$ and $\sigma_z$ were just taken over from the TA Luft model. As the equations for determining concentrations are different this leads to the fact that the concentrations calculated according to the VDI guideline are in case of high stacks even higher than those calculated with the TA Luft model (Biniaris, 1989). If the dispersion model of the VDI guideline had been calibrated - which is, of course, not a trivial task, otherwise it would have already been calibrated - this model would present a relatively large improvement as compared to the TA Luft model.

Should, however, within the framework of a new TA Luft or of a harmonization of dispersion models in the EU (see following chapter) an advanced dispersion model be up for discussion, then the Gaussian dispersion model of the VDI guideline could be more useful to power plant operators. The fact that the calculated immission concentration would have been overestimated should not be of importance if you consider the above statements given in connection with table 1.

## DEVELOPMENT IN THE FIELD OF DISPERSION MODELS IN EUROPE

In Europe, there are a series of initiatives for intensifying the cooperation of those organisations which develop advanced models for calculating the atmospheric dispersion of air pollutants.

In 1991, ERCOFTAG (European Research Community for Flow, Turbulence and Combustion) started a scheme pursuing this aim which includes three workshops - in Denmark (1992), in Switzerland (1993) and in Belgium (1994). The aim of these workshops is the harmonization of the dispersion models applied in the various European states. This does not mean that all states need only use one dispersion model. Harmonization means that the models used for a set task should present results which agree within a certain limit.

In the following, several important recommendations of the first workshop hosted in Ris, Denmark, in May 1992, are listed (Olesen and Mikkalsen, 1992):

- An advanced Gaussian model is recommended.
- The dispersion model should consider the following effects:
  - dry deposition,
  - wet deposition,
  - chemical transformation,
  - inversion layer,
  - mixing layer, and
  - sedimentation.

The TA Luft dispersion model does not consider any of these, but the VDI guideline, by contrast, includes all these effects (see table 2).

- A major problem of the simple Gaussian models is that they describe the state of turbulence in the atmosphere by making a division into dispersion categories (the dispersion model of the TA Luft and of the VDI guideline intend six such categories). Therefore, the advanced Gaussian model should not describe the state of turbulence by means of dispersion categories but by boundary layer parameters such as Monin-Obukhov length, friction velocity and mixing layer height.
- Turbulent mixing should be described as a function of the roughness of the surface (the dispersion models of the TA Luft and the VDI guideline apply only to a roughness as it was present during the tracer experiments in which the dispersion parameters $\sigma_y$ and $\sigma_z$ were obtained).
- The meteorological data should be provided as a time series (ideally over ten years) so that the immission concentrations can be related to any period desired, for instance, daily mean values (the dispersion model of the TA Luft and the VDI guideline do not use a time series but note how often a dispersion category occurs at a certain wind direction and wind velocity).
- The computer programme should be optimised so that it can run on a PC.

Further aspects that were discussed during the workshop:

- Advanced dispersion models not only require the meteorological input of routine meteorological data measured but also fundamental parameters which describe the dispersion such as the Monin-Obukhov length, friction velocity and mixing layer height. Recommendation: a harmonization of meteorological input data should be carried out.
- According to the application of the models, advanced dispersion models should provide more than mean values and percentiles of immission concentrations. This implies, for instance, in the case of a time series input, a time series of immission concentrations, in addition.

## OUTLOOK

From the point of view of power plant operators, the TA Luft dispersion model has served well. Uncertainties concerning this easy-to-handle dispersion model can be accepted if one considers how small the additional impact of a power plant unit is in comparison with both the measured background and the present threshold values.

In the Federal Republic of Germany, several advanced dispersion models have been developed in the past years. The model of the VDI guideline 3782 seems to be of interest for power plant operators due to its analytic solution.

If the efforts made by the EU lead to a harmonization of existing dispersion models the VDI guideline dispersion model could serve power plant operators well, provided it is extended as follows:

- description of the state of turbulence by means of boundary layer parameters,
- consideration of the roughness of the ground,
- calibration of the model.

## LIST OF REFERENCES


Biniaris, S., and Wilhelm, M.: Untersuchung über die Richtigkeit der rechnerischen Bestimmung der Kenngrößen für die Immissions-Zusatzbelastung (I1Z und I2Z) nach dem Ausbreitungsmodell der TA Luft 1986. (Investigations on the accuracy of calculating the concentration data for the additional impact on the ambient air (I1Z and I2Z) according to the dispersion model of the TA Luft 1986). Staubreinhaltung der Luft 1988 (Air Quality Control 1988), no. 48, pp. 351-355.

Biniaris, S.: Probleme im Zusammenhang mit dem Gaußschen Ausbreitungsmodell der TA Luft am Beispiel der Berücksichtigung von Mischungsschichthöhen in der Atmosphäre. (Problems in connection with the Gaussian dispersion model of the TA Luft, in consideration of the problem of mixing layer heights in the atmosphere.) Collection of papers presented at the VGB Conference "Kraftwerk und Umwelt 1989" (Power plants and the environment, 1989).

Briggs, G.A.: Some recent analyses of plume rise observation. Proc. 2nd International Clean Air Congress, Washington 1970. Academic Press (1971).

Martens, R., and Maßmeyer, K.: Übersicht über die Grundlagen atmosphärischer Dispersionsmodelle. (Overview of the basics of atmospheric dispersion models.) Umwelt-Meteorologie (environmental meteorology), VDI/DIN-Schriftenreihe (VDI/DIN series of publications), vol. 15 (1990).

Olesen, H. R., and Mikkalsen, T.: Proceedings of the workshop "Objectives for Next Generation of Practical Short-Range Atmonspheric Dispersion Models", May 6-8, 1992 at Riso, Denmark.

Erste Allgemeine Verwaltungsvorschrift zum Bundes-Immissionsschutzgesetz (Technische Anleitung zur Reinhaltung der Luft - TA Luft)
First General Administrative Regulation pertaining to the Federal Immission Control Law (Technical Instructions on Air Quality Control - TA Luft)
Gemeinsames Ministerialblatt 37 (1986), Nr. 7, S. 93-144
Joint Ministerial Paper 37 (1986), no. 7, pp. 93-144.

Dreizehnte Verordnung zur Durchführung des Bundes-Immissionsschutzgesetzes (Verordnung über Großfeuerungsanlagen - 13. BImSchV), BGBl. I, S. 719
13th Ordinance for the Implementation of the Federal Immission Control Law (Large Furnace Ordinance - 13. BImSchV), Federal Law Gazette I, p. 719 (1983).

VDI guideline 3782, part 3, Calculating plume rise, VDI manual on air quality control, vol. 1 (1985).

VDI guideline 3782, part 1: Dispersion of pollutants in the atmosphere. Gaussian dispersion model for air quality management, October 1992, VDI manual on air quality control, vol. 1 (1992).

FACT-Vol. 18, Combustion Modeling, Scaling and Air Toxins
ASME 1994

# MOST ADVANCED COMBUSTION TECHNOLOGY FOR PRIMARY $NO_X$ REDUCTION

**Hans-Karl Petzel**
Management Power Plant Fenne
Saarbergwerke AG
Völklingen, Germany

**Alfons Leisse**
Process Engineering of Firing Systems
Babcock Lentjes Kraftwerkstechnik GmbH
Oberhaussen, Germany

## ABSTRACT

One of the requirements exacted in obtaining operational approval for large-scale firing systems in the Federal Republik of Germany is keeping - in line with the current state of technical engineering - to a nitrogen oxide emission figure of under 200 mg/m³.

Furnace air staging has proved a success in boiler plants with pulverized lignite firing systems in allowing the standard to be respected without deployment of additional secondary-sided processes.

With the objective in mind also in the case of a pulverized hard coal firing system with dry ash discharge, in attaining an acceptable NOx emission figure for operational approval solely from specific firing steps, the existing firing system of the Völklingen prototype power plant was converted to newly-developed pulverized coal burners and furnace air-staging equipment.

## 1. INTRODUCTION

The Völklingen prototype power plant is a bituminous coal-fired combined-cycle plant with integrated fluidized bed combustor. It has a maximum capacity of 230 $MW_{el}$. At the time of commissioning in 1982 the selected concept was still a technological novelty. This concept is mainly characterized by

- a gas turbine which supplies oxygen to the steam generator,
- the combination of the gas turbine with the air-cooled submerged heating surfaces of two fluidized-bed combustors.

That means 3 possible variants for the operation of the plant /1, 2, 4/:

1. Conventional primary air operation with forced-draft fan and steam generator firing system
2. So-called partial combined-cycle operation with gas turbine and steam generator firing system without forced-draft fan
3. Full combined-cycle operation with gas turbine fluidized-bed and steam generator firing systems.

From the very beginning this advanced steam generator concept has stood out for comparatively low $NO_x$ emission levels /3/. But it was not possible to comply with the emission standards fixed in the Federal Republic of Germany at a later date. Therefore, SAARBERG made it its business to find a suitable $NO_x$ reduction process.

In addition to technological advances aimed at enhancing the generation of electricity, air pollution control devices were installed to decrease $SO_2$ and $NO_x$ emissions. This power station was equiped with one of the first S-H-U flue gas desulfurization systems and has served as a model for expansion of this technology throughout the world.

Today, conventional bituminous coal-fired power plants are normally denitrified by means of the SCR technique; the DENOX reactor is arranged upstream of the electrostatic precipitator or downstream of the flue gas desulfurization system. Due to the existing technical conception and the given space at the Völklingen prototype power plant it would only be possible to install a SCR system between the electrostatic precipitator and the desulfurization plant.

But so far no satisfactory operating results are available for such a variant /5/.

Therefore, intensive efforts were made to reduce $NO_x$ by primary measures. These were backed by the conviction that it should be possible with advanced combustion technology to denitrify highly volatile bituminous coal by primary measures to such an extent that subsequent denitification of the flue gases does not appear reasonable.

After experimental and theoretical advance studies the firing system was converted. So far, results and experiences are available from 2 1/2 years of operation with the possible operating variants of the boiler plant.

The following outline describes the experiences gained in the primary-air mode. This operating mode corresponds to the operating mode of usual, conventional firing systems based on bituminous coal. Thus, the results are important not only for the Völklingen prototype power plant but for power plants in general.

## 2. THE FIRING SYSTEM BEFORE TAKING NOx PRIMARY MEASURES

The steam generator of the Völklingen prototype power plant is designed for a steam flow of 529 t/hr with usual steam conditions (Figure 1).

The coals fired come from the Saar with calorific values of abt. 20 MJ/kg, ash contents of up to 30 %, water contents up to 15 % and volatiles between 38 and 42 % daf. The pulverized coal is conditioned by means of 4 Babcock beater mills. Each mill delivers coal for 2 pulverized coal burners arranged at the same level. Boiler full load can be achieved both with four and with three mills. The maximum thermal output of the burners is about 80 MW.

The arrangement of the pulverizing and firing system is shown in Figure 2 in a very simplified form. The scheme explains the primary-air and the partial combined-cycle mode. It also shows the opposed burners arranged at different levels.

In the primary-air mode the combustion air required for combined pulverizing and drying is heated to 350°C by a steam air heater and a subsequent duct burner and then it is fed to the booster fans of the mills. The classifier temperature is abt. 100 °C, the cold air flow required for controlling this temperature is supplied by the forced-draft fan. The secondary air downstream of the steam air heater is heated to a temperature of 200°C by means of a duct burner.

In the partial combined-cycle mode the exhaust gas of the gas turbine flows - with a maximum temperature of 440°C - to the booster fans of the mills. The gas not needed for the primary side flows to the burners as secondary air. At part load, some gas is delivered to the convection pass of the steam generator. In this operating mode, cold air is provided by a separate fan to control the classifier temperature.

The burners are swirl burners with a core air tube and a concentric primary gas/pulverized-coal tube. Two separate steps are provided for the secondary air because in this way it is possible to respond to the greatly varying volumetric gas flows for the primary-air and the combined-cycle mode. Thus, to observe optimum speed conditions in all operating modes, in the primary-air mode, the secondary air was channeled only over the inside step, in the combined-cycle mode, over both steps.

The design of the entire plant may be learned from Figure 3, it shows a section through the boiler house. One sees the combustion chamber with the paired burners installed at 4 levels, the downstream flue gas passes and the various separate components of the steam generator.

## 3. CONVERSION OF THE FIRING SYSTEM

### 3.1 Mechanism Of NOx Production And Reduction

The knowledge of the chemo-physical mechanisms of $NO_x$ production and reduction begins at the sources of the nitrogen oxides. During the combustion of bituminous coal nitrogen oxides arise from fuel-bound nitrogen (nitrogen $NO_x$) and air nitrogen
(thermal $NO_x$). Thermal $NO_x$ is formed only at temperatures clearly above 1,300°C. In most advanced bituminous coal firing systems with dry ash removal thermal $NO_x$ is therefore less important than fuel $NO_x$.

The formation of fuel $NO_x$ takes place in multiple ways and mainly depends on the fuel properties, the reaction rates of the various reaction partners and reaction conditions.

During the pyrolysis of the fuel which takes place in parallel with primary oxidation (ignition) most of the volatiles are released; through intermediate reactions the nitrogen is available either in atomic form or as fuel nitrogen in conjunction with other outgassing products (HCN, NHx).

NO is formed either by direct oxidation with atomic oxygen or indirectly through reactions with OH radicals which equally arise during the gasification of fuel with high formation rates and which are of great importance for the whole combustion process.

Due to the high reaction rates which are independent of temperature in a wide range, $NO_x$ is primarily produced in the flame core in the direct vicinity of the burners.

A small part of the fuel nitrogen remains in the residual coke and can be involved in the oxidation only in the further course of combustion.
But since this process takes place in parallel with the combustion reaction properly speaking or with residual burnout, the oxidation to NO is hindered due to similarly high reaction rates between carbon and oxygen products.

A reduction of the thermal $NO_x$ portion can only be achieved by reducing the furnace temperature and the flame temperature at the burner itself.

Further $NO_x$ reduction is caused by decreasing the oxygen quantity available, with preference at the pyrolysis stage of the combustion process. In the presence of atomic nitrogen or outgassing fuel nitrogen products the decay reaction of nitrogen oxides already formed is triggered; in that process molecular nitrogen is formed which is no longer involved in further oxidation.
Such a NO reduction is only feasible to the extent to which the production of OH radicals, which is necessary for maintaining the combustion reaction, does not come to a standstill.

### 3.2 Requirements Of Low-NOx Burners

Taking into account the $NO_x$ formation and reaction mechanisms the following demands are placed on a pulverized coal firing system with low-$NO_x$ combustion:

1. Intensive outgassing of coal at high temperature and low oxygen partial pressure at the earliest possible stage.

2. Safe supply of the heat required for pyrolysis by backing the ignition stability in a fuel-enriched oxygen deficiency zone at the burner outlet.

3. Improvement of mixing prerequisites at the burner outlet by steadying the fuel and air distribution in the outlet cross-section in the burner nozzles.

4. Staged and delayed oxygen supply to the fuel products.

5. Adjustment of air mass flows, air ratios and swirling intensity.

The residence time of gas and pulverized coal in the primary fuel-rich flame increases - as is well-known - with the thermal output of the burners.

In addition, the distances of the various air ducts increase with increasing burner output so that it becomes easier to control and delay the mixing processes. By contrast, it is difficult to influence mixing of the reaction partners in smaller burners.

The burners, with a maximum thermal output of almost 80 MW, installed at the Völklingen model power plant present ideal prerequisites for $NO_x$ reduction measures.

## 3.3 CONVERSION WORK

### 3.3.1 Low-NOx Burner

On the basis of these requirements a novel burner type was designed and tested on a large scale within the framework of the denitrification concept.

The installed swirl staged burner is represented in Figure 4.

The oil pilot burner is guided inside a central core air tube. This tube is enclosed by the pulverized coal tube. At a certain distance from the pulverized coal nozzle mouth there is a swirling element which causes the primary air/pulverized coal flow to rotate. In this way steadying of the flow inside the pulverized coal tube and simultaneous enrichment of pulverized coal at the external periphery of the pulverized coal nozzle are achieved. A stabilizing ring is fitted at the nozzle outlet; it splits the pulverized coal flow in this area into various jets and thus creates good conditions for intensive mixing with part of the combustion air.

The secondary air tube is fitted coaxially around the pulverized coal tube and is connected with a spiral-shaped admission housing.
Installed baffle plates and adjustable radial swirl dampers produce a stable swirl flow and even air distribution over the whole nozzle cross section.

In the area of the pulverized coal nozzle a secondary air deflection throat, which is shiftable in axial sense, has been arranged on the pulverized coal tube of the burner.

This element is designed so that the secondary air mass flow is deflected directly at the burner outlet at the time of the primary reactions.
Thus, it is safeguarded that the pyrolysis of the fuel and ignition of the volatiles can take place in an undisturbed atmosphere without additional oxygen enrichment.

Directly at the secondary air nozzle outlet there is the inside burner throat which is formed by the helical-tube evaporator.

The tertiary air nozzle encloses the secondary air tube of the burner and ends in the outside burner throat.
This nozzle has a helical-shaped inlet casing with baffles and adjustable radial swirl dampers just as the secondary air tube.
The form and configuration of the inlet casing and the double throat arrangement result in a stable tertiary air swirl flow which encloses the inside flame zone and thus causes an oxygen admission with local and time delay while the boundaries of the flame extension are enriched with $O_2$.

In the zone between the tertiary air nozzle and the secondary air tube 12 gas burner lances are distributed on the circumference of the burner to alternately inject coke oven gas into the secondary air flow.

The secondary air and the tertiary air flows can be adjusted by separate dampers. The combustion air flow per burner is controlled in dependence on the fuel heat output in an advanced control circuit with the involvement of the primary air flow.

The burners can be operated with air ratios between 0.8 and 1.3.

To observe the boiler set point the differential air flow is admitted to the combustion chamber through 8 burnout air nozzles which are arranged at a single level above the 4th burner level.

### 3.3.2 Nozzle Arrangement Inside The Combustion Chamber

The arrangement of the 8 burnout air nozzles is represented in Figure 5. The combustion chamber has a cross-section of 9.97 x 10.0 m and a height of 33 m. The 8 burners are installed in pairs, at a single level, in vertically staggered opposed arrangement. The vertical distance between the burners is 8 m. At the left and the right side of each burner one sees one lateral air nozzle each.

There are 4 burnout air nozzles each in the front and the rear wall, 11.3 m below the first convection heating surfaces. At the front side, the nozzles are located 4.2 m above the last burner level, at the rear wall, 8.2 m. This arrangement means a compromise between the necessary residence times for nitrogen oxide reduction and residual burnout. The compromise naturally had to give priority to the requirements of burnout in order not to risk the salability of fly ash. Residual burnout is ensured although only a residence time of less than 1.5 seconds is available at full load.

The operating mode of this firing equipment has to take into account the risk of wall corrosion, i.e. it must always safeguard a minimum oxygen concentration at the walls.

A certain number of $O_2$ measuring points is necessary for its adjustment and control. That is why overall 92 sampling points were installed on all walls in the burner belt zone for the commissioning stage. In Figure 5 the points in question have been marked for the rear wall and the right side wall. The two other walls are equipped analogously.

## 4. OPERATING RESULTS

After commissioning of the plant in August 1991 the $NO_x$ level was very low. The initial assumption that it would be necessary to operate the burners at air ratios between $\lambda = 0.8$ and 0.9 in order to reach the guaranteed $NO_x$ value of less than 350 mg/m$^3$, could be dropped because this value was clearly lower.
The guaranteed value is already reached in excess air burner operation with air ratios of
$\lambda = 1.2 - 1.25$.

Thus, commissioning and optimization of the plant were accomplished with comparative ease and quickly. Additional tests were conducted to further exhaust the potential of the new firing concept. The multitude of tests can be subdivided into 4 categories:

1. Testing of burner parameters such as air ratios, air splitting, air temperature, swirl and output capacity
2. Tests regarding the protection of the urnace walls against corrosion
3. Long-time tests regarding the possible utilization of ash
4. Tests regarding the stability of the emission values.

### 4.1 NOx emission As A Function Of The Burner Operating Mode

Obviously the $NO_x$ emission is mostly influenced by the efficiency of the producible reduction zones. The burner air ratio $\lambda$ is a criterion for the furnace reduction zone.

The reduction zone in the flame core near the burner depends on the local oxygen quotient achievable which is greatly influenced by the burner configuration.
The dependencies on the burner air ratio in Figure 6 show that the influence of the reduction zone in the flame core on the $NO_x$ emission is clearly more dominant in the converted firing system than the reduction potential in the combustion chamber.
At boiler full load with 4 mills, the $NO_x$ emission is 370 mg/m$^3$, in excess air operation with a burner air ratio of $\lambda = 1.3$. Relative to the initial value of the system of 800 mg/m$^3$ before the conversion, this means a reduction of the nitrogen oxide emission of
430 mg/m$^3$ or 54 %.

If the burner air ratio is lowered to $\lambda = 0.8$, the emission value decreases by another
150 mg/m$^3$ to 220 mg/m$^3$.

The CO emission behaves reciprocally to the $NO_x$ emission, it rises continuously with decreasing air ratio and reaches a maximum value of abt. 115 mg/m$^3$.

The constant increase of the CO emission with decreasing burner air ratios is attributable to the unfavorable conditions for the residual burnout. The available mixing and burnout time is so short that, in case of a great delay of the combustion, certain residual burnout of carbon monoxide between the heating surfaces cannot be avoided. But such carbon monoxide burnout is less effective due to the unfavorable temperature and mixing conditions.

The shown results represent the possible operating range of the firing system. The emission levels are stable, ignition at the burner outlet is constant under all operating conditions and the residual carbon contained in fly ash is clearly lower than 5%. Thus, the system might be operated with an air ratio of 0.8 and $NO_x$ values of 220 mg/m$^3$ unless there were the risk of furnace corrosion.

Protection of the furnace walls becomes easier by the special burner design. The defined deflection of secondary and tertiary air from the flame core permits the formation of an oxygen enclosure around the flame. This protection can be influenced by means of the swirl and the tertiary air mass flow.

Figure 7 shows the emission behavior with changed air distribution in the burner, for 2 differing air ratios ( $\lambda = 1.25$ and $\lambda = 1.0$) at boiler full load.

The distribution of burner air was varied by changing the position of the tertiary air damper from the 100 % open position to loss of ignition, which occurred at $\lambda = 1.25$ and in the 20 % open position, while the secondary air damper was constantly fully opened.

In a wide range, namely up to the abt. 40 % position of the tertiary air damper, trimming of burner air for higher secondary portions has no influence on the primary reaction and on the emission values not either.

If the secondary air mass flow is increased beyond this point by closing the tertiary air dampers, the flow and pressure conditions at the burner outlet change considerably with $\lambda = 1.25$. In this way the addition of secondary air at the pyrolysis and/or ignition stage is intensified.

Moreover, the ignition point is shifted from the burner mouth towards the combustion chamber as a result of the increase in velocity.

The fact that the burners behave neutral in a wide setting range, relative to the emission result, is attributable to the effect of the secondary air deflection throats. These elements cause identical ignition conditions at the burner for almost all operating conditions, with minimum oxygen quotient.

The flow and reaction conditions at the burner and in the vicinity of the burner were simulated by means of the FLUENT program system.

Figure 8 shows the typical flow pattern for the new swirl staged burner in the combustion chamber. The criterion for the reduction zone in the flame core is the turnover rate of the volatile fuel components, i.e. the progress of pyrolysis and the simultaneous oxidation in the zone near the burner.

### 4.2 NOx Emission As A Function Of The Burner Zone Heat Release Rate

The optimum setting of a firing system was so far mostly dependent on the output capacity. For stability considerations the burner air ratio was, for example, increased with decreasing load. Shutdown of the mills caused another increase of excess air. The resulting higher $NO_x$ emission was compensated by the fallen furnace temperature at part load so that there were often $NO_x$ emissions irrespective of the load.

The operating behavior of the new burners is so stable that their air ratio is kept constant versus load.

In Figure 9 the $NO_x$ emission is plotted versus the burner load range between 66 % and 133 %. The load point of 133 % is defined in such a way that, with the firing system conceived for the four-mill mode, boiler full load can also be achieved with 3 operating mills.

Thus, the burner zone heat release rate rises from 0.75 $MW/m^2$ to 1.0 $MW/m^2$. The combustion temperatures in the respective combustion zone available change in parallel with the change of the characteristic heat release rates.

In spite of the anticipated influence of the so-called thermal $NO_x$ forming mechanism, there is only a very flat rise of the $NO_x$ emission versus the load range.
Overall, the change of emission is abt. 60 $mg/m^3$ with a burner air ratio of $\lambda = 1.0$.

These changes of the measured $NO_x$ values are primarily attributable to the changed flow and pressure conditions in the primary reaction zone of the burners. The firing system of the Völklingen prototype power plant does not show any signs of stronger influence of the $NO_x$ values through the thermal effect.

### 4.3 Protection Of The Furnace Walls Against Corrosion

The above outline has shown that with the new burners $NO_x$ emissions are achievable whose level is drastically lower than the figures known so far. But long operation with these settings is ultimately only admissible if there is excess oxygen near the furnace walls.

The related measuring results for full load of the boiler are given in Figure 10 for a burner air ratio of 0.8. It shows the values measured at the various sampling points of the right side wall and the rear wall. At most points there are sufficient oxygen concentrations of more than 0.1 %. Only at 5 points on the rear wall no oxygen was detected. The lateral air was in the cooling air position.

Therefore, extended operation with these firing parameters can lead to local corrosion. Here the limits of the oxygen enclosure which, with low air ratio, is unable to prevent the recirculation of oxygen-free flue gas between the burners, become evident.

Figure 11 shows that, with a slightly higher air ratio of 0.95, a high oxygen concentration of more than 1 % is already ensured everywhere. At the same time the $NO_x$ emission has risen from 220 to 260 $mg/m^3$.

### 4.4 Possible Reuse Of Ash

To permit permanent operation of a firing system, ultimately also the residues from the combustion process have to comply with the specified requirements. Thus, for example, the carbon content of fly ash is to be kept under 5 % in order to ensure salability of the ash.

During the operating period so far no additional efforts were necessary in order to reach this objective. The grain size distribution also complies with the requirements of reuse.

### 4.5 Stability Of Emission Values

The emission data presented so far are results of defined trial operation. To demonstrate the behavior in normal boiler operation, the half-hour averages of $NO_2$ and CO emissions of March 3.1994 have been plotted in Figure 12 as provided by the emission computer. Almost all NOx half-hour averages were on that day between 230 and 260 $mg/m^3$, the CO emission varied between 20 and 85 $mg/m^3$.

Figure 13 gives the emission values over an extended period. It shows the daily averages of $NO_2$ emissions as provided by the emission computer for the months of February and March 1994. The variation range of the daily $NO_2$ averages is very narrow, it lies between 235 and 275 $mg/m^3$. For the CO emissions the computer presents daily averages in the range of 35 to 85 $mg/m^3$.

Overall, these operating results show that the extremely low $NO_x$ emissions can be achieved with the new firing equipment also in continuous operation.

## 5. OUTLOOK

The firing results achieved with novel pulverized coal burners have furnished proof in the Völklingen prototype power plant that, with an appropriate burner design, the chemo-physical reaction mechanisms can be utilized more intensively for $NO_x$ reduction than in the past.

The special form and configuration of important burner parts cause in the flame a state which otherwise can only be achieved with greatly lower stoichiometry at the burners (Figure 14).

If the potential of these novel burners is used in an optimum manner, it may be assumed that low-$NO_x$ firing systems can operate without burnout air facilities and without downstream denitrification plants.

With an exact definition of the gasification and ignition conditions of the fuel at the burner outlet and consistent deflection and delayed addition of secondary and tertiary air to the primary flame, the effects of furnace air staging are projected into the direct burner area.
Thus, low $NO_x$ emission values and safe residual burnout become possible in burner operation with excess air.

For the combustion of bituminous coal the principle of the opposed burner system has proved very efficient. The fact that the extremely low $NO_x$ emissions were achieved by swirl staged burners with a thermal output of almost 80 MW should be considered in the design of future plants.

## 6. LITERATURE


1. H.K. Petzel: Betriebserfahrungen mit dem Modellkraftwerk Völklingen. Der Maschinenschaden 57 (1984), 5, p. 163 - 172.

2. H.K. Petzel: Modellkraftwerk Völklingen (Grundkonzeption und Einfahrerkenntnisse), Jahrbuch der Dampferzeugungstechnik, 5th edition 1985/85, p. 36 - 47, Vulkan-Verlag, Essen.

3. G. Scholl und L. Stadie: Prozeßtechnische Untersuchungen an dem ersten Kombi-Kraftwerk auf Kohlebasis - Modellkraftwerk Völklingen. Final report on BMFT research project 03 E 1065 B.

4. F. Mayr: Modelllkraftwerk Völklingen - Auslegung und Konstruktion der kohlebefeuerten Dampferzeugeranlage. VGB-Kraftwerkstechnik 57 (1984), No. 5.

5. G. Scholl, H.K. Petzel, F. Bleif, L. Stadie and H. Hückel: Das Modellkraftwerk Völklingen, VGB-Kraftwerkstechnik 72 (1992), 5, p. 427 - 432.


## FIGURES

Figure 1

## Boiler

| | | | |
|---|---|---|---|
| Steam capacity | SH | t/h | 529 |
| Steam capacity | RH | t/h | 495 |
| Steam pressure | SH | bar | 198 |
| Steam pressure | RH | bar | 40.5 |
| Steam temperature | SH | °C | 535 |
| Steam temperature | RH | °C | 535 |
| Flue gas temperature | | °C | 180 |

## Firing system

| | | |
|---|---|---|
| Number of mills | | 4 |
| Number of mills for total capacity | | 3 |
| Mill type | | HS 260/300 |
| Number of burners per mill | | 2 |
| Hot air temperature in primary air operating mode | °C | 200 |
| Hot air temperature in GT operating mode | °C | 440 |
| Classifier temperature | °C | 100 |

## Fuel

| | | |
|---|---|---|
| Calorific value | MJ/kg | 19.3 - 26.4 |
| Volatiles, daf | % | 38.0- 40.0 |
| Ash | % | 6.0 - 27.0 |
| Moisture | % | 8.0 - 18.0 |
| Nitrogen | % | 1.03 - 0.66 |
| Sulphur | % | 0.51 - 0.53 |

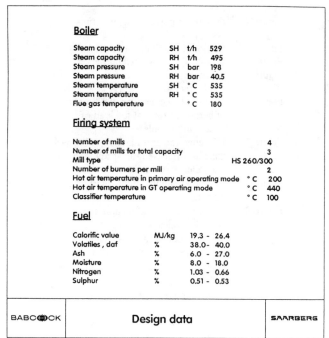

BABCOCK | Design data | SAARBERG

Figure 3

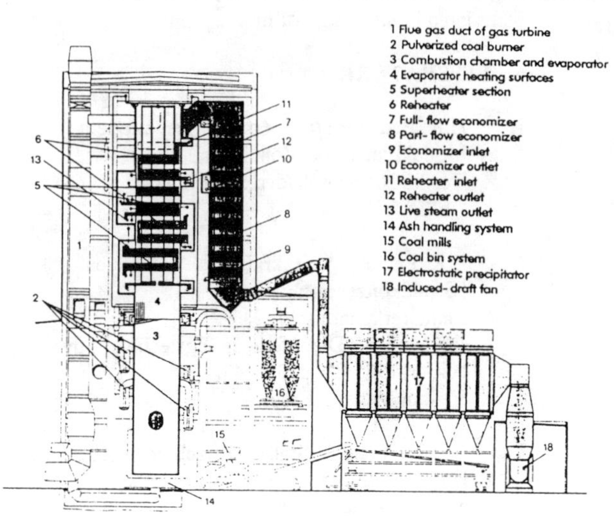


BABCOCK | Section through boiler house | SAARBERG

Figure 2

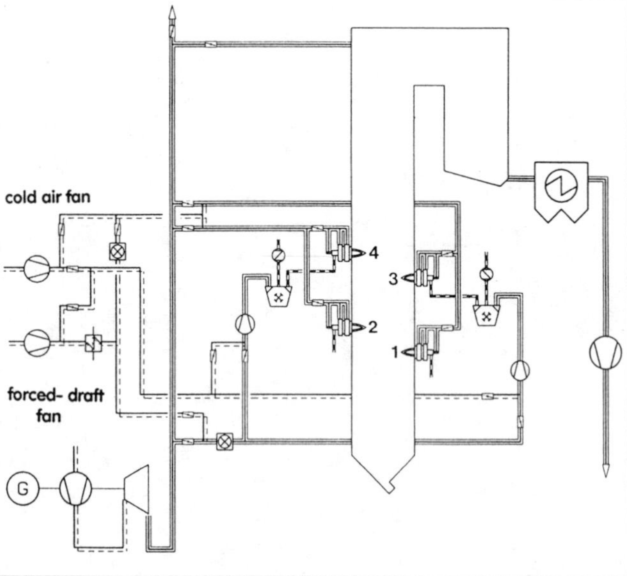


BABCOCK | Scheme of pulverizing and firing system | SAARBERG

Figure 4

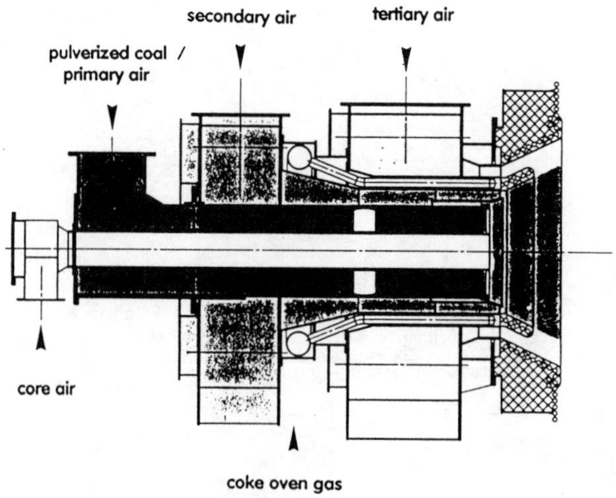


BABCOCK | Babcock Low - NOx swirl staged burner ( DS )

Figure 5

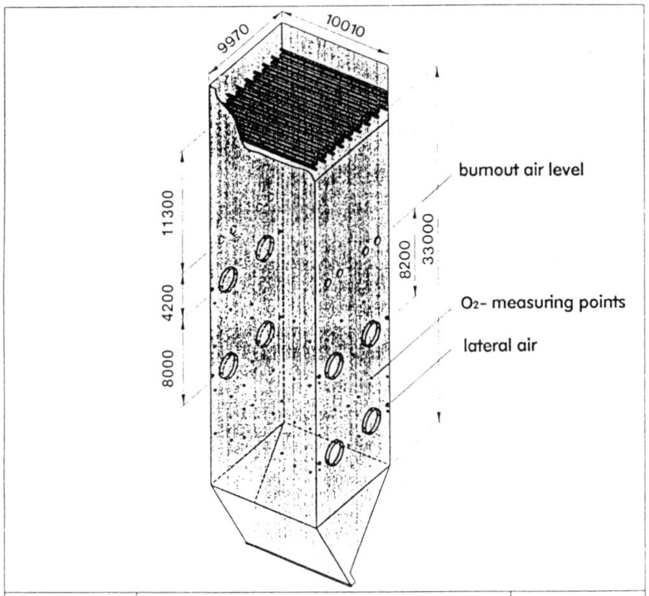


BABCOCK | Arrangement of air ports and $O_2$- sampling points | SAARBERG

Figure 7

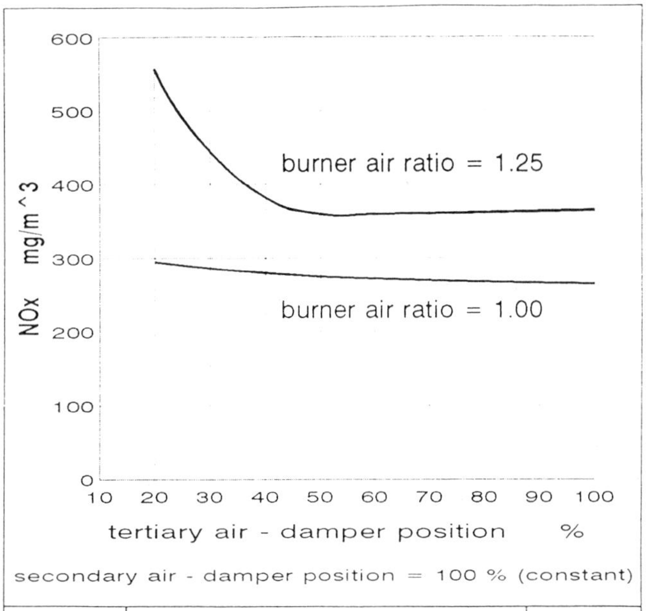


BABCOCK | NOx - emission as a function of burner air distribution | SAARBERG

Figure 6

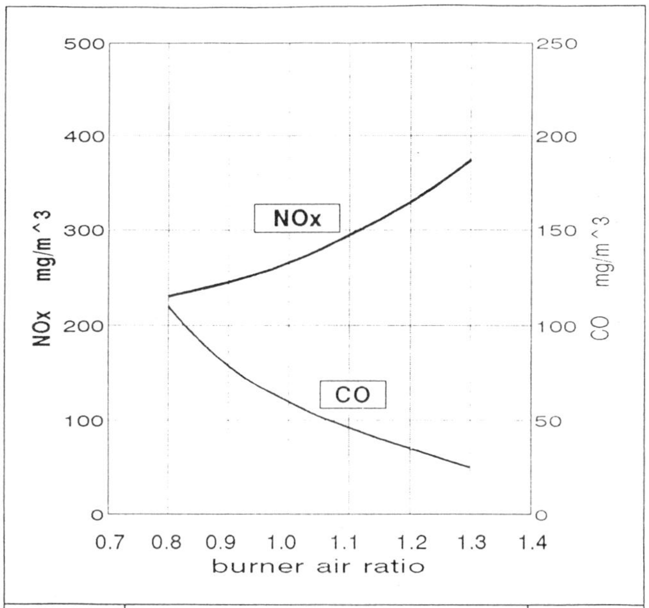


BABCOCK | NOx - and CO - emissions | SAARBERG

Figure 8

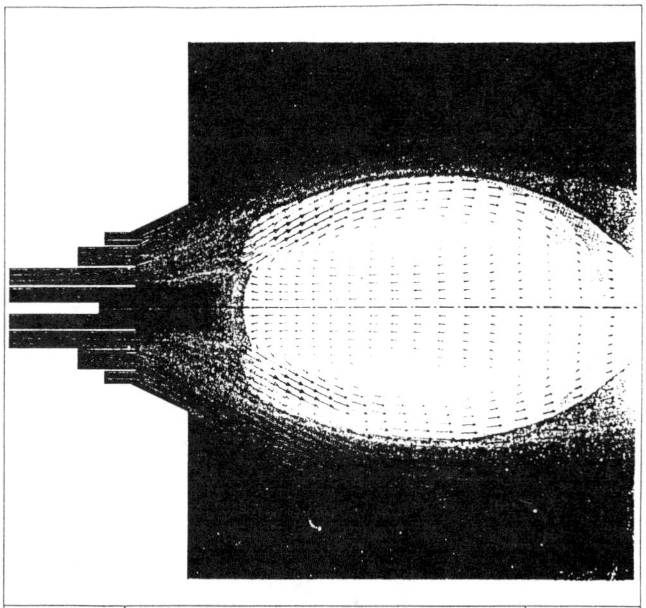

BABCOCK | Computer simulation of the swirl staged burner | SAARBERG

Figure 9

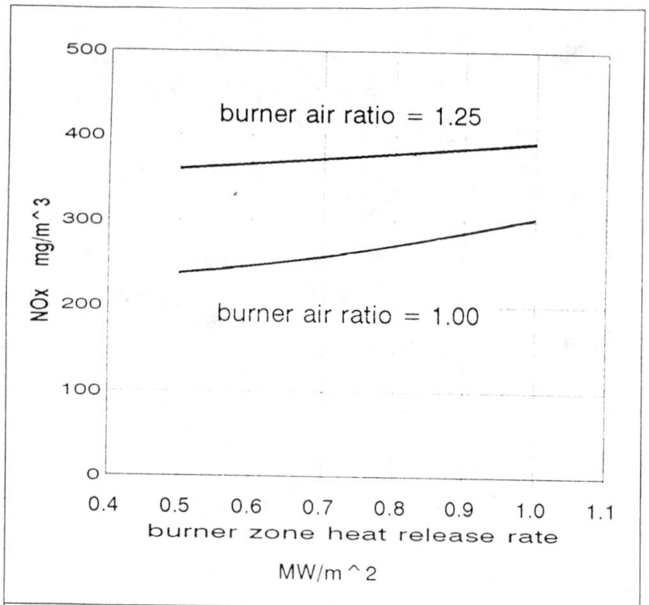


BABCOCK | NOx - emission as a function of burner zone heat release rate | SAARBERG

Figure 10

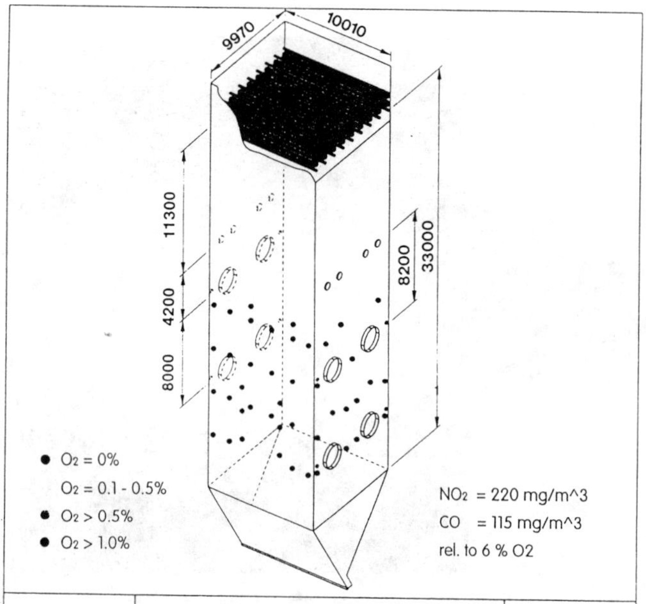


BABCOCK | O2- concentration with a burner air ratio of 0,8 | SAARBERG

Figure 11

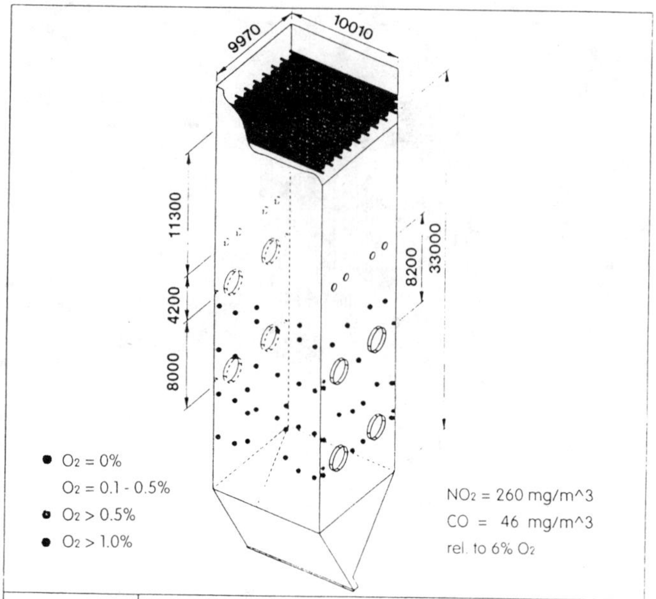


BABCOCK | O2- concentration with a burner air ratio of 0,95 | SAARBERG

Figure 12

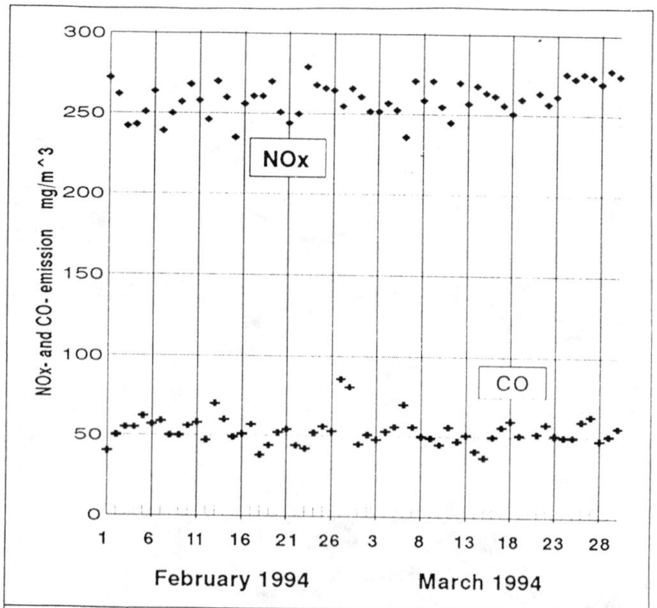


Daily averages of NO2 - and CO - emissions | SAARBERG

Figure 13

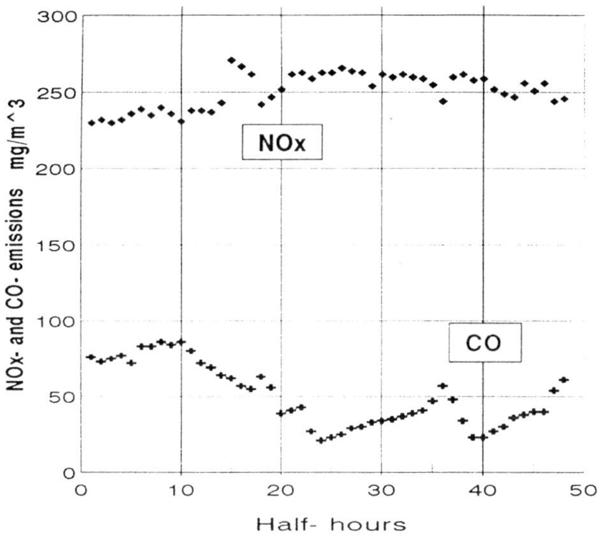


Half- hour averages of $NO_2$ - and CO - emissions

SAARBERG

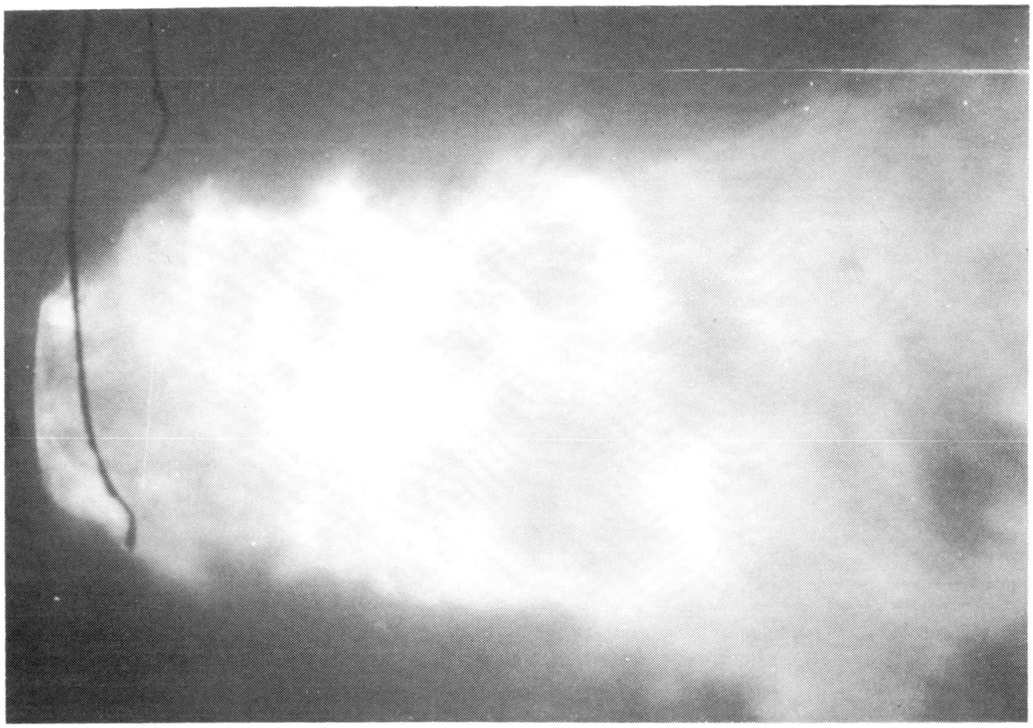

Figure 14 Flame pattern of the swirl staged burner

FACT-Vol. 18, Combustion Modeling, Scaling and Air Toxins
ASME 1994

# LOW $NO_x$ EMISSION FROM AERODYNAMICALLY STAGED OIL-AIR TURBULENT DIFFUSION FLAMES

A. L. Shihadeh, M. A. Toqan, J. M. Beér, P. F. Lewis,
J. D. Teare, J. L. Jiménez, and L. Barta
Combustion Research Facility
Massachusetts Institute of Technology
Cambridge, Massachusetts

## ABSTRACT

An experimental investigation on the reduction of nitrogen oxide emission from swirling, turbulent diffusion flames was conducted using a prototype multi-annular burner. The burner utilizes swirl-induced centrifugal body forces to dampen turbulent exchange between the fuel and air streams, allowing an extended residence time for fuel pyrolysis and fuel-N conversion chemistry in a locally fuel-rich environment prior to burnout. This aerodynamic process therefore emulates the conventional staged combustion process, but without the need for physically separate fuel-rich and -lean stages.

Parametric studies of swirl intensity and external air staging were carried out to investigate the feasibility of aerodynamic staging for low NOx combustion with No. 6 heavy fuel oil. NOx emission was reduced from an uncontrolled 300 ppm (3% O2) to 91 ppm in the optimal configuration. A further reduction from 91 ppm to 53 ppm was realized by external staging (primary stage fuel equivalence ratio = 1.13). A detailed flame structure investigation was carried out for a parametrically optimized, staged flame.

## INTRODUCTION

Combustion air staging has proven to be a highly effective method for reducing NOx emission in a number of practical systems. Typically, these systems rely on physically separate fuel-rich and fuel-lean combustion zones, between which "overfire" air is injected. Drawbacks to this method include higher operating costs, corrosion of the heat transfer surfaces in the fuel-rich first stage, and difficulty in retrofitting existing systems. As an alternative to relying on physically separate zones for staging, the Radially Stratified Flame Core (RSFC) burner was developed at MIT to implement staging by aerodynamic means, so that all of the combustion air is introduced at the burner.

Analogous to the conventional overfire air systems, the RSFC burner aerodynamically creates two combustion zones, one in which fuel-air mixing is suppressed by radial density stratification, and the other characterized by a high degree of mixedness. In the density stratified zone, relatively cool, highly swirling combustion air surrounds a hot, fuel-rich flame core, which is created by injecting a small portion (~15%) of the combustion air near the central fuel jet. Under certain conditions, the centrifugal body forces associated with the swirl damp turbulent mixing between the fuel-rich core and the surrounding air, effectively containing the fuel within a locally fuel-rich "first stage" for an extended residence time, during which the NO reduction chemistry is active. Further downstream, vortex breakdown occurs, and the peripheral combustion air mixes with the products of the fuel-rich core, creating a fuel-lean burnout stage. A schematic of the internal staging process is shown in Figure 1.

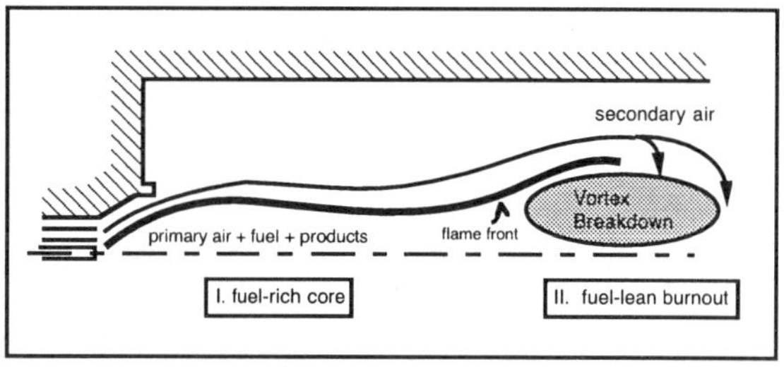


**FIGURE 1** INTERNAL STAGING SCHEMATIC.

In previously published studies of natural gas flames (Toqan et al., 1992), the RSFC burner achieved 70 ppm $NO_X$ emission at 3% $O_2$ (56 ppm CO) without flue gas recirculation, and 15 ppm $NO_X$ (< 10 ppm CO) with 32% of the flue gas recirculated. Using detailed velocity, species concentration, and temperature measurements, those studies demonstrated the role of swirl-induced radial stratification in producing aerodynamically-staged low $NO_x$ flames. In the current work, the applicability of the RSFC burner to No. 6 heavy fuel-oil (0.3 wt % N) flames was investigated. Because the nitrogenous species (fuel-N) in No. 6 fuel-oil are readily converted to $NO_x$ in the presence of $O_2$, particular attention was given to the residence time available in the fuel-rich core to ensure maximum fuel-N conversion to $N_2$. The principal source of NOx emission when burning fuels containing chemically bound nitrogen is the conversion of these nitrogen species (Pershing and Wendt, 1977).

## BURNER AERODYNAMICS

Figure 2 illustrates the important aerodynamic features of a 'typical' low NOx RSFC burner flame. An initial fuel-rich flame core is created by mixing initiated within the burner between the central fuel jet and the primary air. The secondary air, typically constituting 85% of the total burner air, is introduced through a radially displaced annulus. The fuel jet penetrates through an annular internal recirculation zone (IRZ) that extends into the burner quarl. Hot combustion products and some fuel peel off the fuel jet during its passage through the reverse flow zone, and are carried back into the quarl. In the emerging fuel-rich flame, fuel mixes slowly, due to turbulence damping, with the surrounding air. Further downstream of the burner (~ 5 burner diameters), stratification ends, largely due to the decay in tangential velocity, and the remaining secondary air mixes rapidly with the fuel-rich core, completing the combustion.

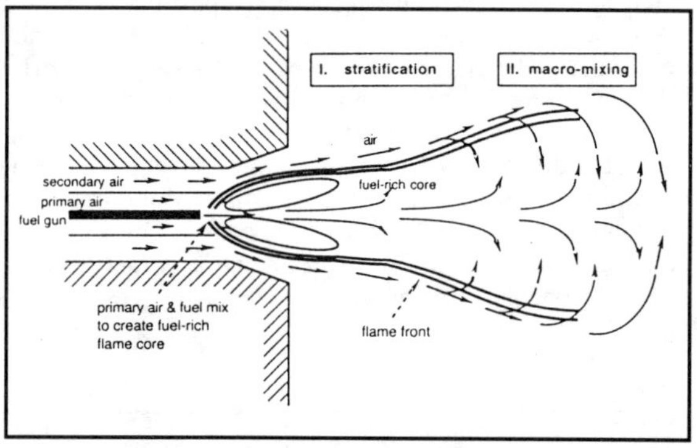


**FIGURE 2** TYPICAL LOW $NO_X$ RSFC FLOW FIELD.

Turbulence damping in the density stratified zone results from a combination of the centrifugal force field and the positive radial density gradient (created by the density difference between the hot burning core and the surrounding air). When an individual fluid eddy displaces a parcel of the relatively dense combustion air toward the flame axis, work is expended in the force field, dissipating turbulent energy. The concept is rooted in the work of Rayleigh (1916), which showed that a rotating fluid is stable with regard to radial interchanges if $\rho Wr$, the product of density, tangential velocity, and radial position, increases with radial distance from the axis of rotation. Beér et al. (1971) adapted Richardson's dimensionless group for the characterization of turbulence damping under conditions of atmospheric inversion to flames with swirling air flow around a hot central core. More recently, the adaptation of the Richardson number to a radially stratified natural gas flame in a low $NO_X$ gas burner was presented by Toqan et al. (1992).

## EXPERIMENTAL

Parametric tests were conducted to study the relationship between exit NOx emission and swirl number, $S$, defined as the non-dimensional ratio of angular momentum to axial momentum and burner radius:

$$S \equiv \frac{G_\phi}{G_x R} \quad \text{(Beér and Chigier, 1963).}$$

In addition, axial profiles of centerline species concentration and temperature were measured for zero and maximum swirl settings to demonstrate the role of swirl in mixing suppression. Parametric studies of external staging (using conventional overfire air) were also conducted to give a relative indication of the efficacy of the RSFC burner internal staging process.

Finally, a detailed flame structure study was conducted in which gas composition and temperature measurements were taken at many locations in an optimized flame. The detailed study was used to elucidate the overlapping mixing and chemistry processes, particularly to address the question of fuel-N conversion. The measurements would indicate the extent to which fuel was confined within a fuel rich core, and whether the temperature in the fuel-rich region was high enough to allow fuel-N conversion to N2 in the available fuel-rich residence time.

The thermal input was maintained at 0.9 MW and the fuel was No. 6 fuel-oil with a 1.5 C/H ratio and a 0.3 wt% N content. The air preheat temperature was 555 K. A twin fluid Y-jet atomizer with air as the atomizing medium was used, except for the detailed flame study, in which steam atomization was employed.

### Experimental Burner

A schematic of the RSFC burner is given in Figure 3. The combustion air is introduced through three concentric annular nozzles, of which the positions of the primary and secondary (as well as the fuel gun) can be adjusted to produce a particular flow field. To produce the desired internally staged flame for the experiments, the primary air nozzle was used to introduce

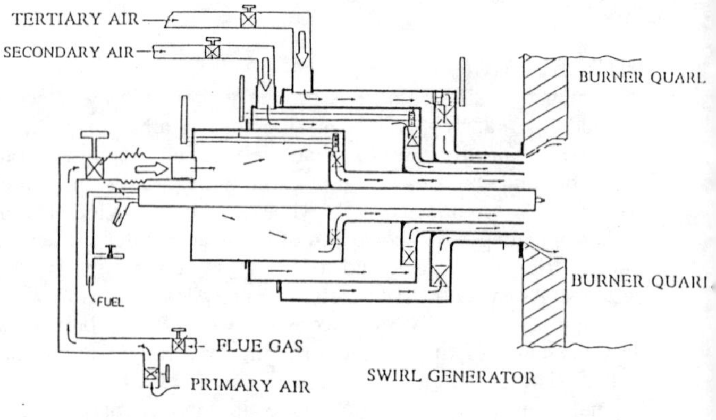


**FIGURE 3** SCHEMATIC OF THE RSFC BURNER.

approximately 15% of the combustion air while the remaining 85% was introduced through the tertiary nozzle (for the purpose of this discussion, the terms "tertiary air" and "secondary air" are interchangeable).

Each nozzle is equipped with moveable block-type swirlers capable of infinitely variable swirl control. For the results reported below, the degree of swirl used in any experiment is indicated by "swirl setting" which is a linear scale of swirler adjustment angle, with 0 representing no swirl and 10 representing maximum swirl. For moveable block swirlers, the theoretical swirl number of the flow issuing from any nozzle is strictly a function of the burner geometry, and can be calculated for any particular swirl setting, as shown in Beér and Chigier (1972). For the burner geometry of the experiments reported here, a swirl setting of 10 corresponds to a swirl number of 0.3 and 0.6 for the primary and tertiary nozzles, respectively.

## Experimental Furnace

The MIT Combustion Research Facility (CRF) was used to conduct the experiments. The CRF is an approximately 10 m long tunnel furnace that consists of several interchangeable water-cooled sections, each with either a bare metal or refractory brick lined surface, and a square (1.2 x 1.2 m) or cylindrical ($\phi$ 0.5 m) cross-section. The thermal capacity of the CRF is 3 MW, though typically it is operated at 1 MW. By varying the sequence of bare-metal and refractory brick sections, the heat extraction along the flame axis can be varied to simulate the temperature history of large scale practical flames. An overfire air injection port is located 3.4 m downstream of the burner.

An access door in each of the furnace sections allows measurement of gas temperature and composition with intrusive traversing probes, including a suction pyrometer for temperature, a water cooled suction probe for major species, and a steam jacketed suction probe for hot cell fourier transform infrared (FTIR) spectrometry for various trace species. The sampling methods and furnace have been described in detail elsewhere (Beér et al., 1985).

## EXPERIMENTAL RESULTS

### Tertiary Air Swirl

Figure 4 illustrates the effect of tertiary air swirl on NOx emission, demonstrating a reduction from 300 ppm at zero swirl, to 120 ppm at the maximum setting, which corresponds to a swirl number of 0.6. The non-zero slope at the maximum setting suggests that further reductions in NOx might be achieved if the burner were modified by increasing the maximum swirl angle. These results are consistent with previous studies of natural gas RSFC flames which demonstrated the presence of radial stratification as a result of swirl (Toqan et al., 1992). As a simple test that the same process was at work in the oil flames, centerline measurements of gas composition and temperature were taken along the flame axis for maximum and zero tertiary swirl cases. The results, shown in Figures 5 through 9 indicate that in the high swirl case, fuel-air mixing is suppressed, while in the zero swirl case, mixing was rapid. Figure 5 shows that a significant amount of oxygen reaches the flame axis in the no-swirl case, whereas practically none is found at the axis when swirl is applied. Similarly, the CO and CO2 profiles shown in Figures 6 and 7 indicate that with swirl, fuel consumption proceeds more slowly, partly accounting for the lower temperatures shown in Figure 9.

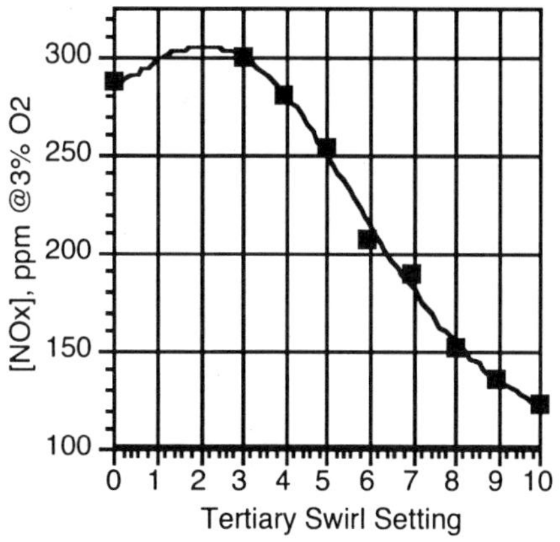


**FIGURE 4** NOx EMISSION VERSUS TERTIARY AIR SWIRL SETTING

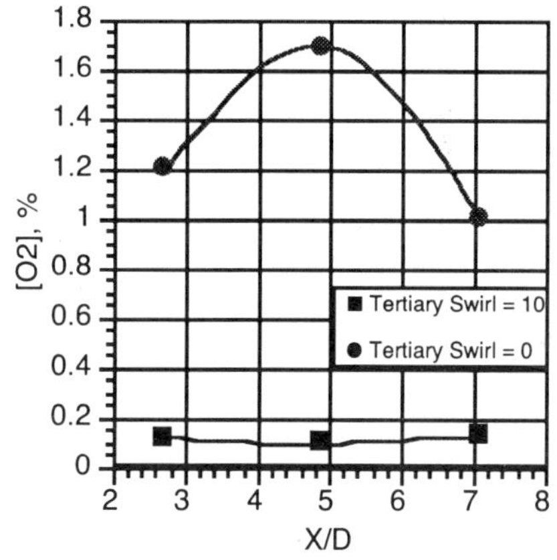


**FIGURE 5** OXYGEN CONCENTRATION AT FLAME CENTERLINE FOR ZERO AND MAXIMUM TERTIARY SWIRL; AXIAL DISTANCE IS INDICATED AS BURNER DIAMETERS FROM BURNER FACE.

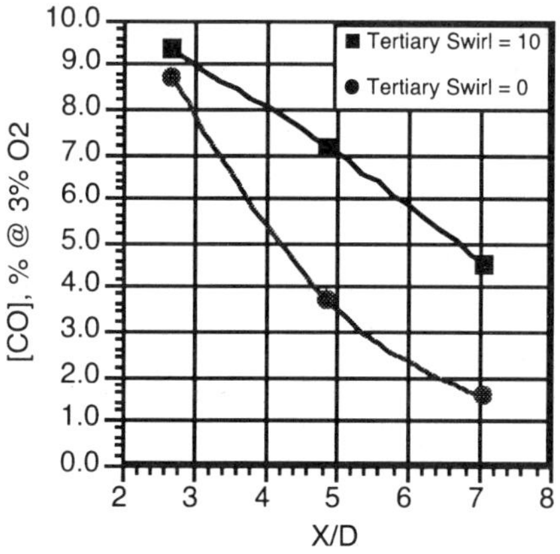


**FIGURE 6** CENTERLINE CO CONCENTRATION.

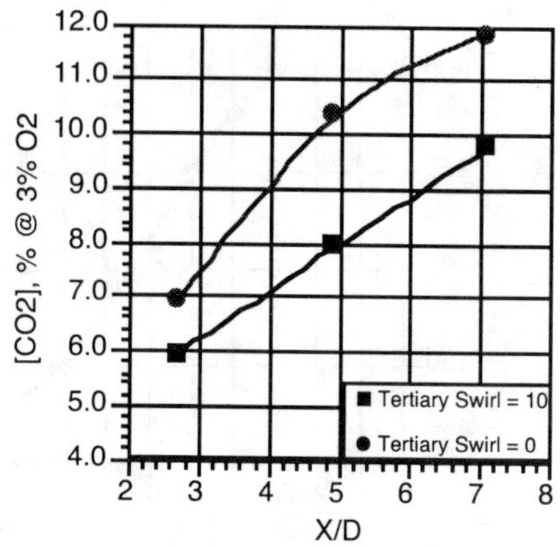


**FIGURE 7** CENTERLINE CO2 CONCENTRATION.

These differences account for the striking contrast in NO evolution shown in Figure 8. In the no-swirl case, NO concentration increases drastically along the axis, as should be expected given the presence of oxygen, whereas in the maximum swirl case, NO continues to decrease along the axis, since both conditions for NO reduction are met: high temperature, and a fuel rich environment. It is interesting to note that at X/D = 2.7, the NO concentration is considerably greater for the high swirl case, possibly due to the fact that the mixture is so rich there that the needed H and OH radicals are present in insufficient concentrations to decompose the NO precursors, such as cyanogen and ammonia species, to N2.

## Primary Air Swirl

Compared to tertiary swirl setting, primary air swirl had the opposite, but quantitatively less significant, effect on NOx emission. As shown in Figure 10, increasing swirl setting from 4 to 10 increased NOx emission from 92 to 120 ppm. The increase in emission is likely due to an increase in the strength of the internal recirculation zone in the near-burner field. As the recirculation zone strengthens, a greater portion of the fuel is diverted around the IRZ (instead of passing through it) to the outer periphery of the flame core, where it readily mixes with the surrounding air, thereby reducing stratification. This increase in mixing rate, however, had the desirable effect of stabilizing the flame; when primary air swirl setting was reduced to below 5, ignition became unstable.

## External Air Staging

Conventional external staging provided a relative indication of how well the internal staging produced by the RSFC burner reduced NOx emission. As shown in Figure 11, external staging reduced NOx emission from 91 ppm to 53 ppm in the best case. Burnout was complete by the furnace exit, and CO emissions were below 50 ppm in all cases. Staging beyond phi = 1.2 did not further reduce emission, likely due to the fact that as fuel equivalence ratio increases, the concentration of oxygen-containing radicals required to decompose the NOx precursors to N2 decreases. The fuel-equivalence ratio at which NOx decomposition is arrested is temperature dependent, with higher temperatures yielding a higher optimal fuel-equivalence ratio and lower NOx concentration (Shihadeh, 1994).

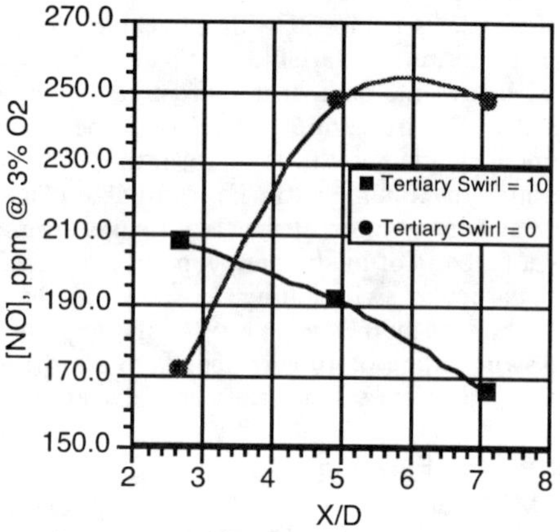


**FIGURE 8** CENTERLINE NO CONCENTRATION.

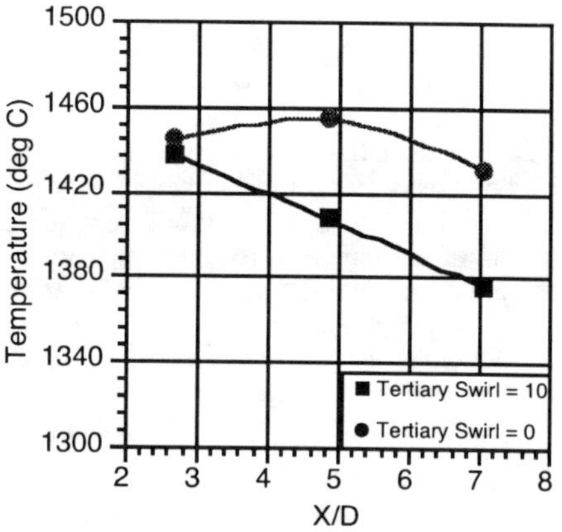


**FIGURE 9** CENTERLINE GAS TEMPERATURE.

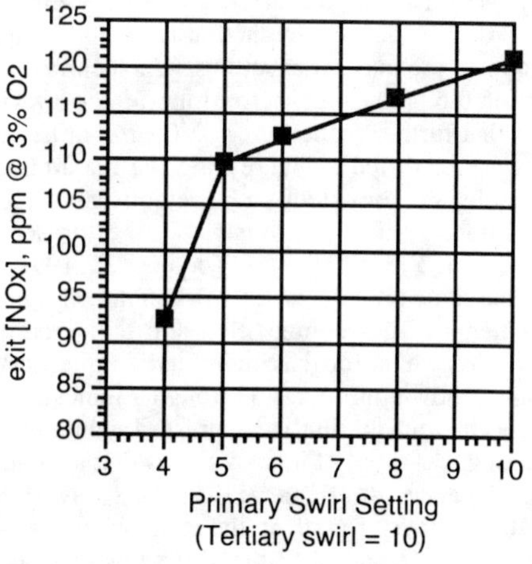


**FIGURE 10** NOx VERSUS PRIMARY SWIRL SETTING.

The success of external staging in reducing NOx indicated that the internal staging process of the RSFC burner had room for improvement, particularly by increasing residence time and/or temperature in the stratified zone to allow the fuel-N to N2 conversion reactions to equilibrate. If stratification ends after all or most fuel-N has been converted to N2, then external staging will have only a marginal effect on NOx emission; at present, its function is to maintain fuel-rich conditions after stratification ends so that the fuel-N reduction reactions may continue.

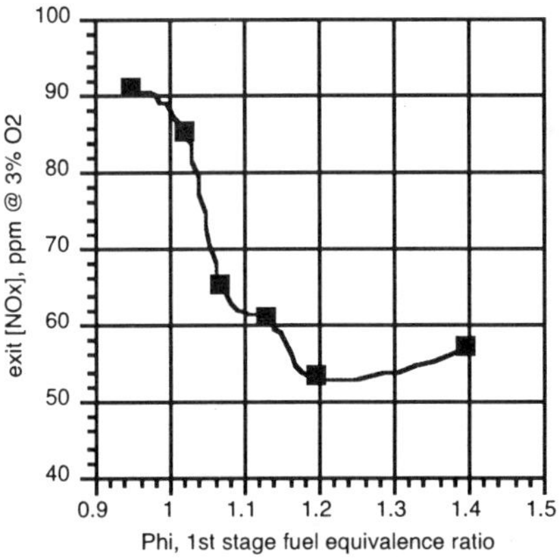


**FIGURE 11** NOx EMISSION VERSUS PRIMARY STAGE FUEL EQUIVALENCE RATIO.

## Flame Structure Study

Detailed in-flame measurements of gas composition and temperature were taken for an externally staged, parametrically optimized flame. The input parameters used to generate this flame are listed in Table 1.

**TABLE 1** BURNER CONDITIONS FOR DETAILED STUDY.

| Nozzle | Flow, % of burner air | Swirl No. | Axial Velocity [m/s] |
|---|---|---|---|
| Primary | 15 | 0.3 | 33 |
| Tertiary | 85 | 0.6 | 50 |

A 0° Y-jet atomizer was used with steam as the atomizing medium. The first stage was operated at a fuel equivalence ratio of 1.13, with the over-fire air injection port positioned at 3.4 m (11 burner diameters) from the burner. Thermal input was maintained at 0.9 MW. These conditions resulted in the following exit emissions:

| | |
|---|---|
| NOx | 58 ppm |
| O2 | 2 % |
| CO2 | 13 % |
| CO | < 30 ppm |

Figures 13 to 18 below show measured concentration and temperature distributions in the near-burner field. They illustrate the ways in which the high-swirl RSFC flame yields low NOx emission. In Figure 12, contours of NO concentration are plotted as a function of axial and radial distance (expressed as burner diameters) from the burner. In the figure, the NO formation and destruction chemistry regimes are apparent in two distinct zones, delineated by the 140 ppm iso-concentration line. In the formation zone, fuel-N is oxidized in an oxygen-rich environment, as evidenced by the O2 contours shown in Figure 13, and the absence of CO in the same region, shown in Figure 14. In a fuel-lean environment fuel-N will be converted to NOx, and thermal NOx will be formed, explaining the observed increase in NO concentration in this zone.

Within the NO reduction zone, there is a particularly intense region of NO destruction near the burner axis, indicated in Figures 13 to 18 by the shaded area. In this region, it can be seen that the O2 concentration is low, while CO (Figure 14) and CH4 (Figure 15) concentrations are relatively high, indicating a fuel-rich local atmosphere. In this context CH4 is important as an indicator of the presence of hydrocarbon fragments, such as CH and CH2. These hydrocarbon fragments play a major role in NOx reduction via the NO "reburn" mechanism (CHi + NO --> HCN + ...) and are likely responsible for the particularly intense NO destruction seen here. It should also be noted that within this region, the rate of destruction increases along the axis, reaching a maximum at a distance between 2.5 and 3.2 burner diameters. This rate increase can be partly explained by the peaking of gas temperature in the same location, as shown in Figure 17, since the fuel-N and NO decomposition reaction rates increase with temperature. The remainder of the NO reduction zone results from the high temperature, oxygen deficient conditions where "reburn" also occurs, though at a slower pace because of the lower concentration of CHi fragments there.

Returning to the question of external staging, it can be seen in the NO contour plot that if the over-fire air (OFA) injection port had been positioned anywhere within the plotted axial domain (instead of at X/D = 11), higher NOx emission would have resulted since the NO reduction reactions had not yet been completed (i.e., taken the mixture to the equilibrium NOx concentration). For example, if the OFA were positioned at X/D = 3.9, it is likely that the exit NOx emission would have been closer to 120 ppm, rather than the achieved 58 ppm.

## CONCLUSIONS

The experiments demonstrated that aerodynamic staging can be highly effective in reducing NOx emission from No. 6 fuel-oil flames, from and uncontrolled 300 ppm to 91 ppm. By imposing external staging on the low NOx RSFC flame, a 40% reduction in NOx emission was obtained (from 91 to 53 ppm), indicating that the internal staging realized in the experiments can theoretically be improved. To some extent, improvements could be achieved by delaying vortex breakdown, and by further optimizing the near burner flow field to create conditions for turbulence damping earlier in the flow - by accelerating ignition, for example.

Even with these improvements, however, it may be difficult to match the NOx emission reduction that can be achieved with conventional external staging, mainly due to the fact that with internal staging, the NO reduction chemistry is limited by the wide distribution of local fuel/air ratios (in the stratified zone) from zero to infinity. As a result, some portion of the mixture is too rich for converting NOx to N2, while elsewhere it is too lean. With external staging, a well-stirred flow pattern can be utilized so that the distribution about the mean equivalence ratio is narrower, allowing the majority of the combustion to occur at an optimal fuel/air ratio, but at the expense of higher corrosiveness in the fuel-rich stage.

One promising strategy, then, may be to combine the two methods of staging, in which case a deeply-staged condition could be achieved (in effect) without as high a first-stage overall fuel-equivalence ratio, and therefore without the normally associated degree of corrosiveness. In cases where physical constraints make staging via over-fire air impossible, staging by aerodynamic means alone is an effective low NOx alternative.

## REFERENCES


Beér, J.M. and N.A. Chigier, "Swirling Jet Flames from an Annular Burner," *5 me Journeé d'Études sur les Flammes*, Paris, 1963.

Beér, J.M. et al.: "Laminarization of Turbulent Flames in Rotating Environments," *Combustion & Flame*, 16, 39-45, 1971.

Beér, J.M. and N.A. Chigier: *Combustion Aerodynamics*, Robert Krieger Publishing Co., Malabar, Fl., 1972.

Beér, J.M., Farmayan, W.F., Teare, J.D., and Toqan, M.A.: "Laboratory-scale study of the combustion of coal derived liquid fuels," EPRI AP4033 Report, p. 182, 1985.

Pershing, D.W. and J. Wendt: "Pulverized Coal Combustion: The Influence of Flame Temperature and Coal Composition on Thermal and Fuel NOx," *Sixteenth Symposium (International) on Combustion*, The Combustion Institute, p. 389-399, 1977.

Rayleigh, L. 1916. *Proc. Roy. Soc.* A93. 148-154

Shihadeh, A.S., "NOx Control and Environmental Regulation: Techniques and Discriminatory Social Outcomes," S.M. Thesis, Massachusetts Institute of Technology, Cambridge, MA.

Toqan, M. A. et al: "Low NO Emission from Radially Stratified Natural Gas - Air Turbulent Diffusion Flames," *Twenty-Fourth Symposium (International) on Combustion*, The Combustion Institute, p. 1391-1397, 1992.


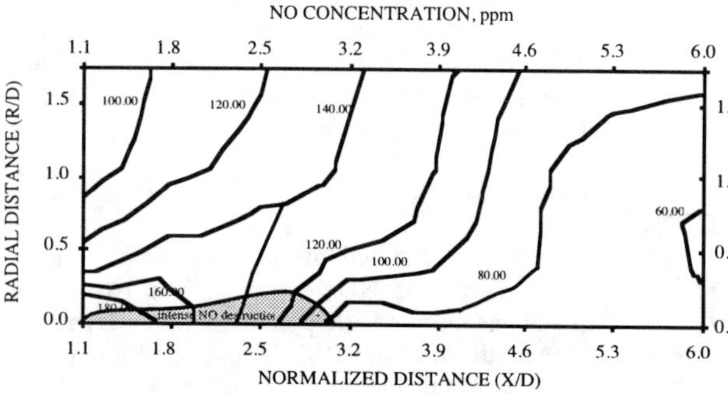


**FIGURE 12** MEASURED NO CONCENTRATION.

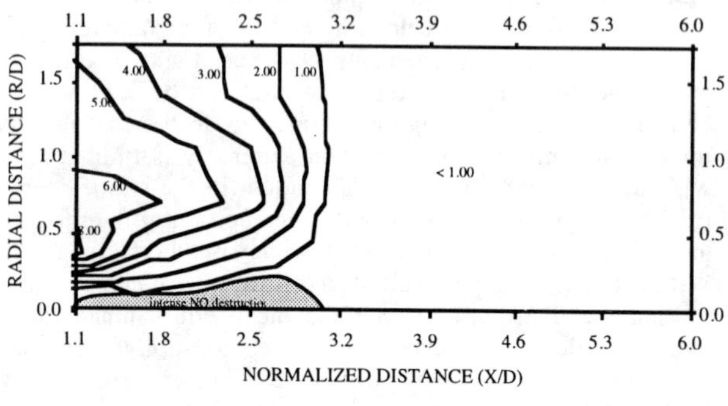


**FIGURE 13** MEASURED O2 CONCENTRATION.

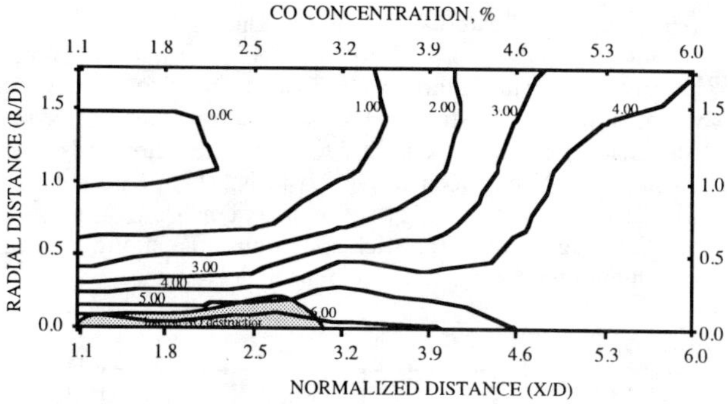


**FIGURE 14** MEASURED CO CONCENTRATION.

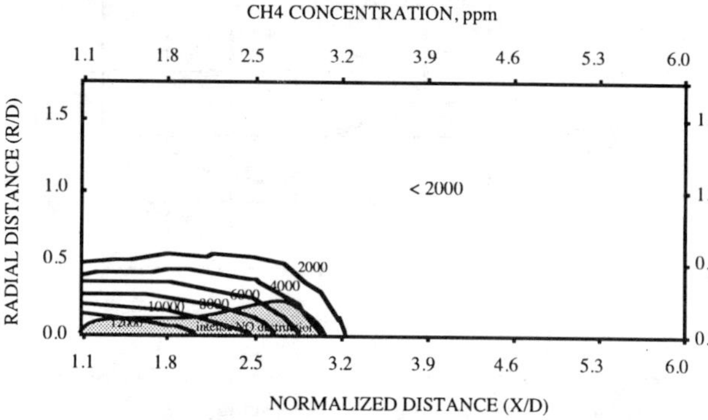


**FIGURE 15** MEASURED CH4 CONCENTRATION (FTIR MEASUREMENT).

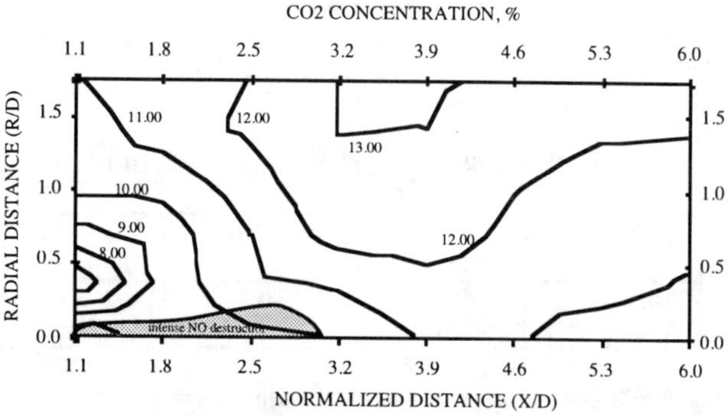


**FIGURE 16** MEASURED CO2 CONCENTRATION.

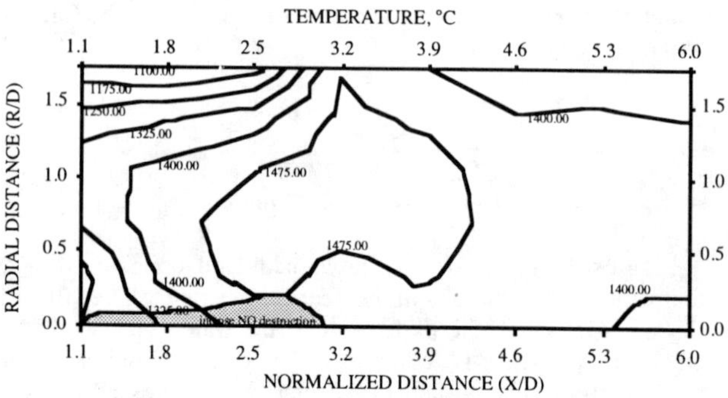


**FIGURE 17** MEASURED GAS TEMPERATURE.

FACT-Vol. 18, Combustion Modeling, Scaling and Air Toxins
ASME 1994

# ENGINEERING ANALYSIS OF RECOVERY BOILER SUPERHEATER CORROSION

**John F. La Fond, Arie Verloop, and Allan R. Walsh**
Jansen Combustion and Boiler Technologies, Inc.
Woodinville, Washington

## ABSTRACT

The occurrence of fire-side corrosion in kraft recovery boiler superheaters has increased in recent years due to the higher demands placed on recovery boilers. Recent research has led to new fundamental understanding of the mechanisms of corrosion in recovery boiler superheaters. However, there has been a need for development of engineering tools that combine fundamental data on superheater deposit chemistry, corrosion mechanisms, and heat transfer analysis to allow practical solutions to this problem.

Factors that play an important role in superheater corrosion include superheater design and boiler operating parameters. These factors are reviewed thoroughly upon initiating an engineering analysis effort. The focal point of the superheater corrosion analysis is a comprehensive computer-based heat transfer analysis. This paper describes the engineering analysis process that has been developed and illustrates its application through three case studies.

## INTRODUCTION

The fire-side corrosion of superheater tubes in recovery boilers has become increasingly more common and troublesome as recovery boilers are fired above their design levels. This often leads to excessive superheated steam temperatures and elevated superheater metal temperatures. These conditions can accelerate corrosive metal loss if proper modifications are not made to the boiler combustion system or superheater design.

Efforts to reduce salt cake make-up by improving chemical capture and recycle efficiencies have contributed to higher potassium and chloride levels in the as-fired black liquor. The presence of these two elements in elevated amounts can contribute to a higher susceptibility to deposit formation and fire-side corrosion in the superheater area.

In addition, the power cycle economics of some recent cogeneration projects require higher steam temperatures from the recovery boiler than previously considered practical. Operation at higher steam temperatures raises superheater tube metal temperatures and corrosion rates.

For these reasons, a better understanding of fire-side superheater corrosion has been of recent interest. Along with need for a better understanding of the mechanisms that promote superheater corrosion, there has been a need for the development of analytical tools to help engineer solutions to corrosion problems. In cases where fire-side superheater corrosion is occurring, engineering analysis can offer the following:

- Prediction of tube metal temperatures to identify areas most vulnerable to corrosive attack
- Identification of limits to boiler operating parameters to maintain tube metal temperatures at safe levels
- Assistance in redesign of superheater tube arrangement and metallurgy to avoid exposure of tube metal to excessive temperatures and heat fluxes.

This paper describes the ingredients of such an engineering analysis tool and examples of its application on three recovery boilers.

## SUPERHEATER DESIGN

The physical arrangement of recovery boiler superheaters plays an important role in their susceptibility to fire-side corrosion. Several factors are critical in superheater design, including the location of the superheaters, the tube geometry and arrangement, the number and location of attemperators, the materials of construction, and the direction of steam flow.

In recovery boiler design, the location of the superheater platens often differs greatly relative to the furnace nose, screen, and furnace cavity (see Figure 1). Superheater location plays an important role in determining radiant heat flux to the superheaters from flue gases and the exposure of the tubes to corrosive compounds from the furnace cavity.

The number of tube passes in a superheater platen (i.e., the number of times that a superheater tube is looped from top to bottom between the feed and discharge headers) affects corrosion rates. With each pass, more surface area is exposed to the flue gases, and the higher the final steam temperature becomes. Figure 2 illustrates how one tube circuit is often routed to reduce its number of passes in comparison to the other tube circuits. The lower flow resistance associated with the shorter tube circuit leads to a higher steam flow in the short circuit and reduced steam flow in the other tube circuits.

Tight radius tube bends cause higher friction losses than large radius bends. These higher friction losses may also result in an imbalance of flow and steam temperatures in some circuits. Tight tube bends in the tube circuits not only cause unbalanced pressure loss to the steam flow, but also prestress areas in the tube material. Corrosion is accelerated in regions of residual stresses [1]. Also, the lower superheater tube bends are often exposed to a larger radiant heat flux than the vertical tube sections. Consequently, the lower tube bends must be carefully inspected during outages, and special attention should be paid to tube bends when performing engineering analyses.

The distance between tubes, the tube arrangement (in-line or staggered tubes), and the tube diameter will impact heat transfer rates. These factors also influence the formation and accumulation of furnace-side deposits.

The header size and feeder tube arrangement partially determine steam flow distribution to the superheater banks. A lack of adequate steam flow through some circuits can lead to high steam temperatures in localized regions of the superheater. In addition, the absence of collection headers between superheater sections may result in large variations in steam and tube metal temperatures since there is no opportunity for interstage mixing of steam.

Many different materials of construction have been used for superheaters in recovery boilers. These include carbon steel, SA-213 T-11, SA-213 T-22, Type 347 and 310 stainless steel, and composite tubing (Type 310 stainless steel over T-22). The use of high grade steels or other metals has helped to alleviate some corrosion problems, but it can be very costly and has not proven to be a panacea.

A common approach to solving mill corrosion problems is to use a more corrosion resistant material in troubled areas. This approach is effective in regions where oxidation corrosion is dominant. However, in cases where molten deposits are present at the tube surface, a change in metallurgy is often ineffective. Tran et al. [2] have investigated corrosion rates for a standard 2-1/4 chromium, 1 molybdenum steel superheater tube material (SA-213 T-22), and a nickel-based alloy (Inconel 800H). Once a critical temperature is reached, in this case between 550 and 600°C (1020-1110°F), the corrosion rate increases to unacceptably high levels. Although the Inconel performs better than the SA-213 T-22, the metal loss remains excessive when a molten deposit is present.

The number, location, and capacity of attemperators will affect the local steam and tube metal temperatures. Inadequate attemperation capacity can limit steam temperature control in superheater banks. Superheaters with more than two banks often require more than one attemperator to assure adequate steam temperature control in all superheater banks.

The steam flow direction relative to the flue gas flow will determine the difference between the steam and flue gas temperature, which affects convective heat transfer. Superheater designs are based on counter-flow (Figure 1a), co-flow (Figure 1b), or some combination of the two. Although a counter-flow configuration will result in more efficient use of superheater surface area, its use places the hottest steam in proximity with the hottest flue gases. The high metal temperatures resulting from the counter-flow configuration could lead to high corrosion rates.

## EFFECT OF BOILER OPERATION ON SUPERHEATER CORROSION

In addition to the design of the superheater, boiler operation will also affect superheater corrosion rates and patterns. In particular, operation of the boiler above the design loading rate will often lead to excessive levels of char particle and smelt carry-over, in conjunction with high flue gas and superheated steam temperatures. Corrosion rates of superheater tubes are determined by the physical state and chemical make-up of the deposits. The physical state of the deposit material depends on the tube metal temperature, which is in turn determined by the local flue gas and steam temperatures. In addition, the composition of the flue gas will affect corrosion rates.

### Nature of Corrosive Deposits

The material that deposits on recovery boiler superheaters can be highly corrosive. The rate of corrosion is accelerated when molten deposit material is in contact with the tube metal surface [2,3]. This may be due to increased contacting at the surface and increased mass transfer rates. Therefore, a key to avoiding severe superheater corrosion is to avoid conditions which lead to a molten deposit at the tube metal surface.

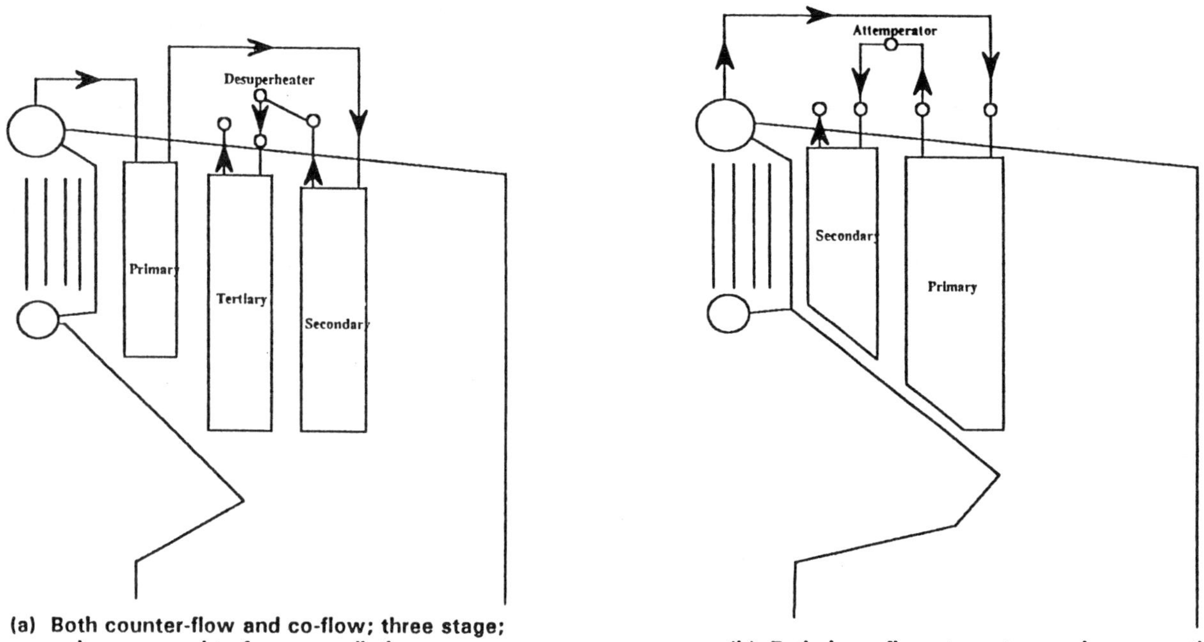


(a) Both counter-flow and co-flow; three stage; secondary exposed to furnace radiation; no collection headers between drum and discharge of secondary.

(b) Entirely co-flow; two stage; primary mostly protected from furnace radiation by nose.

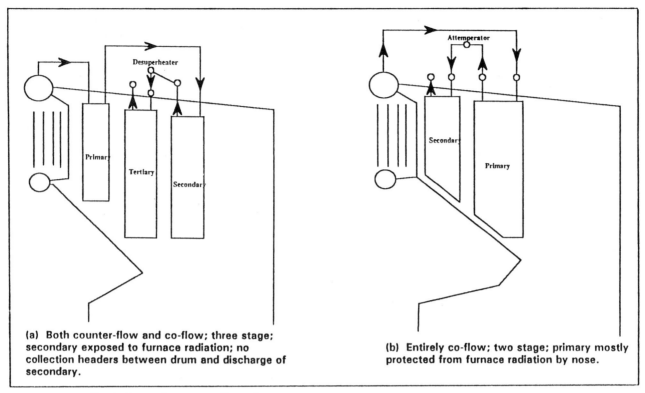

Figure 1. Typical superheater configurations.

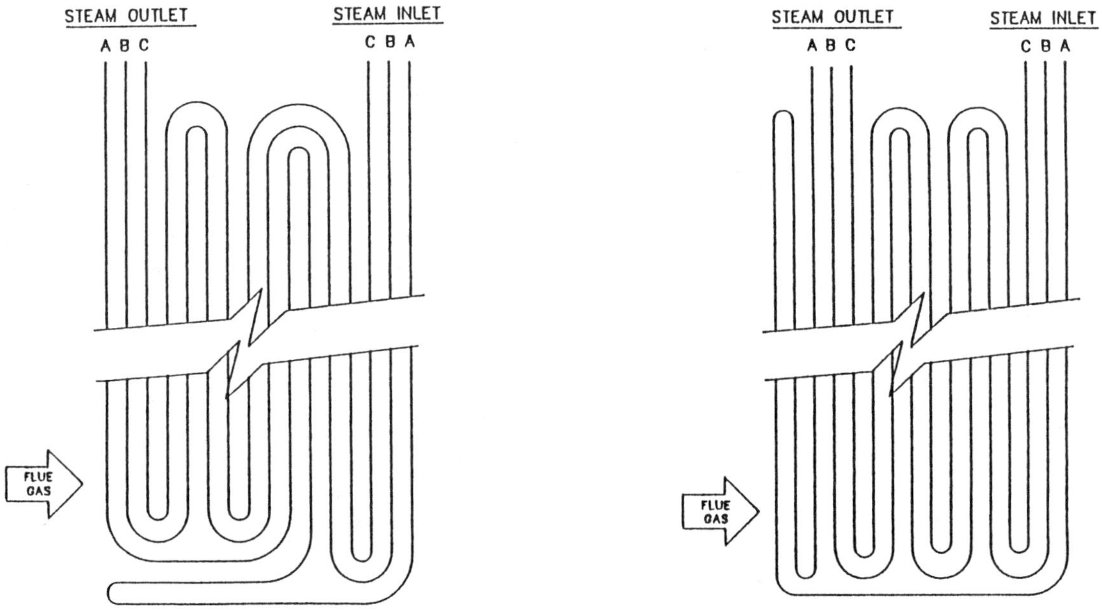


(a) "Hockey stick" and short circuit in A circuit.

(b) A circuit designed to protect tubing at steam discharge.

Figure 2. Examples of two superheater tube arrangements. In both cases, A circuit has four passes, B and C circuits have six passes.

Another factor related to corrosivity of deposits is the presence of significant amounts of unburned carbon, as will occur when char particle carry-over is high. When a boiler is being operated close to its maximum firing capacity, char particles may be carried over into the upper furnace region and deposited on the superheater tubes. Excess carbon can cause local reducing conditions, leading to the formation of $Na_2S$. Without the presence of carbon, the $Na_2S$ would be oxidized to $Na_2SO_4$. The $Na_2S$ can further suppress the melting point of a deposit and can support sulfidation corrosion. Sulfidation corrosion is more severe than oxidation because a passive oxide layer is not formed to reduce the metal wastage rate.

Recent research [2-4] has been instrumental in defining the influence of deposit chemistry on the first melting temperature of recovery boiler deposits. In particular, the role of potassium and chloride in establishing molten conditions has been defined, and this information can be used in corrosion analysis efforts. Increased potassium in the deposits can cause a strong suppression of the first melting point. The decrease in the melting point makes the tube surface more susceptible to corrosive attack.

In particular, the data generated by Tran et al. [2,4] has provided engineers with the ability to approximate the first melting point of the deposits, based on the as-fired black liquor composition. The first melting point serves as a limit to tube metal temperature in analyzing the potential for superheater corrosion in recovery boilers.

Once the metal temperature limit has been established by estimating the first melting temperature of the deposit, an estimation of the actual metal temperature in the superheater is required.

Through conductive, convective, and radiative heat transfer analysis, it has been found that the two most important factors that influence metal temperatures are the local steam temperature and the radiation heat flux from the hot flue gases.

## Steam Temperature

Due to the relatively high heat transfer coefficient on the steam-side versus the fire-side, the tube metal temperature is most closely tied to the steam temperature. Therefore, accurate tracking of steam temperature throughout a superheater is critical in establishing areas that are susceptible to high corrosion rates.

Steam temperatures in recovery boilers depend on superheater design and boiler operation. In general, as recovery boilers are operated at higher loads, the steam temperature increases. The trend is to design and operate recovery boilers at higher pressures. This increases the saturation temperature and the temperature of the steam throughout the superheaters. Increases in the solids loading to the recovery boiler also generally increase the flue gas temperatures in the upper furnace, if the proper countermeasures are not taken. Improved combustion in the lower furnace can allow higher liquor solids loading without increases in flue gas temperature at the

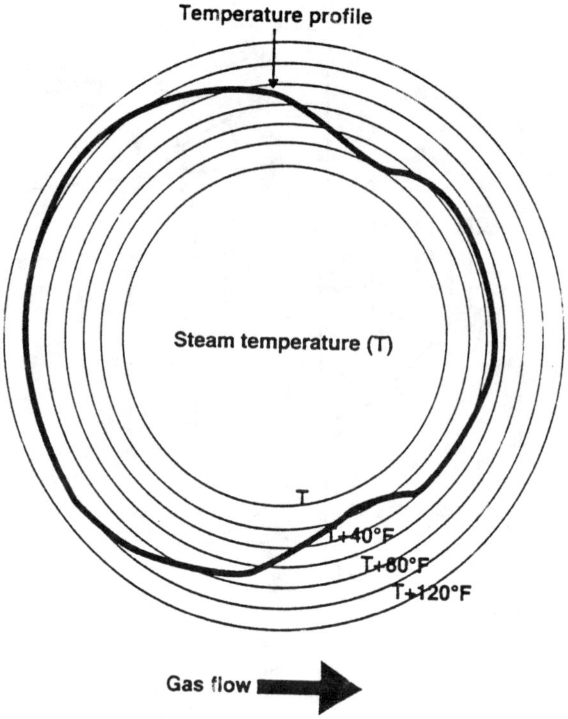


Figure 3. Typical superheater tube circumferential metal temperatures in a recovery boiler.

superheater due to higher radiative heat fluxes in the lower furnace [5].

## Flue Gas Temperature

Flue gas temperatures dominate the rate of radiant heat transfer in the furnace cavity since radiation is a function of the gas temperature raised to the fourth power. High radiant heat flux from the furnace cavity can produce locally higher tube metal temperatures [6]. Figure 3 presents a typical circumferential temperature profile of a recovery boiler superheater tube. In a utility boiler under a large radiant load, the tube metal that faces the furnace cavity (windward) is over 40°C (70°F) higher in temperature compared to the back side of the tube (leeward). A similar effect is present in recovery boilers, although cavity radiation levels are typically not as high as in utility boilers.

Radiant heat flux can be particularly high in recovery boilers that are being pushed beyond the capabilities of their combustion system. In these cases, an extended combustion zone will result in higher temperatures, char particles, and combustible gases in the screen and superheater region.

## ENGINEERING ANALYSIS

The previous sections have highlighted the important features of superheater design and the operating parameters that have the largest impact on corrosion in recovery boiler superheaters. Engineering analysis applies research data and fundamental heat transfer calculations to develop viable solutions to corrosion problems. The elements involved in the engineering analysis include review of superheater design, collection of field data, chemical analysis of black liquor and superheater deposits, and a heat transfer analysis of superheater tubes.

## Design Review

The first step in the engineering analysis is to obtain accurate design drawings of the boiler and superheater in question. In particular, tube diameter, tube wall thickness, tube metallurgy, tube spacing, tube weld locations, header design, attemperator capacity and location, pendant location, and furnace design operating conditions all have an impact on the heat transfer analysis and engineering approach.

## Collection of Field Data

A critical element in the engineering analysis is to collect the required operational information. A superheater corrosion analysis is typically one part of a comprehensive boiler performance evaluation [7]. Included in the evaluation are flue gas temperature measurements, determination of inlet and outlet steam temperatures and flows, and calculation of flue gas flow rates through overall mass and energy balances. This data is used as a cross-check in the overall energy and mass balances around the superheater, developed as part of the heat transfer analysis. Consequently, the uncertainties in the heat transfer analysis can be reduced.

## Chemical Analysis of Process Streams

As part of the boiler performance evaluation, samples of black liquor, smelt, and fly ash are collected for chemical analysis. As mentioned earlier, the potassium and chloride levels in the superheater deposits have a large influence on the first melting temperature and liquid fraction characteristics of the deposits.

Tran et al. have shown from a survey of 49 recovery boilers that relatively little enrichment of potassium and chloride occurs in the superheater deposits as compared to the as-fired black liquor [8]. This allows the use of the as-fired black liquor chemical analysis to calculate an approximate first melting temperature for the superheater deposits without actually retrieving a representative sample from the superheater area. If a more accurate determination of the first melting temperature is required, in-situ probing can be performed to actually measure the temperature where corrosion is accelerated.

With an estimate of the first melting temperature in hand, the engineer next needs to predict tube metal temperatures to determine where the "hot spots" may lie.

## Heat Transfer Analysis

The heart of the engineering evaluations of superheater corrosion is a computer-based heat transfer analysis program. The heat transfer analysis has been developed in a spreadsheet format to allow flexibility in the analysis of different superheater designs. Conductive, convective, and radiative heat transfer rates are calculated for each tube row to provide tracking of steam, metal, and flue gas temperatures throughout the superheater.

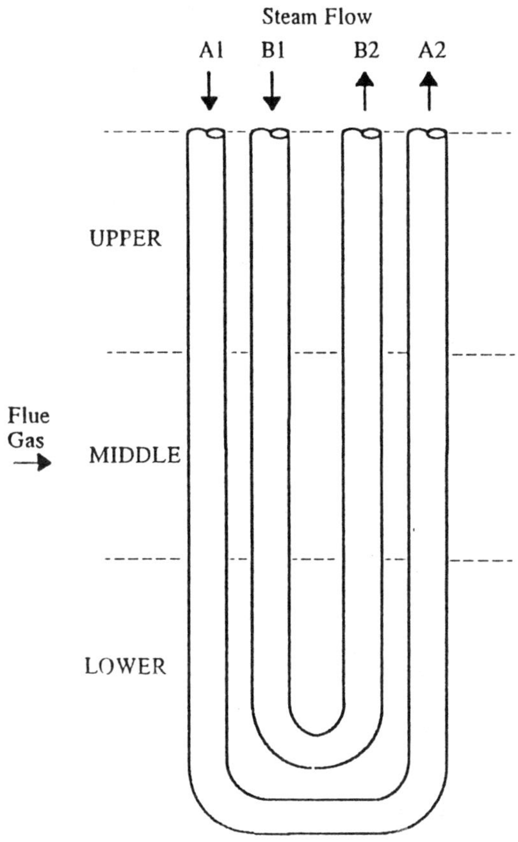


Figure 4. Simplified superheater tube arrangement to illustrate heat transfer calculation scheme.

Since the designs of superheaters can be so different, the calculation steps must be customized for each analysis. The spreadsheet format of the program allows for relatively rapid and straight-forward reconstruction of the program to match the physical arrangement of the superheater being analyzed.

Figure 4 presents a sketch of a simplified two-tube circuit, two-pass superheater to illustrate the calculation methodology. The superheater bank can be broken into any number of vertical zones (in this example, three are shown). Flue gas conditions, preferably based on field measurements, are input to the program for each zone, which gives a variation in temperature, flue gas flow rate, and radiation heat flux in the vertical plane of the superheater.

**Inputs.** The required inputs to the program include flue gas conditions and properties, tube metallurgy and geometric parameters, furnace geometry, and steam conditions and properties.

**Calculation Methodology.** Heat transfer calculations are performed on each tube row of each vertical zone in a progressive step-wise fashion so that the result of one block of calculations can feed into the adjacent block. An open-loop iteration is required for some configurations since there is not always full closure of the underlying equations.

The heat transfer equations used in the program were based on similar equations contained in the literature for recovery boilers [9]. For example, arrangement factors and intertube radiation conductance factors are used in accordance with previously described methods.

Fouling factors are not utilized. Instead, a more fundamental equation is used, allowing for variation in deposit layer thickness throughout the superheater. The deposit layer thickness is determined either through comparison of the deposit radical deformation temperature to the calculated external temperature, or is prescribed in areas that have an average deposit thickness based on soot blower effectiveness. The thermal conductivity of the deposit layer is based on a value contained in the literature [3].

**Outputs.** The outputs from the program include flue gas, steam, and tube metal temperatures; heat absorption profiles; mass and energy balances; and average deposit thicknesses.

In a complicated analysis of this type, some simplifying assumptions must be made. These include:

- No side-to-side variations in flue gas and steam conditions
- Deposition on the tubes is symmetrical and circumferential
- Flue gas over the tube banks is in pure cross-flow
- No mixing between vertical zones
- Distribution of furnace cavity radiation is estimated based on arrangement geometry.

Considering the variability in recovery boiler operations and the general level of uncertainty in the heat transfer equations, these simplifications are not considered detrimental to the practical utility of the heat transfer program.

The usefulness of the program has been demonstrated through the analysis of several recovery boiler superheaters. To illustrate its use, three case histories are described in more detail.

## RESULTS OF CASE HISTORY STUDIES

The following case studies provide examples of the engineering analysis in application and the recommendations that resulted. Results from three studies are presented. Each is from a different boiler manufacturer.

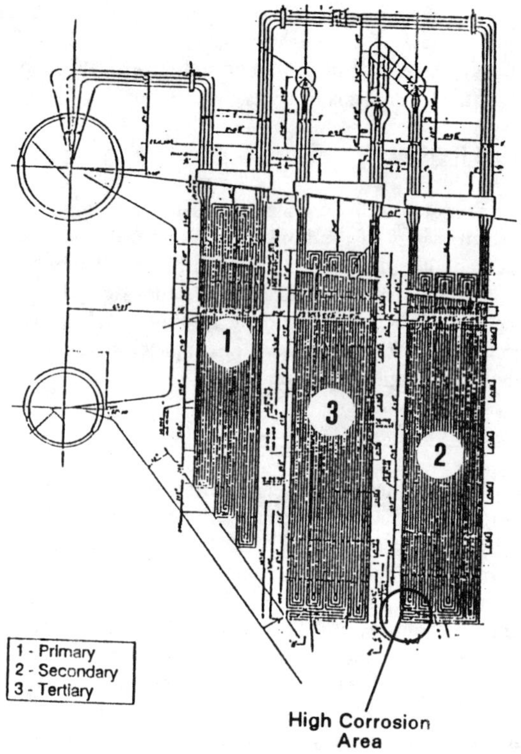


Figure 5. Superheater tube arrangement and area of high corrosion for Case A.

### Case A

**Boiler description.** Boiler A is a unit located in southeastern United States, designed to burn 1,360 metric ton/d (3.0 million lb/d) of liquor solids. During testing, it was fired at 1,550 metric ton/d (3.4 million lb/d), resulting in a heat input of 105% of the original design. Steam flow during testing was 204,000 kg/hr (449,000 lb/hr), and the heat transfer surface area of the superheater was 1,100 $m^2$ (11,900 $ft^2$).

**Problem.** The superheater was experiencing severe corrosion of the lower bends, as located in Figure 5. The corrosion was limited to the right hand side of the furnace in the secondary superheater sections.

**Analysis.** The corroded areas consisted of SA-213 T-11 metal. Analysis showed average steam temperatures of 463°C (865°F) exiting the superheater, with some excursions in temperature above 510°C (950°F). Testing also showed that there was uneven gas flow across the superheater, and there were high levels of char particle carry-over combined with low excess oxygen concentrations.

**Results.** Subsequent heat transfer analysis of the secondary superheater section confirmed high tube metal temperatures in the regions where corrosion was occurring (Figure 6). The first melting temperature of the deposits was determined to be 550°C

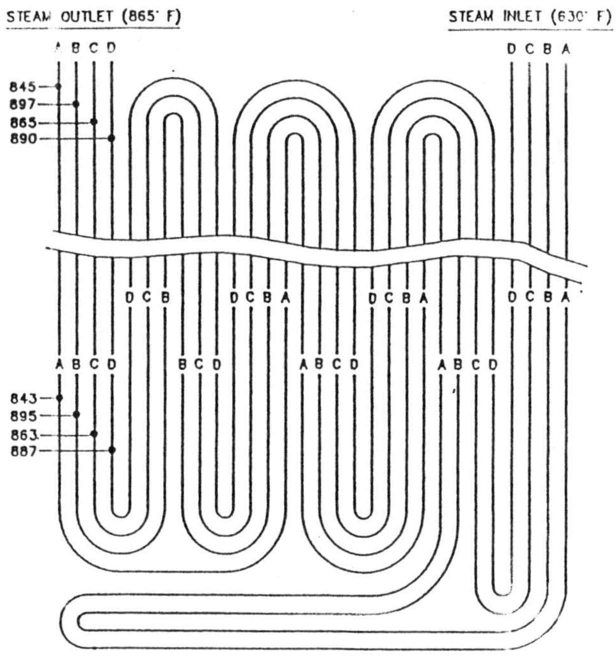


Figure 6. Calculated average tube metal temperatures (°F) for the secondary superheater described in Figure 5 for Case A.

(1020°F), which led to the recommendation of a maximum outlet steam temperature of 440°C (825°F).

**Recommendations.** It was recommended that the combustion conditions be improved in the lower furnace to reduce flue gas temperatures entering the superheater and to reduce carry-over. The combustion system was upgraded by eliminating the tangential flow conditions and modifying the forced draft fans to promote combustion lower in the furnace. In addition, a second attemperator was installed and monitoring of the steam conditions was improved.

The boiler upgrade allowed the mill to increase the liquor solids firing rate by 17% to 1,820 metric ton/d (4.0 million lb/d), while eliminating the superheater corrosion problems.

## Case B

**Boiler description.** Boiler B is a unit located in the lower midwestern region of the United States. It was designed to burn 955 metric ton/d (2.1 million lb/d) of liquor solids, but during testing, it was fired at only 682 metric ton/d (1.5 million lb/d). This resulted in a heat input of 70% of the original design. Steam flow during testing was 102,000 kg/hr (225,000 lb/hr), and the superheater heat transfer surface area was 1,190 m² (12,800 ft²).

**Problem.** The unit was experiencing severe corrosion of the lower bends of the tertiary superheater, as located in Figure 7.

**Analysis.** The failed areas consisted of SA-213 Type 347 metal. Analysis showed average steam temperatures of 432°C (810°F) exiting the tertiary superheater.

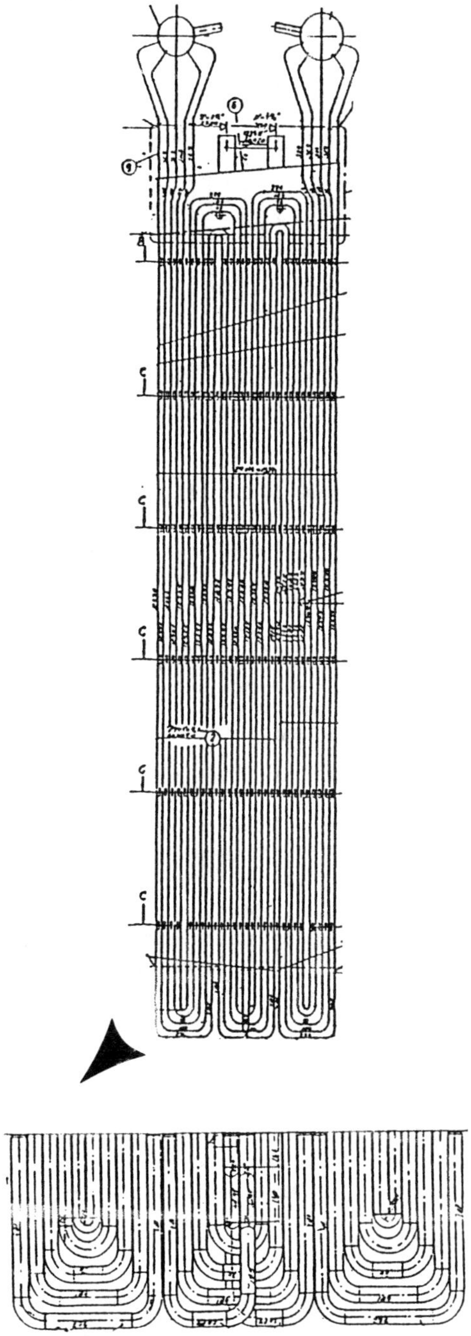

Figure 7. Tertiary superheater arrangement for Case B.

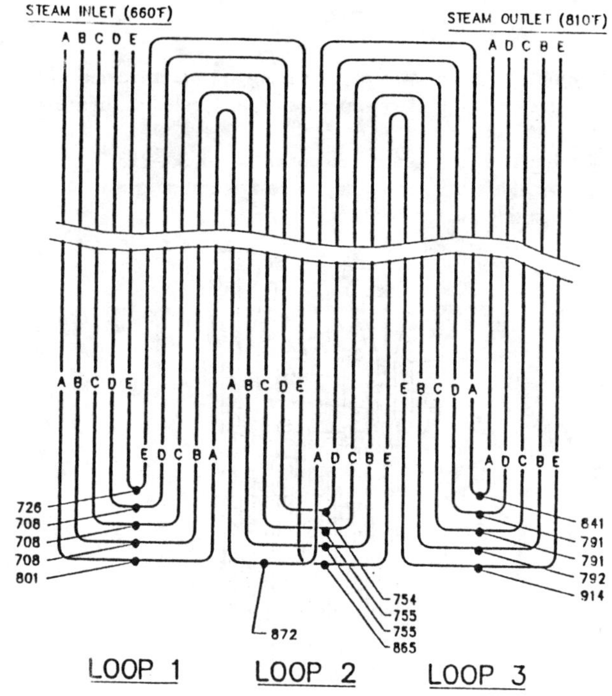


Figure 8. Calculated average tube metal temperatures (°F) for the superheater described in Figure 7 for Case B.

**Results.** Heat transfer analysis of the tertiary superheater section confirmed high tube metal temperatures in the regions where corrosion was occurring (Figure 8). The first melting temperature of the deposit was determined to be 520°C (970°F).

**Recommendations.** It was recommended that more complete steam temperature data and deposit thermal characteristics be obtained through additional field testing. This data would be used to make recommendations on modifications to the superheaters.

The client has used the results from the study to work with the boiler manufacturer on additional testing and design modification. The superheater was eventually replaced with a different design.

## Case C

**Boiler description.** Boiler C is a unit located in southeastern United States, designed to burn 2,270 metric ton/d (5.0 million lb/d) of liquor solids. Design steam flow is 342,000 kg/hr (752,000 lb/hr), and the superheater heat transfer surface area is 2,990 m² (32,200 ft²).

**Problem.** The boiler was scheduled to come on-line in 1991, and prior to startup the mill wanted an independent analysis of potential superheater corrosion. The analysis included an evaluation of the attemperation scheme and capacity for the superheater arrangement as shown in Figure 9.

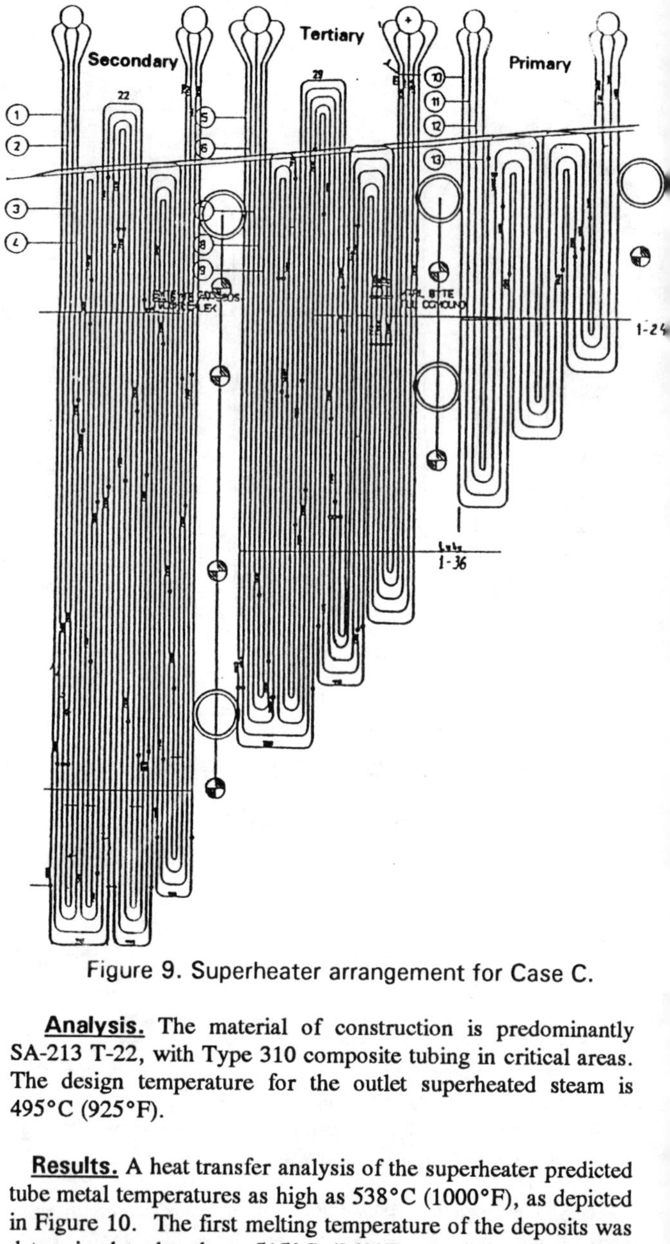


Figure 9. Superheater arrangement for Case C.

**Analysis.** The material of construction is predominantly SA-213 T-22, with Type 310 composite tubing in critical areas. The design temperature for the outlet superheated steam is 495°C (925°F).

**Results.** A heat transfer analysis of the superheater predicted tube metal temperatures as high as 538°C (1000°F), as depicted in Figure 10. The first melting temperature of the deposits was determined to be about 515°C (960°F), partially due to high potassium levels in the liquor. It was also determined that the effectiveness of the primary attemperator may be limited by saturation conditions.

**Recommendations.** It was recommended that the boiler begin operation with a steam temperature limit of 455°C (850°F), with close monitoring of the superheaters. Improvements were to be made on the steam temperature monitoring system to identify potential hot circuits. And finally, it was recommended that the feasibility of reducing the potassium levels in the liquor be evaluated.

The client continued to work with the boiler manufacturer to assure acceptable operation, and a corrosion probe was installed in the boiler at start-up.

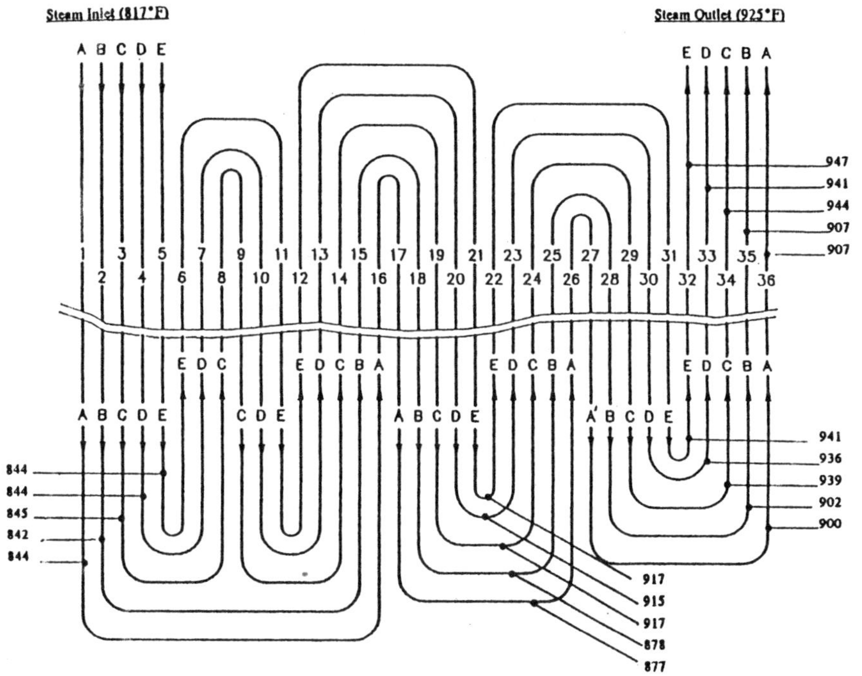


Figure 10: Calculated average tube metal temperatures (°F) for the tertiary superheater described in Figure 9 for Case C.

## CONCLUSIONS

The methods described in this paper have been successfully used for analysis of the superheaters of seven recovery boilers with the following general conclusions:

- Energy balances on the steam- and flue gas-side close within 5%.
- Areas most likely to suffer from severe corrosive attack can be identified.
- Steam attemperation schemes can be redesigned and easily evaluated.
- Superheater metallurgy and tube arrangement designs can be evaluated prior to operation to determine critical areas.

To date, the analysis has been applied to units supplied by Ahlstrom (1), B&W (3), CE (3), and Gotaverken (1). The program will continue to be updated as more information becomes available.

## REFERENCES


1. Fontana, Mars G., Corrosion Engineering, McGraw Hill Book Company, 1986.

2. Tran, H.N., D. Barham, and M. Hupa, Corrosion 1988, NACE, "Fireside Corrosion in Kraft Recovery Boilers. An Overview of the Effect of Deposit and Flue Gas Chemistry," St. Louis, Paper No. 437, 1988.

3. Backman, R., and M. Hupa, Tappi Journal, "Fouling and Corrosion Mechanisms in the Recovery Boilers Superheater Area 11," 70 (6): 123 (1987).

4. Tran, H.N., TAPPI Kraft Recovery Operations Short Course, "Kraft Recovery Boiler Plugging and Prevention," Orlando, pp. 181-190, 1991.

5. La Fond, J.F., A. Verloop, and A.R. Walsh, Tappi Journal, "Recovery Boiler Fireside Capacity: An Update of Theory and Practice," 76(9): 107, 1993.

6. Combustion - Fossil Power Systems, Combustion Engineering, Inc., Windsor, CT, (1981).

7. Verloop, A., T.W. Sonnichsen, and O. Strandell, Tappi Journal, "An Overview of Recovery Boiler Performance Evaluations," 73 (3): 145, 1990.

8. Tran, H.N., D. Barham, and D.W. Reeve, International Chemical Recovery Conference, "Chloride and Potassium in the Kraft Chemical Recovery Cycle," Ottawa, Ontario, Canada, pp. 23-28, 1989.

9. Steam - Its Generation and Use, 37th ed., The Babcock and Wilcox Company, New York, 1963.



Originally presented in the 1991 *Tappi* Engineering Conference, September 30 - October 3, Nashville, TN,

FACT-Vol. 18, Combustion Modeling, Scaling and Air Toxins
ASME 1994

# ALKALIS IN ALTERNATIVE BIOFUELS

**Thomas R. Miles and Thomas R. Miles, Jr.**
Consultants
Portland, Oregon

**Richard W. Bryers**
Foster Wheeler Dev. Corp.
Livingston, New Jersey

**Larry L. Baxter**
Combustion Research Fac.
Sandia National Labs.
Livermore, California

**Bryan M. Jenkins**
Biological and Agricultural Engineering
University of California
Davis, California

**Laurance L. Oden**
Bureau of Mines
U.S. Department of the Interior
Albany, Oregon

## ABSTRACT

The alkali content and behavior of inorganic material of annually produced biofuels severely limits their use for generating electrical power in conventional furnaces. A recent eighteen-month investigation of the chemistry and firing characteristics of 26 different biofuels has been conducted. Firing conditions were simulated in the laboratory for eleven biofuels.

This paper describes some results from the investigation including fuel properties, deposits, deposition mechanisms, and implications for biomass boiler design, fuel sampling and characterization.

Urban wood fuels, agricultural residues, energy crops, and other potential alternate fuels are included in the study. Conventional methods for establishing fuel alkali content and determining ash sticky temperatures were deceptive. The crux of the problem was found to be the high concentration of potassium in biofuels and its reactions with other fuel constituents which lower the "sticky temperature" of the ash to the 650° C to 760° C (1200° F-1400° F).

## ALKALI DEPOSIT INVESTIGATON

High levels of alkali in annual crop biomass fuels promote serious fouling of convection surfaces and slagging of fluidized beds and grates in combustion boilers. Increased competition for woody biomass has forced power plants to burn lower quality residue fuels or annual crop fuels. Biomass plants built in the 1980s have found that they can only burn limited quantities of some of the offset agricultural residues required by their operating permits. The National Renewable Energy Laboratory (NREL) of the US DOE recognized the problem in 1991 and sponsored the Alkali Deposit Investigation. (Miles, 1993, Baxter, 1993a)

The objective of the project was to organize and implement a joint industry and NREL investigation of possible solutions to the alkaline deposit problems experienced with biomass fuels in conventional boilers. NREL funding was fully matched by nine industry co-sponsors.

The project consisted of sampling, analysis and database development supported by the investigation. The biomass electric power industries contributed their experience and made their facilities available for testing. An advisory team of experts worked with the industries to identify problems, to assess the suitability of current solutions and to suggest future directions.

## ALKALI DEPOSITS FOUND IN BIOFUEL BOILERS

Power plants firing biofuels in conventional boilers have experienced fireside deposits and convection pass fouling that have reduced plant availability and efficiency, caused costly unscheduled outages and accelerated tube wear and corrosion. Deposits include:

- fused glassy deposits on refractory surfaces
- agglomeration of grate ash
- agglomeration of sand media in fluidized beds
- agglomeration and fouling of flyash in hot gas ducts.

Fouling of the upper furnace and convection passes include:

- reflective flyash ash coating waterwalls
- buildup on screen and superheater tubes
- bridging between tubes causing irregular flow and erosion
- hard calcium deposits from excess lime added to prevent agglomeration in bubbling beds.

While sootblowing can sometimes remove buildup, coverage is often not adequate or possible. Corrosion under deposits contributes to tube wastage and unscheduled outages. Corrosion in economizers and air preheaters becomes most evident in the fifth year of operation.

Alkali and alkaline earths in combination with other elements such as sulfur and chlorine are common constituents of these deposits. The term "alkali" is used here to describe the sum of potassium and sodium compounds, generically expressed as the oxides $K_2O$ and $Na_2O$. High levels of alkali, particularly potassium, distinguish biofuels from coal or peat.

## BIOFUELS AND ALKALI CONTENT

Tables 1-3 list several fuels fired in biomass plants. These include woody fuels from urban wood waste residues and urban construction, woody fruit and nut chips, pits, nuts, shells and hulls, field crop residues including straw, stalks, fibers, and manure. The fuels are listed according to the concentration of alkali oxides. (Miles, 1993, Mylchreest and Butler, 1991) This index can be used by industry to identify those fuels that have sufficient alkali compared with other inorganic constituents to signal a potential for slagging. All fauna and flora are dependent on potassium and some sodium ions for their growth processes.

Table 1 shows that the biofuel containing the least alkali is the stemwood or trunk of a tree and the larger limbs and branches. The sap in the cambium layer is rich in alkali but does not deposit it in the wood itself. (Marschner, 1986) The small branches, twigs and foliage that consist of high percentages of annual growth and cambium tissue are quite rich in potassium with sodium salts and organic complexes. Table 1 illustrates the low levels of alkali in the pine stemwood furniture trim, hogged wood sawmill and forest residues compared with wood with high annual growth. Wood has low intrinsic ash containing low silica and high calcium relative to potassium. In the case of willow or hybrid poplar, the combination of a small stem with its higher cambium-to-bole ratio plus the branches and twigs results in higher alkali levels. Some woods, like white oak, contain a high percentage of alkali in the ash, but have very little ash, illustrating the point that all the values shown must be considered to properly evaluate a fuel.

Clean urban wood waste consists of ground pallets, crating and construction lumber. While residues from construction are cleaner than from demolition, Table 2 shows that contaminants in both fuels account for higher levels of silica and aluminum than in ash from wood industry residues. Sample of fines removed from dirty urban wood waste with a 3mm (1/8th in) screen contains high levels of ash and alkali similar to the mill yard waste in Table 2.

Table 1. Fuel and Ash Properties of Woody Biomass

| **Fuel Type** | **Hog Fuel** | **Furniture Waste** | **Hybrid Poplar** | **Forest Residue** | **Willow, Top** |
|---|---|---|---|---|---|
| **Ash, %** | 1.0 | 3.61 | 2.70 | 3.97 | 2.34 |
| **Chlorine %** | | <0.01 | .04 | .04 | <0.01 |
| **HHV, Mj/kg** | 20.95 | 20.15 | 18.95 | 20.18 | 19.81 |
| **Composition** | | | | | |
| **$SiO_2$** | 35.18 | 57.62 | .88 | 17.78 | 2.05 |
| **$Al_2O_3$** | 2.31 | 12.23 | .31 | 3.55 | 1.97 |
| **$TiO_2$** | .01 | .50 | .16 | .50 | .03 |
| **$Fe_2O_3$** | 4.41 | 5.63 | .57 | 1.58 | .35 |
| **CaO** | 25.37 | 13.89 | 44.4 | 45.46 | 34.18 |
| **MgO** | 7.62 | 3.28 | 4.32 | 7.48 | 2.98 |
| **$Na_2O$** | 5.64 | 2.36 | .23 | 2.13 | 2.67 |
| **$K_2O$** | 9.26 | 3.77 | 20.08 | 8.52 | 18.40 |
| **$SO_3$** | 3.03 | 1.00 | 3.95 | 2.78 | 2.92 |
| **$P_2O_5$** | 5.68 | .50 | .15 | .44 | 7.10 |
| **$CO_2$** | | | 19.52 | | 22.64 |
| **Undet.** | 1.58 | -.78 | 5.43 | 9.78 | 4.71 |
| **Total, %** | 100.00 | 100.00 | 100.00 | 100.00 | 100.00 |
| **$Na_2O+K_2O$, kg/GJ** | .07 | .10 | .16 | .20 | .23 |

Table 2. Fuel and Ash Properties of Urban Wood Fuel Blends

| **Fuel Type** | **Urban Wood-Ag** | **Urban Wood** | **Urban Wood-Almond** | **Urban Wood -Straw** | **Mill Yard Waste** |
|---|---|---|---|---|---|
| **Ash, %** | 2.50 | 5.54 | 6.78 | 8.19 | 20.37 |
| **Chlorine %** | .05 | .06 | .03 | .13 | .30 |
| **HHV Mj/kg** | 19.50 | 19.46 | 18.46 | 18.82 | 16.32 |
| **Composition** | | | | | |
| **$SiO_2$** | 28.81 | 55.12 | 45.60 | 55.50 | 59.65 |
| **$Al_2O_3$** | 8.47 | 12.49 | 10.75 | 9.37 | 3.06 |
| **$TiO_2$** | .83 | .72 | .54 | .50 | .32 |
| **$Fe_2O_3$** | 3.28 | 4.51 | 4.06 | 4.77 | 1.97 |
| **CaO** | 27.99 | 13.53 | 18.96 | 11.04 | 23.75 |
| **MgO** | 4.49 | 2.93 | 4.22 | 2.55 | 2.15 |
| **$Na_2O$** | 3.18 | 3.19 | 3.08 | 2.98 | 1.00 |
| **$K_2O$** | 8.86 | 4.78 | 6.26 | 6.40 | 2.96 |
| **$SO_3$** | 2.00 | 1.92 | 2.06 | 1.80 | 2.44 |
| **$P_2O_5$** | 2.57 | .88 | 1.47 | 1.04 | 1.97 |
| **$CO_2$** | 6.07 | | | | |
| **Undet.** | 3.45 | -.07 | 3.00 | 4.05 | .73 |
| **Total, %** | 100.00 | 100.00 | 100.00 | 100.00 | 100.00 |
| **$Na_2O+K_2O$ kg/GJ** | .14 | .21 | .37 | .38 | .46 |

While many power plants were intended to burn agriculture residues most now fire only the woody stumps and prunings, with small quantities of nuts, shells and agricultural residues. Agricultural residues in Table 3 illustrate the very wide range of alkali content in these fuels. Any annual crop contains substantial percentages of alkali deposited throughout the plant. In addition to a species' characteristic alkali and ash content, growth site, sensitivity to soil content, rainfall and the variety all contribute to the variation in concentration. Also there is the wide spread in wheat straw according to the growing site. Bagasse from sugar cane has a low alkali content because the soluble alkali is rinsed out with the sugar after crushing. The very high ash content plus moderate alkali in rice straw presents a very serious deposit and agglomeration problem. Almond hulls and shells are convenient to burn but have very high potassium contents.

Few plants now burn plantation or "energy" crops, like switchgrass, or straw. Some special dedicated furnaces in Denmark are burning 100% straw with partial success. There are none in the U.S. It is notable that the Danish plants that fire 100% straw are still considered to be first generation facilities by their owners. There are frequent outages for maintenance. Since steam is their primary product, electrical power production is subject to process steam and district heating demands.

Table 3. Fuel and Ash Properties of Agricultural Fuels

| **Fuel Type** | **Switch-grass** | **Wheat Straw Lo-alk** | **Wheat Straw Hi-alk** | **Rice Straw** | **Almond Hulls** |
|---|---|---|---|---|---|
| **Ash, %** | 8.97 | 4.32 | 9.55 | 18.67 | 6.13 |
| **Chlorine %** | .19 | .14 | 1.79 | .58 | .02 |
| **HHV Mj/kg** | 18.08 | 18.48 | 16.78 | 18.91 | 18.90 |
| **Composition** | | | | | |
| **$SiO_2$** | 65.18 | 46.07 | 37.06 | 74.67 | 9.28 |
| **$Al_2O_3$** | 4.51 | 1.69 | 2.66 | 1.04 | 2.09 |
| **$TiO_2$** | .24 | .09 | .17 | .09 | .05 |
| **$Fe_2O_3$** | 2.03 | 1.85 | .84 | .85 | .76 |
| **CaO** | 5.60 | 9.95 | 4.91 | 3.01 | 8.07 |
| **MgO** | 3.00 | 2.45 | 2.55 | 1.75 | 3.31 |
| **$Na_2O$** | .58 | 1.18 | 9.74 | .96 | .87 |
| **$K_2O$** | 11.60 | 25.20 | 21.70 | 12.30 | 52.90 |
| **$SO_3$** | .44 | 4.92 | 4.44 | 1.24 | .34 |
| **$P_2O_5$** | 4.50 | 3.32 | 2.04 | 1.41 | 5.10 |
| **$CO_2$** | | | | | 20.12 |
| **Undet.** | 2.32 | 3.28 | 14.32 | 13.89 | -2.89 |
| **Total, %** | 100.00 | 100.00 | 100.00 | 100.00 | 100.00 |
| **$Na_2O+K_2O$, kg/GJ** | .56 | .58 | 1.66 | 1.22 | 1.62 |

Biomass plant experience with deposits has been related to alkali concentrations in the fuel. Some plants found that alkali concentrations in the range of 0 to 0.17 kg/GJ (0.4 lb./MMBtu) are manageable, 0.17 to 0.34 kg/GJ (0.4 to 0.8 lb./MMBtu) questionable and 0.34 kg/GJ (0.8 lb/MMBtu) and higher unmanageable in conventional biomass boilers. Most wastewood fuels are manageable, having high calcium compared with potassium or silica. Urban wood residues contain high silica and low potassium provided they are processed. They become troublesome when fired with high alkali pits, nuts, hulls and shells. Unmanageable fuels include straws, stalks and 100% pits, nuts and shells like those shown in Table 3.

Knowing the alkali values of each fuel it is possible and sometimes practical to blend fuels with some confidence and adjust according to experience, although blending may not always result in a better fuel. Some blends in Table 2 are high in alkali and silica and contain sulfur, chlorine and phosphorus. Deposits result from the interaction of alkali with other elements such sulfur, phosphorus, and iron. Chlorine is not found in slags but has serious corrosion effects in convective areas where it deposits as chlorides and it is responsible for the release of alkali from the fuel. (French et al., 1994)

However, the slagging tendency of a boiler cannot be anticipated in terms of fuel properties alone. While the tendency to slag generally increases with increasing alkali content, the form of the alkali and other inorganic constituents as well as boiler operating conditions and boiler design have large impacts on deposit properties. This project delineated the major variables causing slagging. A conceptual description of deposit generation was proposed and both field and laboratory data were gathered to test it. The description remains to be formalized in the form of a computer based program that incorporates fuel properties, boiler operation, and boiler design in predictions of ash deposit properties throughout the boiler.

## DEPOSITS FROM FUEL TESTS

Eight field tests were conducted at five host power plants operated by industrial sponsors. Plants ranged in capacity from 5 MWe to 25 MWe. Boilers included spreader stokers with traveling grates, circulating fluidized beds, and bubbling fluidized beds. Five boilers were in California (Marysville, Woodland, Mendota, Delano, El Centro), two were in Denmark (Haslev, Slagelse) and one in England. Samples and plant data were collected by project personnel. All deposit analyses were performed by Hazen Research Inc. Additional studies were conducted by Sandia National Laboratories and the Bureau of Mines.

Laboratory tests simulating boiler conditions were also conducted in the Sandia Multi Fuel Combustor (MFC) on two of the field fuels and eleven representative individual fuels taken from the field tests. (Baxter, 1993a, Jenkins et al., 1994)

## Deposits From Highly Alkaline Fuels

Deposits from highly alkaline fuels were observed with straw fired in stokers with traveling grates at Haslev, Slagelse, El Centro, and in the suspension fired Multi Fuel Combustor. The Danish stokers push straw gently onto a cooled grate. El Centro is a spreader stoker with a traveling grate and the Sandia Multi Fuel Combustor burns finely divided fuel in suspension. Reports and analyses were also received from a spreader stoker traveling grate firing manure with wood.

Table 4. Deposits From Straw in a Danish Stoker

| Location | Fuel Ash | Grate Ash | Upper Wall | Super-heater | Dust Collector |
|---|---|---|---|---|---|
| Form | | slag | powder | agglom | flyash |
| Composition | | | | | |
| $SiO_2$ | 63.42 | 62.26 | 18.62 | 20.92 | 8.64 |
| $Al_2O_3$ | 1.95 | 1.94 | 1.12 | 4.20 | 1.41 |
| $TiO_2$ | .02 | .07 | .02 | .03 | .02 |
| $Fe_2O_3$ | .66 | .48 | .32 | * | 2.80 |
| CaO | 4.20 | 10.59 | 14.41 | 16.45 | 5.82 |
| MgO | .46 | 2.15 | 2.45 | 1.38 | .56 |
| $Na_2O$ | .83 | .47 | .47 | 1.31 | .79 |
| $K_2O$ | 13.10 | 17.70 | 33.40 | 40.13 | 49.81 |
| $SO_3$ | 1.95 | .04 | 8.67 | 10.90 | 8.86 |
| $P_2O_5$ | 4.96 | 3.74 | 3.46 | 5.41 | 2.12 |
| $CO_2$ | | .12 | .30 | .44 | .64 |
| Undet. | 8.45 | .44 | 16.76 | -1.17 | 18.53 |
| Total, % | 100.00 | 100.00 | 100.00 | 100.00 | 100.00 |
| Chlorine % | 3.40 | .04 | 15.20 | 22.85 | 21.80 |

* Iron in shot used for cleaning removed.

Table 5. Deposits From 80% Wood-20% Straw(Spreader Stoker)

| Location | Fuel Ash | Grate | Stoker brick | Upper wall | Super-heater |
|---|---|---|---|---|---|
| Form | blend | slag | fused | powder | wedge |
| Composition | | | | | |
| $SiO_2$ | 57.58 | 60.75 | 61.75 | 5.41 | 33.77 |
| $Al_2O_3$ | 10.16 | 10.72 | 11.65 | 1.63 | 9.47 |
| $TiO_2$ | .48 | 0.56 | .54 | .07 | .50 |
| $Fe_2O_3$ | 3.98 | 3.79 | 3.81 | 2.74 | 3.57 |
| CaO | 11.29 | 11.25 | 10.89 | 4.97 | 14.68 |
| MgO | 2.96 | 3.23 | 2.83 | 1.26 | 3.79 |
| $Na_2O$ | 3.04 | 2.32 | 2.50 | 9.05 | 4.09 |
| $K_2O$ | 6.89 | 6.01 | 6.00 | 27.90 | 11.80 |
| $SO_3$ | 2.26 | .13 | .05 | 41.90 | 16.30 |
| $P_2O_5$ | 1.07 | 1.28 | 1.08 | .76 | 1.12 |
| $CO_2$ | | | | | |
| Undet. | .29 | -.04 | -1.10 | 4.31 | .91 |
| Total | 100.00 | 100.00 | 100.00 | 100.00 | 100.00 |
| Chlorine % | 2.24 | .01 | .01 | .21 | .29 |

Deposits on or near the grate or near high temperature flames were either glassy fused coatings or agglomerates of ash bonded together with a glassy material. Deposits on walls parallel to gas flow were white crusty reflective ash or, where flyash impinged on furnace walls, filamentous accumulations of loosely bonded flyash. Deposits on superheater tubes perpendicular to flow were mostly granular, extending in wedge shapes from the tube surface into the gas stream. Flyash that collected on air heaters, economizers and dust collectors was granular when wood was fired with straw and powder in the Danish straw plants. Corrosion was evident in the dust collection systems of the straw plants.

## Blends Of Wood And Agricultural Residues

Deposits from firing blended wood and agricultural residues were observed with urban wood waste and straw, prunings and pits, nuts and shells in: bubbling fluidized beds, circulating fluidized beds, in semi suspension in a spreader stoker and in the Multi Fuel Combustor.

There were fewer fireside deposits when firing wood blends. Most of the plants have learned how to avoid agglomerates by adjusting fuels and operating conditions. Refractory was coated with slag in spreader stokers, but the furnace portion of the fluidized beds was relatively clean. Deposits accumulated on feed tubes and other hot protuberances in the furnaces. Deposits on walls parallel to gas flow were coated with varying accumulations of flyash in the stoker fired boilers. Fluidized bed waterwalls were kept relatively clean by the high sand velocities. Granular deposits caked on hot gas ducting and cyclones. Deposits on superheater tubes perpendicular to flow were solid but granular in nature. Corrosion was evident on superheater tubes. Salt crystals appeared in deposits where straw was fired. Flyash collected from air heaters, economizers and dust collectors was sandy or soft where limestone was added.

Tables 5 and 6 show deposits found in the conventioanl traveling grate, fluidized bed and circulating fluidized bed boilers. These are discussed in detail by Jenkins et al. (1994). These boilers all operated at similar bed temperatures and flue gas exit temperatures. Deposits collected from similar temperature zones in differnt boilers had similar compositions. Grate and wall deposits were similar to the fuel ash. Potassium and calcium silicates and sulfates deposited on screen tubes and tertiary superheaters. Chlorides and carbonates appeared in the cooler zones.

Table 6. Deposits From a Bubbing Fluidized Bed Wood-Ag Fuel

| Location | Fuel Ash | Furnace wall | Screen tubes | 3° SH | 2° SH front |
|---|---|---|---|---|---|
| Form | | agglom | wedge | wedge | deposit |
| **Composition** | | | | | |
| $SiO_2$ | 28.27 | 62.76 | 9.43 | 8.97 | 14.53 |
| $Al_2O_3$ | 8.28 | 11.49 | 4.57 | 3.16 | 3.76 |
| $TiO_2$ | .84 | .53 | .32 | .27 | .54 |
| $Fe_2O_3$ | 3.33 | 2.94 | 1.48 | 4.29 | 2.18 |
| CaO | 28.99 | 1.15 | 31.74 | 19.34 | 25.78 |
| MgO | 4.51 | 1.40 | 3.44 | 2.03 | 3.27 |
| $Na_2O$ | 2.81 | 2.26 | 1.83 | 3.14 | 2.91 |
| $K_2O$ | 9.58 | 4.72 | 13.20 | 25.90 | 17.20 |
| $SO_3$ | 2.14 | .36 | 27.70 | 17.90 | 13.00 |
| $P_2O_5$ | 2.95 | 0.74 | 2.35 | 1.40 | 2.33 |
| $CO_2$ | 4.98 | .26 | 1.38 | 2.33 | 4.91 |
| Undet. | 3.32 | 11.39 | 2.56 | 20.24 | 9.59 |
| Total | 100.00 | 100.00 | 100.00 | 100.00 | 100.00 |
| Chlorine % | 1.74 | .13 | 2.99 | 14.20 | 9.74 |

Figure 1 is a SEM cross section of a deposit sample showing a layer of calcium silicate on the surface of the flyash particles, binding adjacent quartz ($SiO_2$) particles . These are individually analyzed by electron microprobe techniques for elemental composition. A glassy or fused layer binding flyash particles was the most common form of deposit found in superheaters. Binders were identified by x-ray diffraction as potassium or calcium aluminum silicates, calcium or potassium sulfates and calcium carbonates.

Figure 1. Eutectic Bonding in Superheater Deposit Wood Ag Blend. Note $CaSiO_2$ bond between $SiO_2$ particles.

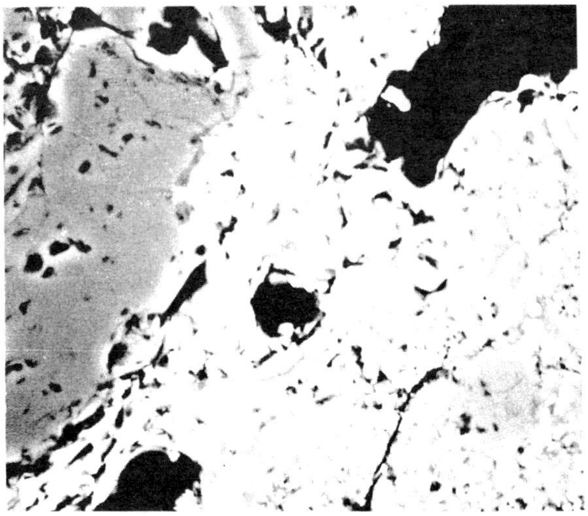

## Laboratory Combustion Tests

Combustion tests in the pilot-scale Multi Fuel Combustor showed that silica from straw absorbed potassium from gases at 900° C . Over time silica nodules deposited on combustor walls changed from a silica to potassium ratio, Si/K, of 30:1 to 4:1. As the deposit absorbed the alkali the amorphous silica changed to glass. (Baxter et al., 1993)

## DEPOSIT FORMATION

Deposits observed in this project are consistent with all known mechanisms for deposit formation (particle impaction, condensation, thermophoresis and chemical reaction). (Baxter et al, 1993)

### Potassium

On the basis of the fuel characterization thus far, biofuels contain compounds which have relatively low melting temperatures. Besides inorganic compounds, biomass also contains potassium in organic form which will vaporize and decompose during combustion to form oxides, hydroxides, chlorides and sulfates, depending on the local composition and residence time of the products of combustion. These compounds all have very low melting temperatures. Their impact on fireside deposits will depend on their vapor pressure and whether they condense homogeneously on tube surfaces creating a fused scale, or heterogeneously on flyash, causing a sticky surface on particles approaching the tube surfaces.

When potassium condenses on a flyash particle it forms a particle whose surface is enriched with potassium. Consequently, the melting temperature or stickiness of the surface will depend upon the rates of condensation and diffusion of potassium into the core of the particle and will be lower than the silica grains originating in the biomass.

Potassium in straw and annual crops is atomically dispersed and intimately mixed with organic structures, as shown by chemical fractionation. Atomically dispersed material includes thenongranular portin of the inorganic material. It occurs as dissolved salts in inherent moisture, cations attached to carboxilic and other functional groups, complex ions, and chemisorbed material. A fraction (≈ 15-40 %) of this potassium volatilizes during combustion. The silicon in the fuel does not appreciably vaporize under these conditions whether from the wood or the straw. When firing the wood-straw mixture in Table 5, for example, wall and grate deposits resembled the chemistry of the wood fuel ash but deposits in the convective passes reflected the volatile potassium from the straw. (Baxter et al., 1993) Alkali vapor migration was apparent in the fireside deposits in all boilers. Wall deposits in the stokers and fluidized beds were enriched in silica and low in potassium compared with flyash or superheater deposits.

### Sulfur

If sulfur is present in the fuel, or in a supplemental fuel such as oil, fouling and slagging will be a problem as sulfates will form. Huppa and Backman (1983) developed a fouling index for firing wood with oil, gas or peat that is based on sulfation of soluble alkali and alkaline earths. Alkali sulfates are unstable at typical combustion temperatures of 900°C (1650° F). During combustion volatilized alkali condenses either directly on cool surfaces or in the thermal boundary layer near such surfaces. At deposit surface temperatures, sulfates are stable and will form as alkali-containing condensate reacts with gas-phase sulfur species. Alkaline earth sulfates (calcium sulfate or anhydrite, for example) are more refractory and may form as calcium oxide particles react with sulfur at higher temperatures. Calcium sulfates will also form in deposits as gas-phase sulfur species react with calcium species. Neither alkali nor alkaline earth species will react with sulfur if they are in the form of silicates. Sulfates have low vapor pressure and will condense out on tube surfaces and flyash. If no sulfur is present, the hydroxides or oxides formed will have very high vapor pressures and will not condense until after they leave the boiler.

It was observed that most woods, which contain high levels of calcium, significant potassium, and low levels of sulfur, cause few deposit problems when burned alone. When burned as a blend with sulfur-containing species, such as manure or an annual crop, deposits become enriched in sulfur as $CaSO_4$ and $K_2SO_4$.

Sulfur capture by calcium was observed both in fixed and fluidized beds. Lime addition in a circulating fluidized bed resulted in calcium and sulfur concentrations in the cooler recycle loop and convection passes.

## Chlorine

Chlorine volatilizes readily during combustion. Stable chlorine-containing vapors generated during combustion include alkali chlorides, and hydrogen chloride. The propensity of chlorine to facilitate alkali vaporization is significant. For example, chlorine is among the few materials that will react readily with alkali in the form of silicates. The reaction produces a gas-phase alkali chloride. These alkali salt vapors are both volatile and stable, and they tend to condense further downstream in a combustor than non-chlorinated alkali vapors.

Experience indicated that chlorine, rather than alkali, can be the limiting reactant in determining the total amount of alkali vapor produced. Deposit formation and tenacity tend to be increased with increasing amounts of vaporized material. Therefore, both chlorine and alkali content are important in predicting deposit properties. It is not uncommon that fuels high in alkali but low in chlorine present less severe deposition problems than fuels with intermediate concentrations of alkaline chlorine.

## Silica

Potassium and sodium, the alkali metals, as oxides, hydroxides or in metallo-organic compounds, will form low melting eutectics with silicates. A phase diagram showing the melting point of various mixtures of potassium oxides ($K_2O$), with silica ($SiO_2$), which makes up the bulk of the ash in biofuels, shows that a mixture of 32% $K_2O$ and 68% $SiO_2$ melts at 769° C (1420° F). Note that $SiO_2$ alone melts at 1700° C (3100° F). This ratio is very close to the 25% -35% alkali ($K_2O$+$Na_2O$) found in many biomass ashes. The addition of alkali to silica is a well known technique for producing commercial glasses. The glass that formed on refractory when firing straw in a spreader stoker reflects this ratio. We found similar values in the agglomerated flyash and grate ash deposits in the test boilers.

Bed deposits in fluidized bed combustors at temperatures of 760° C to 900° C (1400° F to 1650° F ) are also dominated by the silica portion of the fuel. While the gross analyses of bed deposits appear to be mostly silicates, surface enrichment may be present. Chlorine and sulfur are depleted largely because of the relatively high temperatures in this region of the combustor (sulfates decompose and chlorides vaporize at high temperatures).

## DEPOSITION MECHANISMS

Deposition mechanisms are reviewed elsewhere (Baxter, 1993a). They can be divided into two groups: those dealing with gases and those dealing with particles.

### Condensation and Chemical Reaction

Surfaces oriented parallel to the gas flow and subject to deposition by condensation and diffusion of small particles, but free of impaction by large particles, will collect a thin film of potassium sulfate. This is what happened in spreader stokers firing wood with straw, wood with manure, and straw alone. This deposition is unavoidable. One has to allow for the loss in adsorption due to this fouling in the original design, otherwise the plant must add surface or be derated.

Operating a furnace at partial load on a wood-straw fuel demonstrated the formation of sulfur compounds by condensation. Furnace gas exit temperatures were 760° C (1400° F). Reflective white ash deposits appeared in the upper furnace and in convective passes. High sulfur concentrations, as in Table 5, indicated that alkalis were in the form of sulfates. Granular deposits and some agglomerates formed on the walls and on convection tube passes. No molten deposits were observed. The absence of silicon, aluminum, iron or titanium or other refractory compounds indicated that few particles accumulated in the region. In this case it is likely that atomically dispersed potassium in the fuel originated from the straw and formed a vapor during combustion. Since this region of the furnace has low particle impact, deposits are formed primarily through condensation and chemical reaction.

### Particle Impaction and Thermophoresis: Wall And Grate Deposits

Wall deposits are formed from the inorganic portion of the ash. The flyash impacts on the walls and is the major contributor to the total biomass deposit. Table 6.

Since potassium deposited on the flyash surface will make it sticky, deposition will be primarily by inertial impaction of large particles and diffusion of the small particles. Deposition will occur primarily on surfaces immersed in the gas stream perpendicular to flow. Screen tubes and conventional convection surfaces placed in the gas stream at temperatures a few degrees below the melting temperature of eutectic mixtures formed by the condensation of potassium compounds on silicates should be relatively free of fouling. Below 730° C to 760° C (1350° F to 1400° F) fouling is apparently minimal.

At temperatures above the melting point, the surfaces will be subject to fouling and sootblowing will be required. The effectiveness of sootblowing depends on the deposit tenacity which can only be predicted empirically. Without this experience one can only rely upon phase diagrams and one can not accurately predict the surface conditions of the flyash.

### Combinations of Mechanisms

Deposits on slag screens at the entrance to the convection passes exhibit a combination of alkali condensation followed by sulfation and particle impaction as shown in Tables 5 and 6. The composition of these deposits is intermediate between the particle-dominated wall deposit and the white condensate-dominated ceiling deposits.

## BOILER DESIGN AND OPERATION

Deposition of biomass ash is an extremely complex process, making the design and operation of steam generators using biomass as a fuel a very difficult task. Successful utilization of biomass as a source of energy requires some understanding of the physical and chemical behavior of the compounds formed by the inorganics during combustion and the mechanism by which the flyash impacts on heat recovery surfaces. Since biomass is an extremely variable fuel, furnaces should be liberally designed to handle the worst-case scenario.

The formation of a molten phase appears to be the key factor influencing retention of ash impacting on a surface. Since flyash is subjected to a quenching process during heat recovery, the solidification temperature becomes the most important factor in selecting gas temperatures suitable for immersing heat-transfer surfaces. The solidification temperature may be many degrees below the melting temperature of the biomass ash determined in the laboratory since it is dependent upon eutectics formed as a result of condensation of volatile alkalis on silicates. Melts may form at temperatures as low as 760°C (1400° F). To avoid serious fouling, convective heat-transfer surfaces should not be immersed in the flue gas stream perpendicular to the direction of flow at temperatures in excess of ≈732°C (1350° F). Designs meeting this type of specification have been successfully adapted to biomass combustion, as illustrated in Figure 2. To do otherwise would reduce the cost of the steam generator but increase the cost of maintenance and operation. Depending upon the level of alkalis present in the fuel, the problem may or may not be controlled by soot blowing.

Flue gas exit temperatures can be used in combination with fuel properties to develop design criteria for burning low quality, high alkali fuels. Temperature control might be used to reduce the amount of liquid in the deposit or on particles passing convection surfaces. In one particular instance the increase in furnace exit gas temperature from 790° C (1450° F) to 850° C (1560° F) in a stoker-fired steam generator firing wood and straw increased the rate and quantity of deposits. This increase in furnace temperature was accompanied by a rapid accumulation of ash on the grate, and molten deposits on fireside refractory. Deposition in fluidized beds often coincided with local high temperatures caused by combustion of dry fuels or an imbalance in secondary air. Temperature regulation to 850° C (1560° F) or less has reduced deposits and agglomeration in Danish fluidized beds firing straw with coal. (Miles, Jr., 1993)

Corrosion is a companion to alkali deposits, occurring under deposits especially where alkali chlorides or sulfur are present. The Danish plants have found that superheaters with flue gas temperatures above 750° C (1382° F) experience high temperature erosion and corrosion due to reactions between the carbon steel tubes, chlorides and flue gas particles. (Miles, Jr., 1993) A melt of alkaline chlorides with sulfur makes ideal conditions for high temperature corrosion. Since high temperatures are important for power generation, corrosion should be the focus of further alkali deposit investigations.

Fluidized beds offer a better opportunity to utilize inhibitors such as magnesium (MgO) because the material is intimately mixed as combustion proceeds. Plants have preferred the low cost lime, limestone ($CaCO_3$) or dolomite ($CaCO_3.MgCO_3$), which are added for sulfur emission control, to also prevent bed agglomeration. In some cases high alumina (25% to 35% $Al_2O_3$) bed media such as kaolin, local clays or mullite have been used where erosion can be controlled. Gasification of high alkali fuels at lower temperatures in auxiliary fluidized beds with inhibitors and burning the gas over the existing high temperature beds or grates offers a possible practical solution .

## FUEL SAMPLING AND CHARACTERIZATION

### Fuel Sampling

Many fuel and ash analyses provided by participating plants at the outset of the project were of limited use due to improper sampling and inappropriate methods of analysis. Good sampling is especially important because of the natural variability of biomass and the apparent variation that can result from concentrations of fines in the fuel. Biomass power plants that use systematic sampling and analysis have improved their plant operation and avoided excessive deposits. The project used the guidelines of the Biomass Boiler Test Procedure proposed by the TAPPI steam and power committee in order to define sampling methods and use consistent procedures. (TAPPI, 1992) This procedure also specifies methods of analysis for fuel size and ash chemistry.

### Fuel Characterization

In this study many analytical methods were used to further understand the mechanisms of deposit formation and composition. Fuel analyses contributed by participating plants and those sampled during the project were compiled into a reference database. When these analyses were used to interpret

Figure 2. Straw Fired Danish Boiler at Haslev

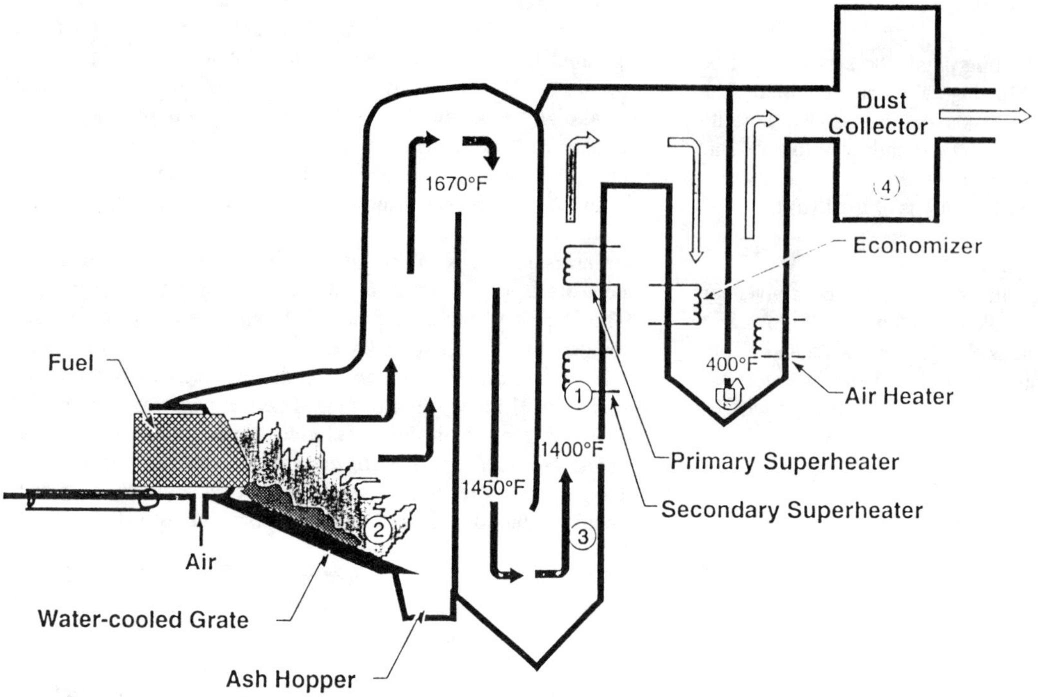


deposit behavior it became evident that a wide variety of methods are used. A common set of analytical methods should be selectede and some methods need to be added to routine fuel analysis.

Particle size as-fired should be included in the analysis. Fines (<10 mesh or <1.7 mm) contain most of the alkali and ash in urban wood waste and should be measured to determine if screening is necessary.

A variety of methods are used for proximate, ultimate and elemental analyses. While several methods are similar some significant differences exist that can produce misleading results. Oxygen should be measured directly. Most laboratories calculate oxygen as the difference between the total sample mass and that of the other major elements. But it is clear that other elements such as chlorine may distort the oxygen value. Chlorine should be a normal part of fuel analysis because of its importance in alkali release.

Samples should be ashed at 600° C with no calcining. (TAPPI, 1992) High temperature (750° C) ashing vaporizes much of the alkali as does prolonged calcining. For example, coal samples are ashed at 750°C and calcined at 1000° C. Measurements of potassium in biofules were found to be 15% to 25% lower when they were ashed at these temperatures.

Ash fusion temperatures are not useful for predicting deposit behavior in biomass for several reasons. First of all, the ASME ash fusion temperature does not simulate melting temperatures resulting from the condensation of alkalis volatilized during combustion on silicates. Secondly, substantial alkalis are lost during the ashing process in preparing pyrometric cones for ash fusion temperatures, thereby artificially raising the measured fusion temperature.

To characterize the fuel it is important to know that potassium is apparently tied up as an organically bound element or in some other form which is volatilized during combustion. Sodium, on the other hand, may appear in various forms since it is not required for plant growth. It is usually present at much lower concentrations, unless available from adventitious materials such as silicates.

Various methods were tested to determine stickiness in biofuels. The stickiness of the flyash in a sulfur/chlorine free fuel can be predicted by DTA or similar techniques in the laboratory, in which case only the endotherms above 950° C - 1000° C may apply. If sulfur is present in any of the fuels being fired, the melting temperatures cannot be predicted by laboratory characterization since they are dependent upon post-combustion reactions and physical state changes. In this case one must know how the elements occur in the fuel and how they react in the boiler.

Three methods of chemical fractionation was used on several fuels in this study to determine what portion of the fuel is readily volatilized during combustion. All methods showed that the major portion of phosphorus, sulfur and potassium in biofuels are water soluble or ion exchangeable and therefore reactive in combustion as shown in Table 5. These results distinguish biofuels from coal in which potassium may be bound in clays such as illite where it is less reactive.

Analysis of water soluble alkali in the fuel, or wet digestion followed by mineral analysis, resulted in more reliable alkali data than thermal ashing methods. A wet ashing method that developed by the Bureau of Mines provided a better account of all elements in the sample compared with standard thermal methods used for coal.

## CONCLUSIONS

This investigation did not discover a "silver bullet" to allow burning 100% agricultural residues in biofuels in conventional boilers. However, it did identify the extent and nature of alkaline deposits. Observation and investigation extended to several field trials. An enhanced understanding has been gained of the physical and chemical mechanisms that produce deposits.

Several mechanisms of deposit formation have been observed with biofuels. Furnaces function as fractionators, producing deposits of varying chemical composition and depositing them in distinctly separate sections of the boiler by combinations of the deposition mechanisms: condensation, inertial impaction, thermophoresis and chemical reaction. Condensation of alkali on flyash and deposit accumulation by inertial impaction appear to be more important than initially expected for biomass deposit formation.

Some general guidelines may be developed for the relationship between fuels and deposits which could be used by industry to: avoid deposits with existing boiler designs; develop techniques for screening fuels; and establish guidelines for fuel blends in a particular plant. Further validation of the mechanisms interpreted from deposit behavior would enable operators to increase boiler availability. In the future, investigators need to measure physical deposit properties such as emissivity, heat transfer rates and coefficients and tenacity, and correlate them with observed mechanisms like sulfation.

A database of fuels has been assembled, including chemical analyses, physical characterization and other pertinent data. Virtually all annual crop residues are too high in potassium to be fired in conventional boilers except in minimal amounts when fired with woody fuels or when leached. Better characterization is needed for more fuels using consistent methods. Levels of alkali and other elements that cause problems can be better identified. The "alkali index" is a useful guide to the tendency of a fuel to slag or foul, especially above about 0.6 kg/GJ (1.5 lb. alkali/MMBtu), but it is not absolute and doesn't account for furnace or boiler conditions. Better measurement of operating conditions is needed to correlate with fuel and deposit characteristics.

New furnaces should be designed with ample surface and with deposits in mind. Gasification or pyrolysis should be tested as alternatives for alkali removal. Alkali behavior and fuel properties should be related to boiler operation and design such as flue gas temperatures. Fuel selection and boiler optimization could result in decreased maintenance and increased availability. A method of prediction for deposit formation based on fuel characteristics and furnace conditions should be developed. This may take the form of a mathematical algorithm expressed in a computer program.

This alkali deposit investigation could contribute to advances in corrosion studies. Corrosion was not within the scope of this deposit work but it is a common problem reported by participating plants and it is recognized that the two are inseparable.

## ACKNOWLEDGMENTS

The authors acknowledge the support and participation by Richard Bain and Ralph Overend of the National Renewable Energy Laboratory (NREL subcontract TZ-2-11226-1) and in particular the patience and cooperation of the industrial sponsors.

The generosity of UC Davis, Sandia National Laboratories and U.S Bureau of Mines in providing scientists as well as laboratory facilities and data has resulted in a new dimension of applied science to the project.

## PROJECT PARTICIPANTS

In August of 1992 a grant of $150,0000 was provided by NREL with the hope of enlisting contributing sponsors from industry. Eventually seven power plant sponsors (one from Denmark) joined the project representing nine power plants: Delano Energy Company, Inc., Woodland Biomass Power, Ltd., and Mendota Biomass Power of Thermo Electron Energy Systems; Hydra-Co Operations; Sithe Energies; Wheelabrator Environmental Systems, Inc.; and Elkraft Power Company, Ltd., with Electric Power Research Institute, Foster Wheeler Corporation and National Wood Energy Association.

Thomas R. Miles, P.E., provided a design engineer's perspective and with Thomas R. Miles, Jr. coordinated the project, locating industry and expert participants and arranging meetings and tests at labs and operating plants.

Dr. Laurance Oden, Bureau of Mines Research Laboratory, Albany, Oregon provided x-ray diffraction (XRD), scanning electron microscopy (SEM) and mineralogical analysis. Dr. Larry Baxter, Combustion Research Facility, Sandia National Laboratories, contributed his experience from coal research, conducted Multi Fuel Combustor tests, chemical fractionation and deposit interpretation as part of a parallel NREL project. Mr. Richard Bryers, Foster Wheeler Corporation, contributed analysis and interpretation of fuels and deposits including low temperature ashing and DTA analysis. Dr. Bryan Jenkins, Biological and Agricultural Engineering Department of the

University of California at Davis, contributed his long experience with biofuels, coordinated field tests, assisted in Multi Fuel Combustor tests, and analysis and interpretation of deposits. Mike Jones, Sithe Industries, was the project field coordinator for plant tests.

Other advisors to the project were George Wiltsee, Appel Consultants; Jerry Radway, EnerChem, Inc.; and Al Duzy as well as many others. Thus the project included a consortium of industry, science and engineering.

## REFERENCES


Baxter, L. L., 1993a, "Ash Deposition During Biomass And Coal Combustion: A Mechanistic Approach," *Biomass and Bioenergy*, Pergamon Press Ltd., Great Britain. Vol. 4, No. 2, pp. 85-102 .

Baxter, L. L. 1993b, "Biomass Combustion for Electric Power Generation," presented at NREL, August 27, 1993.

Baxter, L. L., Miles, T. R., Miles, T. R. Jr., Jenkins, B. M., Richards, G. R. and Oden, L. L. , 1993, "Transformations and Deposition of Inorganic Material in Biomass Boilers". In M. G. Carvalho ed., *Second International Conference on Combustion Technologies for a Clean Environment*, 1 (pp. Biomass II: 9-15). Lisbon, Portugal: Commission of European Communities.

Bryers, R. W., 1992, "Fireside Behaviour of Mineral Impurities in Fuels From Marchwood 1963 to the Sheraton Palm Coast 1992." In S.A. Benson ed., *Inorganic Transformations And Ash Deposition During Combustion*, American Society of Mechanical Engineers, New York, pp. 3-68.

French, R. J., Dayton, D. C., Milne, T. A., 1994, "The Direct Observation of Alkali Vapor Species in Biomass Combustion and Gasification," Report NREL/TP-430-5597, National Renewable Energy Laboratory, Golden, Colorado, January.

Huppa, M., Backman, R., 1983, "Slagging and Fouling During Combined Burning of Bark and Oil, Gas or Peat," in R.W. Bryers, ed., *Fouling of Heat Exchanger Surfaces*, Engineering Foundation, 345 East 47th, New York, NY 10017, pp. 419-432.

Jenkins, B. M., L. L. Baxter, T. R. Miles, T. R. Miles, Jr., L. L. Oden, R. W. Bryers, E. Winther, 1994, "Composition of Ash Deposits in Biomass Fueld Boilers: Results of Full-Scale Experiments and Laboratory Simulations," Paper No. 946007 presented at the 1994 International Summer Meeting, June 19-24, Kansas City, ASAE, 2950 Niles Rd., St. Joseph, MI 49085-9659.

Marschner, H., 1986, *Mineral Nutrition of Higher Plants.* Academic Press, New York.

Miles, T.R., 1992, "Operating Experience With Ash Deposition in Biomass Combustion Systems," Presented to the Biomass Combustion Conference, Reno, Nevada. National Renewable Energy Laboratory, Golden, Colorado.

Miles, T.R. 1993, "Alkali Deposits in Biomass Power Plant Boilers", Proceedings of the conference *Strategic Benefits of Biomass and Waste Fuels*, March 30-April 1, 1993. Electric Power Research Institute, TR-103146, Project 3295-02, December, pp.4-1-4-5.

Miles, T. R., Miles, T.R., Jr., Baxter, L. L., Jenkins, B. M., and Oden, L. L., 1993, "Alkali Slagging Problems With Biomass Fuels," Proceedings of *the First Biomass Conference of the Americas*, August 30-September 2, 1993, Burlington, Vermont. National Renewable Energy Laboratory, Golden, Colorado. NREL/CP-200-5768, DE93010050. Vol 1, pp.406-421.

Miles, T. R., Jr., 1993, "Alkali Deposits in European Straw Boilers," Report to National Renewable Energy Laboratory, August 15, 1993. NREL subcontract TZ-2-11226-1, Alkali Deposit Investigation.

Mylchreest, D., Butler, J., 1991, "Design and Operating Considerations for Biomass Fired CFB's." Presented at: CIBO Fluid Bed II, Indianapolis, Indiana, Council of Industrial Boiler Owners, Burke, Virginia.

Technical Association of the Pulp and Paper Industry, 1992, "TAPPI CA-4967: Biomass Boiler Test Procedure," issued September 30, 1991. Prepared by Power Boiler Subcommittee of the Steam and Power Committee of the TAPPI Engineering Division.

FACT-Vol. 18, Combustion Modeling, Scaling and Air Toxins
ASME 1994

# GAS REBURNING IN TANGENTIALLY-FIRED, WALL-FIRED AND CYCLONE-FIRED BOILERS

**T. James May**
Illinois Power Co.
Decatur, Illinois

**Eric G. Rindahl**
Public Service Co. of Colorado
Denver, Colorado

**Thomas Booker**
City Water Light & Power
Springfield, Illinois

**James C. Opatrny, Robert T. Keen, Max E. Light, Anupam Sanyal, Todd M. Sommer and Blair A. Folsom**
Energy and Environmental Research Corp.
Chicago, Illinois

**John M. Pratapas**
Gas Research Institute
Chicago, Illinois

## ABSTRACT

Gas Reburning has been successfully demonstrated for over 4,428 hours on three coal fired utility boilers as of March 31, 1994. Typically, $NO_x$ reductions have been above 60% in long-term, load-following operation. The thermal performance of the boilers has been virtually unaffected by Gas Reburning.

At Illinois Power's Hennepin Station, Gas Reburning in a 71 MWe tangentially-fired boiler achieved an average $NO_x$ reduction of 67% from the original baseline $NO_x$ level of 0.75 lb $NO_x/10^6$ Btu (323 mg/MJ) over a one year period. The nominal natural gas input was 18% of total heat input. Even at 10% gas heat input, $NO_x$ reduction of 55% was achieved. The demonstration project at Hennepin has been completed and Illinois Power has elected to retain the Gas Reburning system for potential $NO_x$ compliance.

At Public Service Company of Colorado's Cherokee Station, a Gas Reburning-Low $NO_x$ Burner system on a 172 MWe wall-fired boiler has achieved overall $NO_x$ reductions of 60-73% in parametric and long-term testing, based on the original baseline $NO_x$ level of 0.73 lb/$10^6$ Btu (314 mg/MJ). $NO_x$ reduction is as high as 60-65% even at relatively low natural gas usage (5-10% of total heat input). The $NO_x$ reduction by Low $NO_x$ Burners alone is typically 30-40%. $NO_x$ reduction has been found to be insensitive to changes in recirculated flue gas (2-7% of total flue gas) injected with natural gas.

At City Water, Light and Power Company's Lakeside Station in Springfield, Illinois, Gas Reburning in a 33 MWe cyclone-fired boiler has achieved an average $NO_x$ reduction of 66% (range 52-77%) at gas heat inputs of 20-26% in long-term testing, based on a baseline $NO_x$ level of 1.0 lb/$10^6$ Btu (430 mg/MJ).

This paper presents a summary of the operating experience at each site and discusses the long term impacts of applying this technology to units with tangential, cyclone and wall-fired (with Low $NO_x$ Burner) configurations.

## INTRODUCTION

The Energy and Environmental Research Corporation (EER) is conducting field evaluations of Gas Reburning integrated with Sorbent Injection on two utility boilers and with Low $NO_x$ Burners on a third unit, in two U.S. Department of Energy cofunded Clean Coal Technology (CCT) projects. Gas Reburning-Sorbent Injection (GR-SI) is being demonstrated on tangential and cyclone-fired units, while Gas Reburning-Low $NO_x$ Burners (GR-LNB) is being demonstrated on a wall-fired unit (Table 1):

- Hennepin Station Unit 1, a 71 MWe tangentially-fired unit, owned and operated by Illinois Power Company at Hennepin, Illinois. Long-term GR-SI demonstration testing over a one year period was completed in October 1992. Illinois Power has retained the Gas Reburning system for potential $NO_x$ compliance.
- Lakeside Station Unit 7, a 33 MWe cyclone-fired unit, owned and operated by City Water, Light & Power Company, the municipal utility of the City of Springfield, Illinois. Long-term GR-SI demonstration testing was completed in June 1994.
- Cherokee Station Unit 3, a 172 MWe wall-fired unit, owned and operated by Public Service Company of Colorado in Denver, Colorado. Long-term GR-LNB demonstration testing was initiated in April 1993.

Gas Reburning is the primary focus of this paper. The associated technologies will be briefly discussed.

## GAS REBURNING PROCESS

Gas Reburning has been evaluated over the past two decades. It is a proven $NO_x$ emission control technology which can be retrofitted to coal-, oil-, or gas-fired boilers. The reburning process, divides the furnace into three zones (exemplified by a wall-fired boiler in Figure 1):

Primary Combustion Zone: Approximately 75 to 85 percent of the heat is released by coal or another primary fuel in this zone, under a slightly reduced excess air level than normal.

Reburning Zone: The reburning fuel (normally 15 to 25 percent of total heat input) is injected above the burners to create a slightly fuel-rich zone in which $NO_x$ is reduced to $N_2$. The reburning fuel may be injected with or without recirculated flue gas (FGR). Natural gas is an effective reburning fuel, but other fuels including coal, fuel oil, coal-water slurry, or coke oven gas may be used. The quantity of reburning fuel can be significantly reduced when the $NO_x$ control requirement is less than 60%.

Burnout Zone: In the third and final zone, additional combustion air (called overfire air or OFA) is added to oxidize any remaining fuel fragments (hydrocarbons, CO, and carbon) and complete the combustion process.

The stoichiometric ratios, the ratios of air to the stoichiometrtic air required for complete combustion, for the three zones are important to Gas Reburning. The typical stoichiometric ratios in the primary ($SR_1$), reburning ($SR_2$), and burnout ($SR_3$) zones are 1.10, 0.90 and 1.15, respectively, when applied to tangential and wall-fired units. Cyclone-fired units are operated with an $SR_1$ of 1.15.

## $NO_x$ REDUCTION

Figure 2 shows that $NO_x$ emissions decrease with increasing reburning fuel input (expressed as a percent of the total heat input) for all three types of utility boilers. For the tangentially-fired boiler and the wall-fired boiler with LNB, $NO_x$ emissions level off between 12 and 22% gas inputs. It will be desirable to use 12% instead of 22% gas input in these cases because natural gas is currently more expensive than coal at these plants. $NO_x$ emissions are higher in cyclone-fired units due to the high heat release in the cyclones, required for molten slag formation, and higher excess air level (minimum of 15 percent) in the Primary Zone. $NO_x$ emissions were optimized with respect to reburning fuel input, zone stoichiometries, and FGR flow, before initiation of the long-term demonstration testing.

Long-term $NO_x$ data for the three types of boilers are shown in Figures 3, 4 and 5 and summarized in Table 2. The average $NO_x$ reductions ranged from 64 to 67% based on "as found" baseline NOx levels, before the installation of Gas Reburning systems. For the tangentially-fired boiler, $NO_x$ reductions were 67% and 55% at 18% and 10% gas heat inputs respectively (Angello, 1992). For the wall-fired boiler with LNB, the $NO_x$ reduction was as high as 60-65% even at relatively low gas heat inputs (5-10%). Reburning accounted for 35 to 42% $NO_x$ reduction from the LNB $NO_x$ level, in this gas heat input range (Hong, 1993). For the cyclone-fired boiler, $NO_x$ reductions were 67% and 50% at 25% and 15% gas heat inputs respectively (Keen, 1993). Variations in $NO_x$ reductions over the long-term demonstration periods are due to some fluctuations in reburning fuel inputs, zone stoichiometries and other operating parameters during these periods.

The Phase I Title IV $NO_x$ regulatory limits for tangentially-fired and wall-fired boilers are 0.45 and 0.50 lb/$10^6$ Btu (or 194 and 215 mg/MJ), respectively, on an annual average. Any over compliance can be averaged with other affected boilers or the compliance level can be met by lowering the gas input.

## CO EMISSIONS AND CARBON LOSSES

As shown in Figures 6 and 7, Gas Reburning reduces both CO emissions and carbon losses from the wall-fired boiler from levels measured under LNB operation, with or without OFA. Low $NO_x$ Burners stage the mixing of coal and air resulting in a reduction in combustion completion, as evident in high CO emissions and carbon-in-ash. The OFA input under Gas Reburning operation effectively mixes with furnace gas resulting in burnout of combustible matter. Gas Reburning also results in an increase in the upper furnace gas temperature, which promotes combustible matter burnout. While OFA is effective in CO emissions reduction, its input must also be optimized with respect to the requirements of the Gas Reburning process (i.e. zone stoichiometries) and thermal efficiency. The Low $NO_x$ Burners on the wall-fired unit were modified, later in the test program, to enhance combustion efficiency.

## $SO_2$, $CO_2$, OPACITY AND PARTICULATES

Table 3 shows impacts of Gas Reburning on the tangentially-fired boiler emissions. $SO_2$ and $CO_2$ were reduced by 18% and 8%, respectively, at 18% gas input. These are due to differences in fuel qualities; natural gas has no sulfur and a lower carbon to hydrogen ratio. There was essentially no change in opacity or particulate emissions, which were maintained below 0.04 lb/$10^6$ Btu (17 mg/MJ). Gas Reburning operation at the other sites had the same impacts on $SO_2$ and $CO_2$ emissions and stack opacity. A positive effect on particulate emissions is due to a reduction in particulate loadings into the electrostatic precipitator, in proportion to the percent gas heat input.

## BOILER PERFORMANCE IMPACTS

Table 4 shows that Gas Reburning has minimal or no impacts on thermal performance or durability of the tangentially-fired boiler. The thermal efficiency and steam temperature showed only minor changes compared to baseline results. A similar trend has been observed for the thermal performance of the wall-fired boiler (Table 5) and of the cyclone-fired boiler (Figure 8). Gas Reburning had positive impacts on the thermal performance of the wall-fired unit, by increasing the main and reheat steam temperatures and reducing the unit net heat rate. The durability impacts of Gas Reburning on the wall-fired and cyclone-fired boilers will be evaluated upon completion of long-term demonstration testing.

## GAS REBURNING EXPERIENCE HIGHLIGHTS

$NO_x$ level can be varied according to the regulatory requirement by changing the gas input. $NO_x$ emissions averaged from 0.245 to 0.344 lb/$10^6$ Btu (105 to 148 mg/MJ) over the demonstration periods at the three sites. These levels were achieved with average gas heat inputs of 13% (wall-fired), 18% (tangentially-fired), and 23% (cyclone-fired). Since these average emissions are significantly below compliance limits, reduced amounts of gas input may be used to achieve $NO_x$ compliance levels.

The gas input can be decreased from the design point for maximum $NO_x$ reduction without substantial loss in $NO_x$ control. A leveling off in $NO_x$ emissions was evident in the 10 to 20 percent range for the wall and tangentially-fired units.

Gas Reburning can achieve approximately the same $NO_x$ control with and without LNB. $NO_x$ reductions for the wall-fired unit averaged 64% compared to 67% at the tangentially-fired unit. This is due partly to the difference in average gas heat input, but also to a difference in reburning process efficiency when the primary $NO_x$ level is decreased (as from LNBs).

Reciruculated flue gas, injected with natural gas, has a minor impact on $NO_x$ reduction.

CO level can be minimized by optimizing OFA flow in Gas Reburning. This was especially evident in results from the wall-fired unit.

## GAS REBURNING APPLICABILITY

Gas Reburning can be applied to tangentially-fired, wall-fired and cyclone-fired boilers. When more stringent $NO_x$ limits are promulgated by January 1, 1997, Gas Reburning or technologies integrated with Gas Reburning or Selective Catalytic Reduction (SCR) can be applied for all types of boilers (Sanyal, 1993). The integrated technologies include:

- Advanced Gas Reburning (AGR): Gas Reburning (GR) and Selective Non-Catalytic Reduction (SNCR)
- Combi$NO_x$: Gas Reburning (GR) + SNCR + Methanol Injection

These technologies are at various scales of demonstration (pilot to full-scale). Test results indicate the following levels of $NO_x$ control may be achieved: GR and GR-LNB (50 - 70%), AGR (80 - 85%), and Combi$NO_x$ (90%). Therefore, the gas technology to be implemented depends on the required level of $NO_x$ control.

## CONCLUSIONS

The following conclusions are drawn from the three Gas Reburning demonstrations on different types of boilers evaluated over 4,400 hours:

- Effective $NO_x$ control (>60%)
- Consistent and reliable operation
- Minor thermal impacts
- Simultaneous $SO_2$ and $CO_2$ control due to gas firing

## ACKNOWLEDGEMENT

The three Gas Reburning demonstrations are sponsored by:

- U.S. Department of Energy
- Gas Research Institute
- Illinois State Department of Energy and Natural Resources
- Electric Power Research Institute
- Colorado Interstate Gas Company
- Illinois Power Company
- City Water, Light and Power Company of Springfield, Illinois
- Public Service Company of Colorado
- Energy and Environmental Research Corporation

## REFERENCES

Angello, L. C., Engelhardt, D. A., Folsom, B. A., Opatrny, J. C., Sommer, T. M., and Ritz, H. J., "Gas Reburning-Sorbent Injection Demonstration Results," presented at the U.S. Department of Energy First Annual Clean Coal Technology Conference, Cleveland, Ohio September 22-23, 1992.

Hong, C. C., Light, J. M., Moser, H. M., Sanyal, A., Sommer, T. M., Folsom, B. A., Payne, R., Ritz, H. J., "Gas Reburning and Low $NO_x$ Burners on a Wall-Fired Boiler," Second Annual Clean Coal Technology Conference, Atlanta, Georgia, September 7-9, 1993.

Keen, R. T., Hong, C. C., Opatrny, J. C., Sommer, T. M., Folsom, B. A., Payne, R., Ritz, H. J., Pratapas, J. M., May, T. J., and Krueger, M. S., "Enhancing the Use of Coal by Gas Reburning and Sorbent Injection," Second Annual Clean Coal Technology Conference, Atlanta, Georgia, September 7-9, 1993.

Sanyal, A., Sommer, T. M., Hong, C. C., Folsom, B. A., Payne R., Seeker, W. R., and Ritz, H. J., "Advanced $NO_x$ Control Technologies," Tenth Annual International Pittsburgh Coal Conference, September 20-24, 1993.

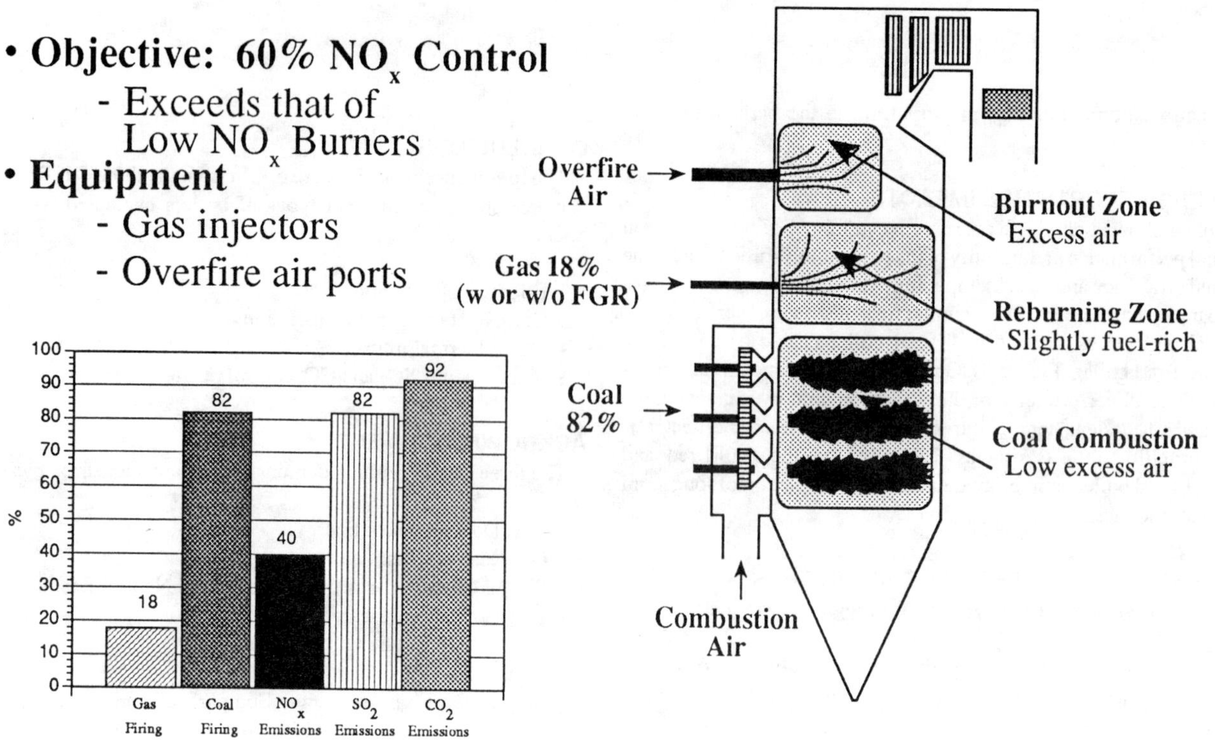


Figure 1. Gas Reburning process.

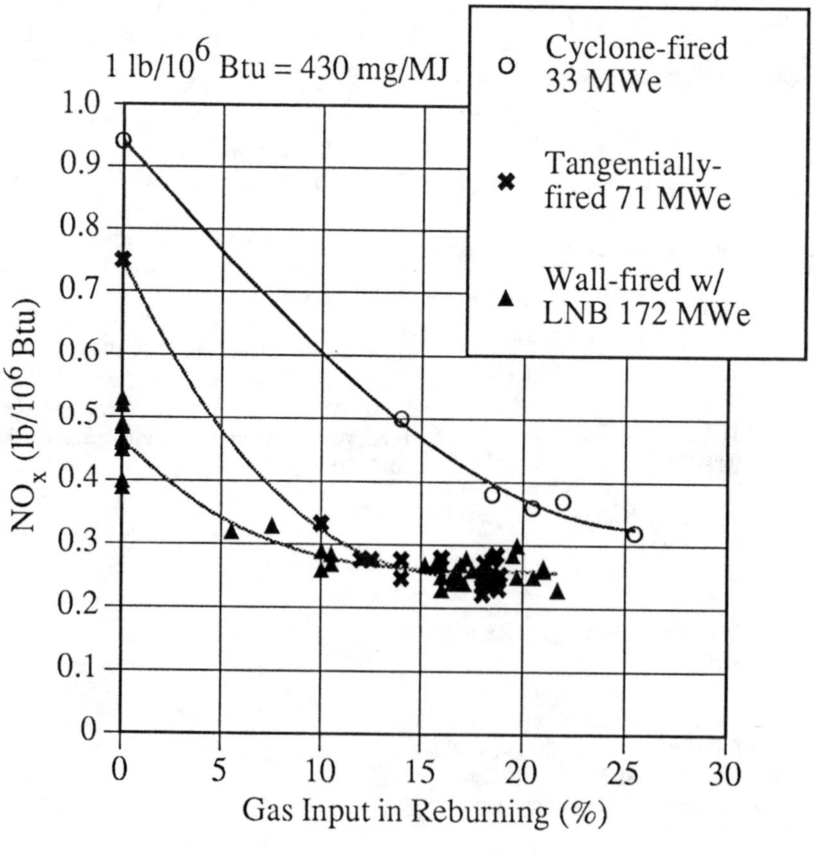


Figure 2. Gas Reburning in three utility boilers.

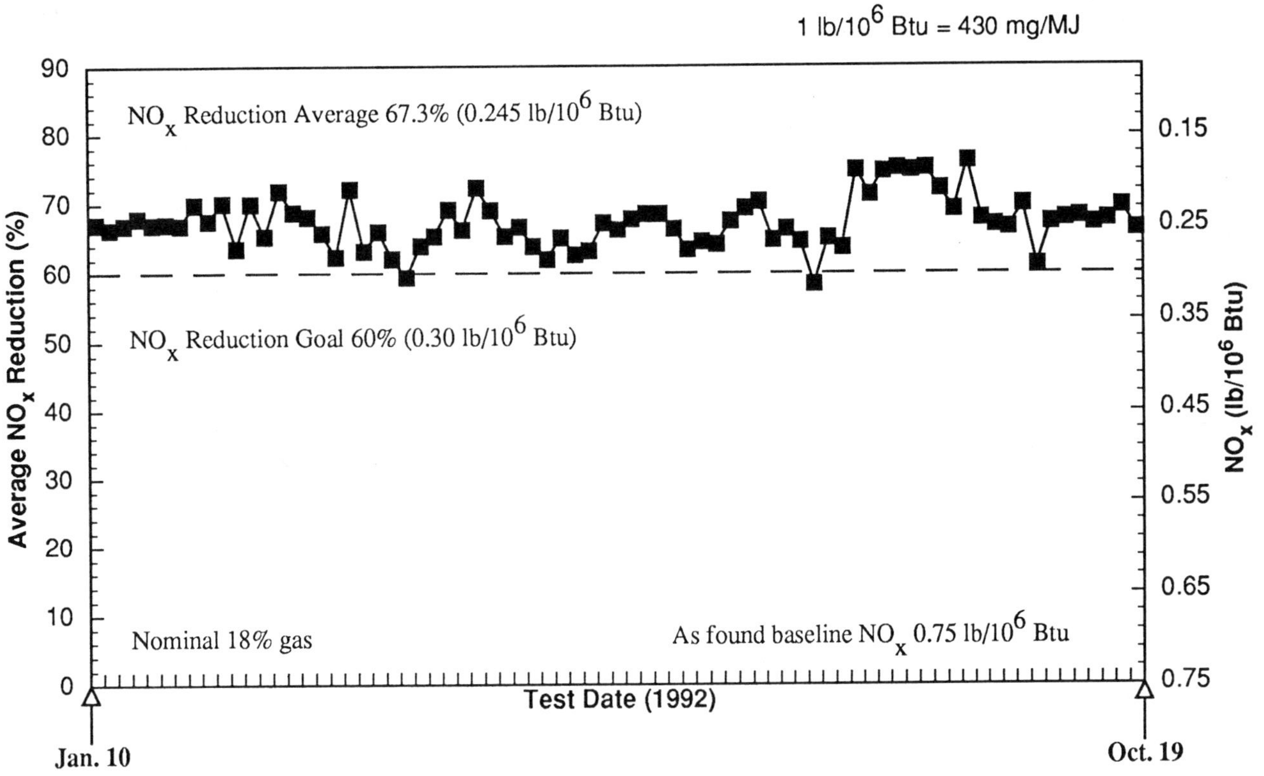


Figure 3. Long term $NO_x$ data for the Tangentially fired boiler.

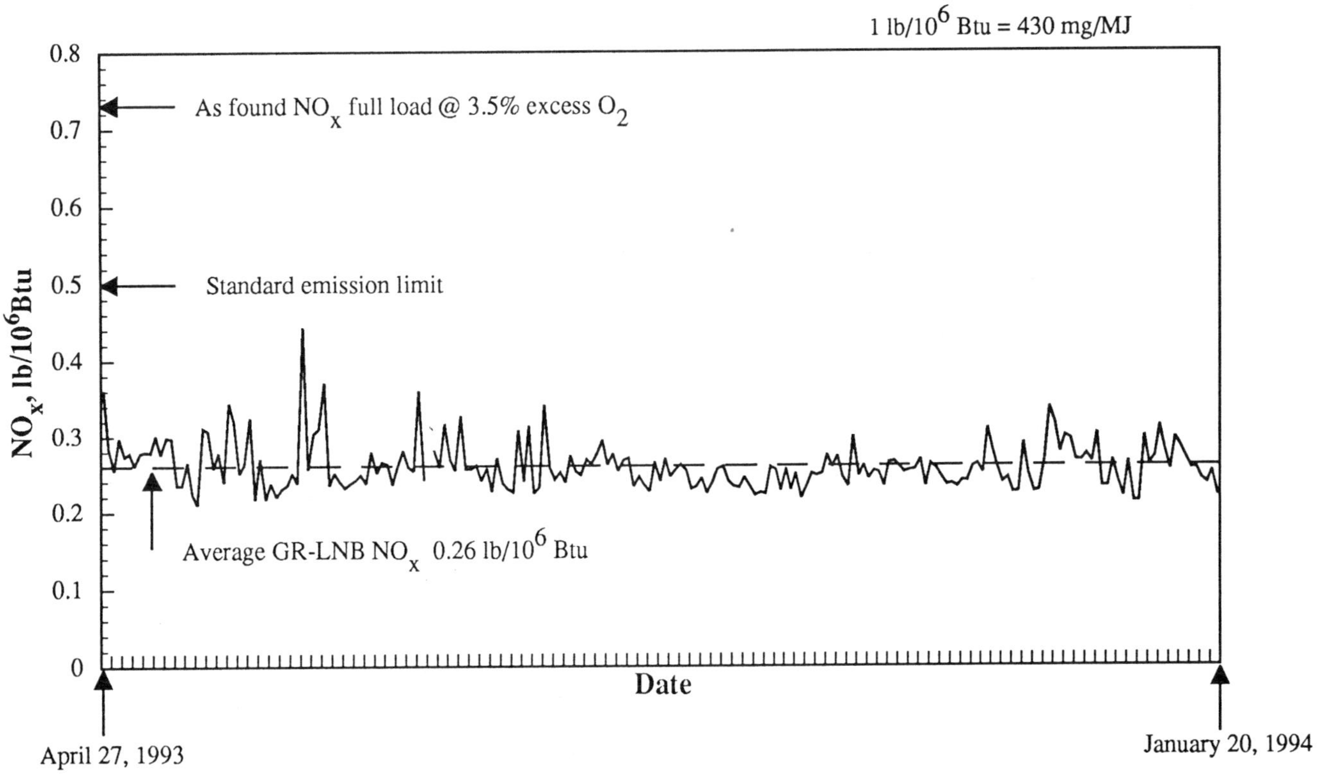


Figure 4. Long term $NO_x$ data for the wall fired boiler with low $NO_x$ burners.

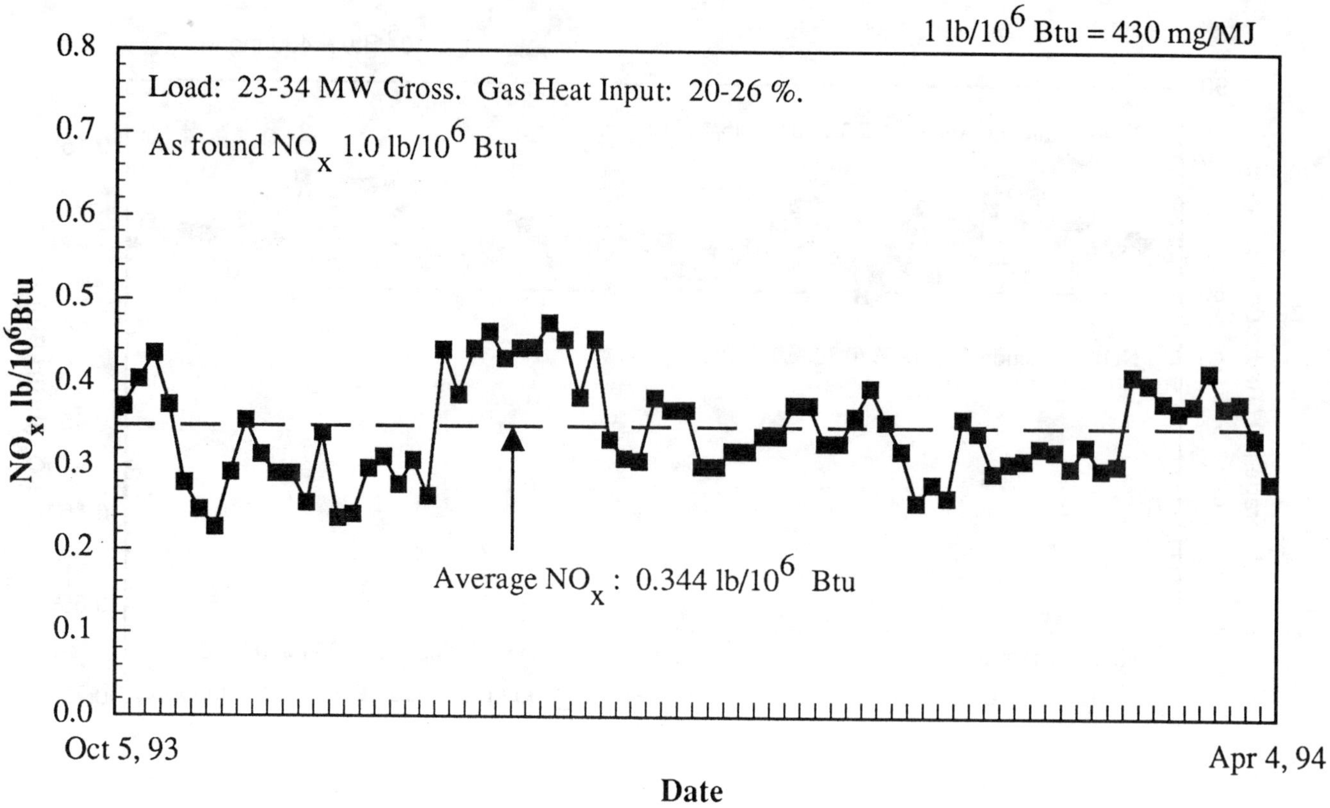


Figure 5. Long-term $NO_x$ data for a cyclone fired boiler.

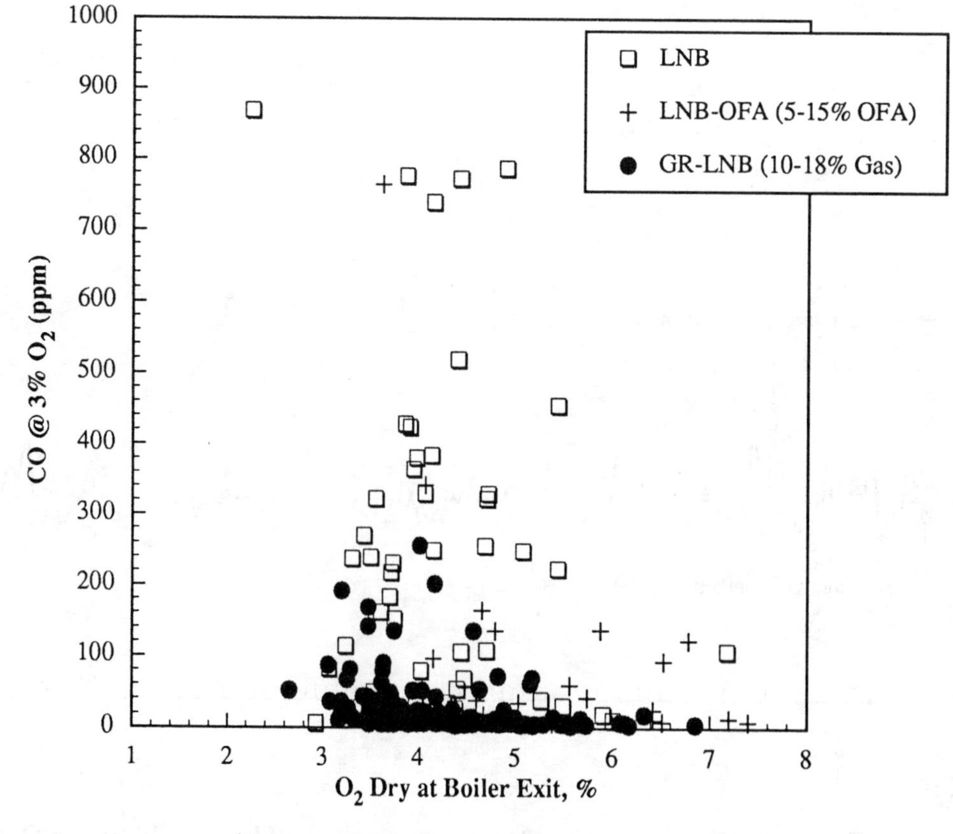


Figure 6. Gas Reburning reduces CO emissions from a wall-fired boiler.

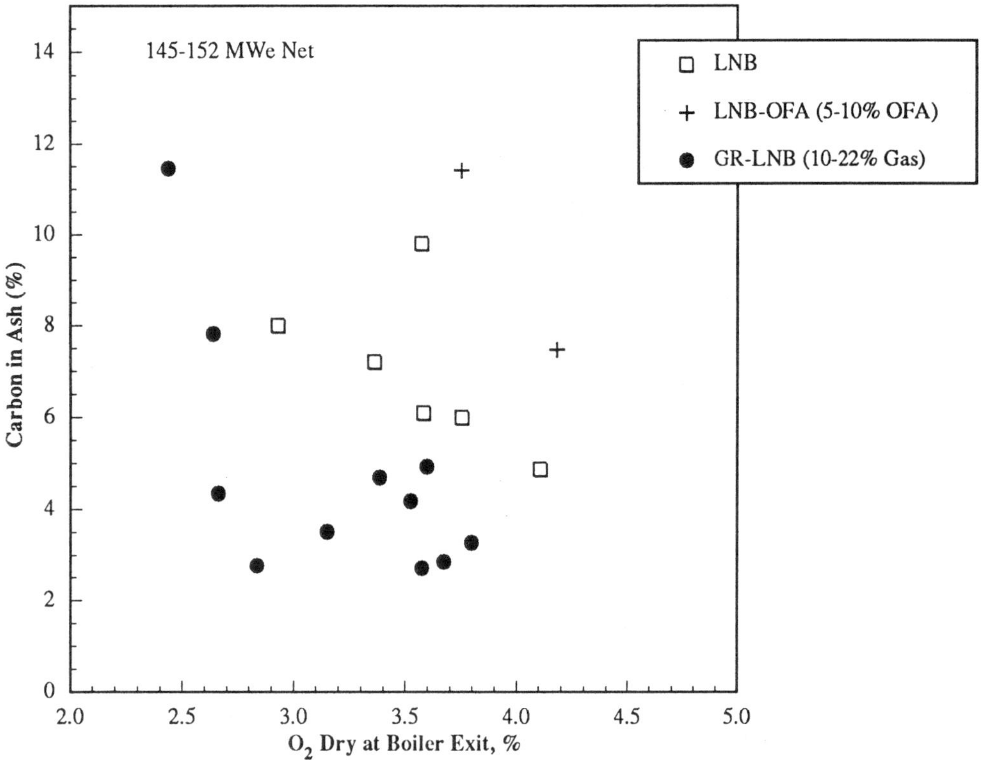


Figure 7. Gas Reburning can reduce carbon loss.

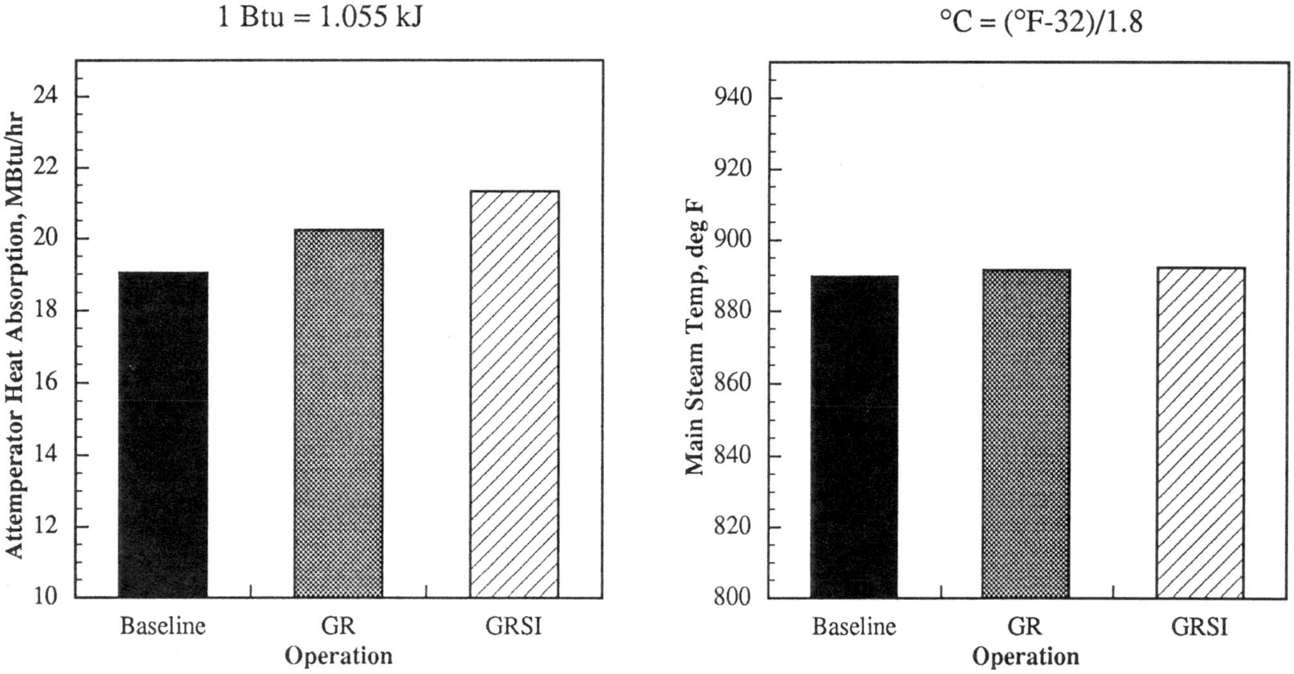


Figure 8. Attemperation impacts from the cyclone-fired boiler.

TABLE 1. GAS REBURNING DEMONSTRATIONS

| Boiler | MWe | Location | Long-Term Testing | Hours |
|---|---|---|---|---|
| Tangentially Fired | 71 | Hennepin, IL | Completed | 1,154 |
| Cyclone | 33 | Springfield, IL | On-going | 255 |
| Wall-Fired | 172 | Denver, CO | On-going | 3,019 |
| | | | | 4,428 |

TABLE 2. LONG-TERM $NO_x$ DATA

| Boiler | % Gas | Average NOx | | Average % NOx |
|---|---|---|---|---|
| | | $lb/10^6$ Btu | mg/MJ | Reduction |
| Tangentially Fired | Nominal 18 | 0.245 | 105 | 67 |
| Cyclone | 20-26 | 0.344 | 148 | 66 |
| Wall-Fired | 5-19 | 0.260 | 112 | 64 |

TABLE 3. TANGENTIALLY FIRED BOILER EMISSION IMPACTS

| Variable | Change from Baseline | Comments |
|---|---|---|
| $NO_x$ | -67% Average | With 18% gas input |
| CO | -1 to +9 ppm | at 70 and 60 MW |
| $CO_2$ | -8% | Less carbon in gas than coal |
| $SO_2$ | -18% | Zero sulfur in gas |
| Opacity | No change | Below 20% |
| Particulates | No change | Below 0.04 $lb/10^6$ Btu (17 mg/MJ) |

TABLE 4. TANGENTIALLY FIRED BOILER PERFORMANCE IMPACTS

| Variable | Change from Baseline | Comments |
|---|---|---|
| *Thermal effects* | | |
| Thermal efficiency | -0.3 to -1.1% | Latent heat losses due to hydrogen |
| Carbon in Ash | +0.5 to 1.7% | Minimal impact on boiler efficiency |
| Steam temperature | No change | Controlled by attemperation |
| Heat distribution | Minor changes | Within normal ranges |
| *Durability* | | |
| Tube wastage | No increase | Extensive ultrasonic surveys |
| Tube metallurgy | No change | Destructive tests of tube samples |
| Projected life | No change | Boiler life unaffected |

TABLE 5. THERMAL PERFORMANCE OF A WALL-FIRED BOILER AT FULL LOAD

| Operation | LNB | GR-LNB |
|---|---|---|
| Process Variables: | | |
| Load, MWe net | 148 | 148 |
| Boiler exit $O_2$, % dry | 3.90 | 3.60 |
| Gas heat input, % | 0.0 | 12.6 |
| Overfire air, % of total air | 0.0 | 19.3 |
| Steam Side: | | |
| Main steam temperature, °F (°C) (Design 1,005°F or 540°C) | 965 (518) | 992 (533) |
| Superheat attemperation, lb./hr (kg/h) | 5,030 (2,280) | 14,200 (6,440) |
| Hot reheat steam temperature, °F (°C) | 942 (506) | 985 (529) |
| Reheat attemperation, lb/hr (kg/h) | 0 (0) | 91 (41) |
| Economizer outlet gas temperature, °F (°C) | 694 (368) | 711 (377) |
| Thermal efficiency, % | 88.00 | 87.55 |
| Gross heat rate, Btu/kWh (kJ/kWh) | 9,487 (10,010) | 9,316 (9,828) |
| Net heat rate, Btu/kWh (kJ/kWh) | 10,261 (10,825) | 10,093 (10,648) |

## AUTHOR INDEX

## FACT-Vol. 18
## Combustion Modeling, Scaling and Air Toxins



**Book Number: G00878**